MW01224956

Global Warming

edited by

Stuart A. Harris

Global Warming
Edited by Stuart A. Harris

Published by Sciyo
Janeza Trdine 9, 51000 Rijeka, Croatia

Publishing Process Manager Jelena Marusic
Technical Editor Teodora Smiljanic
Cover Designer Martina Sirotic
Image Copyright Armin Rose, 2010. Used under license from Shutterstock.com

First published October 2010
Printed in India

A free online edition of this book is available at **www.sciyo.com**
Additional hard copies can be obtained from **publication@sciyo.com**

Global Warming, Edited by Stuart A. Harris
p. cm.
ISBN 978-953-307-149-7

SCIYO.COM
WHERE KNOWLEDGE IS FREE

Contents

Preface IX

Section 1

Chapter 1 **Impact of Global Warming on Tropical Cyclones and Monsoons** 1
K. Muni Krishna and S. Ramalingeswara Rao

Chapter 2 **Greenhouse Gases and Their Importance to Life** 15
Stuart A. Harris

Chapter 3 **Global Warming: CO2 vs Sun** 23
Georgios A. Florides, Paul Christodoulides and Vassilios Messaritis

Section 2

Chapter 4 **Global Warming, Glacier Melt & Sea Level Rise: New Perspectives** 63
Madhav L Khandekar

Chapter 5 **Potential Changes in Hydrologic Hazards under Global Climate Change** 77
Koji Dairaku

Section 3

Chapter 6 **On the Effect of Global Warming and the UAE Built Environment** 95
Hassan Radhi

Chapter 7 **Transport Planning and Global Warming** 111
Pedro Pérez, Emilio Ortega, Belén Martín, Isabel Otero and Andrés Monzón

Section 4

Chapter 8 **Global Warming and Hydropower in Turkey for a Clean and Sustainable Energy Future** 125
I. Yuksel and H. Arman

Chapter 9 **Role of Nuclear Energy to a Low Carbon Society** 141
Shinzo SAITO, Masuro OGAWA and Ryutaro HINO

Chapter 10 **Global Warming** 159
John O'M. Bockris

Section 5

Chapter 11 **Simulating Alpine Tundra Vegetation Dynamics in Response to Global Warming in China** 221
Yanqing A. Zhang, Minghua Song, and Jeffery M. Welker

Preface

This book is intended to be an introduction to the complex effects of climatic change on the activities and life of mankind, particularly in regard to the changes widely known as global warming. It consists of eleven chapters dealing with five different aspects of the problem. The chapters are written by 11 different authors from ten different countries, examining aspects of global warming as it affects their fields of study.

Global warming is a concept developed during a conference in Brazil, at which Maurice Strong was trying to find arguments for obtaining additional funds to aid under-developed countries. This warming is based on the temperature difference between the mean annual temperature (MAAT) in the late 1800s and that existing today (Intergovernmental Panel of Climate Change, 2007). It is usually attributed to anthropogenic changes in carbon dioxide in the atmosphere based on the perceived similarity of the change in MAAT with the increasing carbon dioxide content in the atmosphere. This concept was given tremendous publicity by the media, which swayed public opinion and gave environmentalists a major cause to champion. Quickly, global warming spawned a major industry providing employment for many people, providing extra funds for climate-related research, but skewing that research towards looking for a proof of its presumed veracity. There have been a number of vocal critics among those who could survive without research grants, while the publication of numerous e-mails sent by climate specialists at the University of East Anglia appears to have somewhat tarnished the image of some of the ardent supporters of global warming and its dependence on atmospheric carbon dioxide levels.

It is a fact that carbon dioxide is a greenhouse gas, but it is only one of a great number of factors influencing the MAAT. The latter is the result of the sum of all these influences and it is highly unlikely that it is the only one changing at the present time. Certainly, the regular cyclic controls such as the Milankovich Cycles are believed to be continuing to greatly influence the climate as they have done during the last 3.5 Ma (Imbrie and Imbrie, 1980; Campbell et al., 1998). While there is no question as to whether the MAAT is warmer now than in the late 1800s, the actual cause is still being debated.

The concept of climatic change is well known to geologists, who try to determine the nature of past climates from sedimentary rocks. While these studies show spectacular changes over time at a given location, these changes appear to have remained in a relatively narrow range for over 4 billion years. The sea has never boiled, nor has it frozen solid according to the geological record. There certainly have been substantial changes with time, e.g., the Mesozoic Era was one in which the Tethys Sea lay along the tropics from eastern China, across Asia and southern Europe to the Atlantic Ocean. Since water absorbs approximately five times as much solar radiation as soil, the sea was much warmer than now, and this resulted in higher temperatures everywhere except in Antarctica. When the Himalayan, Alpine and Persian Mountains rose as Africa moved north against the Eurasian Plate, this sea was replaced by land, which probably accounts for the cooling trend that has taken place since the beginning of the Tertiary Era (Harris, 1992).

It is not well known that at least 30% of the heat absorbed in the Tropics must be moved northwards if the land areas in the Northern Hemisphere are to continue with their present temperatures. Surface ocean currents and thermohaline currents transport the heat in the oceans, while changes in the movement of air masses do the same job in continental areas. Changes in these can be very abrupt, e.g., the beginning and end of the Younger Dryas event that is now regarded as being the result of the draining of Lake Agassiz into the North Atlantic Ocean, and resulted in the deflection of the Gulf Stream southwards. The MAAT across northern Europe dropped at least 18 °C (Isarin, 1997) and permafrost with attendant ice-wedges developed throughout the region during a 100-year period with an abrupt beginning and end (Renssen and Vandenburghe, 2003). Changes brought about by movement of the average position of air mass boundaries can occur in about 15 years (Harris, 2009).

Contrary to common assumptions and many published papers, not all areas are experiencing continued warming (Harris, 2009; Krishna and Rao, 2010)). There is enormous variation at a given place from year to year, so only the instrumental records longer than 30 years can reasonably demonstrate these changes. The mechanism of change appears to be variation in the movements of the air masses. At present, these are only starting to be investigated.

The chapters in this book are divided into five sections. Section one consists of three papers on subjects concerning the climate. The first one examines the available instrumental data on the MAAT and precipitation produced by the Monsoon in India between 1880 and 2006. It demonstrates that there are multi-decadal periods of warmer and colder temperatures, as well as variations in rainfall resulting from changes in temperature on the equatorial side of the Tropical Easterly Jet Stream. In chapter 2, the history of the carbon dioxide content in the atmosphere over the last 4 billion years is found not to match the climatic history of the Earth. The dominant greenhouse gas (water vapour) is many times more abundant, though it shows very large variations due to continuous changes in temperature and relative humidity. These dwarf the effects of variations in carbon dioxide, but there appears to be a need to make a choice between aiding desertification by curbing carbon emissions or letting them increase and so partially counteracting reductions in precipitation. The third chapter discusses the relative merits of carbon dioxide and variations in the activity of the sun as factors affecting increasing global temperatures, concluding that the latter may be more important. Both chapters 2 and 3 indicate that allowing the carbon dioxide content of the lower atmosphere to increase may actually aid biological activity.

Section two includes two papers dealing with the effect of the increasing MAAT on natural hazards. In chapter 4, the effect of melting glaciers on stream flow is discussed, together with the problem of predicting the possible rise in sea level this may produce. This is one of the most contentious issues for many low-lying island states in the Pacific Ocean, though all such estimates suffer from the lack of knowledge of possible ongoing changes in the volume of the ocean basins. Chapter 5 discusses the potential changes in hydrologic hazards as a result of the assumed climatic change scenario. Floods have done tremendous damage and caused enormous loss of life in the belt from Pakistan to north-east China this year (2010), so this represents a major problem with or without climate change due to the extreme precipitation events associated with the monsoons and typhoons in Southeast Asia.

Section 3 consists of two papers dealing with planning and climatic change. Chapter 6 describes the effects of global warming on the urban environment in the United Arab Emirates. It summarizes the results of various types of construction on energy consumption in the expanding cities and suggests means of reducing the impact of these. All countries must have means of transportation, particularly in cities, and chapter 7 discusses how this can be achieved with the minimum impact on the climate.

Section 4 examines three alternative energy sources that might supplement energy from conventional sources. Chapter 8 discusses the considerable potential for expanding the amount of energy produced by hydropower in Turkey. Similar possibilities occur in other countries, especially China. Chapter 9 discusses the role of nuclear energy in the future given the depleting supplies of conventional fuels. In chapter 10, the possibilities of transforming hydrogen into the safer fuel, methane, is examined. Hydrogen is a non-polluting fuel, but is very dangerous due to its extreme flammability.

Section 5 provides the results of a study which attempts to predict what will happen to the organic carbon in the alpine meadows of Tibet. Chapter 11 provides the results of modeling at two scales, one for the whole of the Qinghai-Tibetan Plateau, and the other for the area around the Haibei Alpine Tundra Ecosystem Research Station in the north-east of the Plateau.

It is hoped that this group of chapters will offer a good introduction to some of the major issues currently being discussed which relate to global warming.

29 August, 2010

Editor

Stuart A. Harris, Calgary
Department of Geography,
University of Calgary,
Calgary,
Canada

Section 1

1

Impact of Global Warming on Tropical Cyclones and Monsoons

K. Muni Krishna and S. Ramalingeswara Rao
Andhra University
India

1. Introduction

Tropical cyclone is one of the most hampered natural hazard in the North Indian Ocean. The North Indian Ocean is divided by the Indian sub continent into two ocean basin one is Bay of Bengal and the other one is Arabian Sea. Bay of Bengal is the most vulnerable to cyclones than Arabian Sea. Recent studies suggest that cyclone activity over the North Indian Ocean (NIO) has changed over the second half of the 20th Century (Mooley, 1980; Rao, 2002; Knutson & Tuleya, 2004; Emanuel, 2005; Landsea, 2005; IPCC, 2007; Muni Krishna, 2009; Yu and Wang, 2009). General features include a poleward shift in strom track location, increased strom intensity, but a decrease in total storm numbers and also the ocean response in the weak of cyclone. Sea surface temperature (SST) is a fuel to tropical cyclones for their genesis and intensification. Global warming heats both the sea surface and the deep water, thus creating ideal conditions for a cyclone to survive and thrive in its long journey from tropical depression to Category Four or Five superstorm.
SST increasing is so fast and high in the equatorial Indian Ocean compared with other the oceans. It has increased 0.6°C over the NIO since 1960, the largest warming among the tropical oceans. Recent increase in frequency of severe tropical cyclones is related to the increase in SST in response to global warming. Higher SSTs are generally accompanied by increased water vapour in the lower troposphere, thus moist static energy that fuels convection. The large scale thermodynamic environment (measured by Convective Available Potential Energy, CAPE) become more favorable for tropical cyclones depends on how changes in atmospheric circulation, especially subsidence, affect the static stability of the atmosphere, and how the wind shear changes (IPCC, 2007).
Despite an increase in SST over the Bay of Bengal (Sikka 2006), observational records indicate for a decline in the number of depressions over the Bay of Bengal since 1976 (Xavier and Joseph 2000), and various factors are attributed to this trend that includes weakening of the low-level westerly flow over the Arabian Sea (Joseph and Simon 2005), decrease in the horizontal and vertical wind shears as well as in moisture and convection over the Bay of Bengal (Mandke & Bhide 2003; Dash et al., 2004).
Vertical wind shear and high static stability has an adverse influence on tropical cyclone formation and on cyclone strength and longevity (Gray, 1968; Hebert, 1978; DeMaria, 1996; Shen et al., 2000; Garner et al., 2009). Joseph & Simon (2005) indicate that low level jet stream associated with Indian summer monsoon over the NIO is weakening in recent years, which reduces the vertical easterly shear and thus it is favorable for the formation of more

intense tropical cyclones. In the NIO, vertical wind shear is determined by gradients of SST both locally within the ocean basin and remotely from the Indo-Pacific (Shen et al., 2000). High static stability suppresses deep convection during cyclogenesis and educes the potential intensity (Emanuel, 1986; Holland, 1997) of organized cyclones. The contrast between SST and upper tropospheric air temperature is decide the stability.

Tropical cyclones produce significant changes in the underlying ocean thermodynamic structure, which also involves SST changes. SST may decrease by up to 6°C as a result of strong wind forcing. Vertical turbulent mixing within the upper oceanic layer, accompanied by the mixed layer deepening and entrainment of cooler thermocline water to the warm mixed layer, is the primary mechanism of SST decrease during the tropical cyclone passage. The heat fluxes to the atmosphere account for less than 20% of the total SST decrease.

Surface-air-temperature over the world has been warmed by 0.7°C since last 100 years. This is due to both natural and anthropogenic forcing, which result in year-to-year change of temperatures over the globe and there is a drastic change in shooting up of temperature in the last three and half decades due to abrupt increase of Green House Gases (GHGs), which geared up catastrophic climate change over several parts of the globe. Recently climate experts at a monitoring station in Hawaii reported CO_2 level in the atmosphere have reached a record 387 parts per million, which is 40% higher than before the industrial revolution. Tyndall centre for climate change research, for instance suggests that even global cuts of 3% a year starting in 2020, could leave us with 4°C of warming by the end of the century. The Inter governmental Panel on Climate Change (IPCC) has explained the impact of global warming upon mankind with special reference to developing countries of Africa and Asia and alerted the developed countries to reduce GHGs. Of the developing countries, India with its second highest population in the world is mainly affected by way of vagaries of monsoon in terms of floods, droughts and extreme episodes due to climate change. In the fourth assessment report of the IPCC, it is estimated that there will be 2°C enhancement of temperature in the coming 30-years.

Several effects of global warming, including steady sea level rise, increased cyclonic activity and changes in ambient temperature and precipitation patterns are projected in India. Heavy monsoon rains in central India between 1981 and 2000 were more intense and frequent than in the 1950s and 1960s and increased by 10% since the early 1950s and it was attributed to global warming by Goswami et al., 2006. Extreme events like severe drought in the year 2002 and 100cm heavy rainfall on 26th July, 2005 were a few examples during monsoon season. There are some more studies, which indicate that India's long-term monsoon climatic stability is threatened by global warming. Of them, Hingane et al., (1985) studied the long-term trends of surface-air-temperatures of India with a limited data and their analysis showed that the mean annual temperature has increased by 0.4°C during the past century. Later Rupakumar & Hingane (1988) have reported the results of the analysis of long-term trends of surface-air- temperatures of six industrial cities in India. Next, Murthy et al., (2000) estimated costs associated with a low GHG energy strategy in terms of foregone income and welfare of the poor. The impact of climate change on agricultural crop yields in India, GDP and welfare is well studied by Kumar and Parikh (2001a and 2001b) and Rosenzweig and Parry (1994). Lal et al., (2001) concluded that annual mean area-averaged surface warming over the Indian subcontinent to range between 3.5°C and 5.5°C over the region during 2080s, while the DEFRA (2005) suggested that for a warming of 2°C, the yields of both rice and wheat will fall in most places, with the beneficial effect of increased CO_2 being more than offset by the temperature changes over India; similar results have been found for soybean (Mall et al., 2004). Next Battacharya and Narasimha (2005) studied the possible association between Indian monsoon rainfall and solar activity. Ashrit et al., (2005) investigated the impact of anthropogenic climate changes on the Indian summer monsoon and the ENSO-monsoon teleconnection. Later Bhaskaran and Lal (2007) studied the impact of doubling CO_2 concentrations on climate by using UK Met Office models, which is a coupled climate model indicated reasonable simulation of present day climate over the Indian region. In a pilot study, Bhanu Kumar et al., (2007 and 2008) thoroughly investigated increase of surface-air-temperature trends over two states of India namely Rajasthan and Andhra Pradesh with a limited data and they concluded that there is a significant warming trend. In view of the above studies, an attempt has been made in the present study to investigate the relation between weather disturbances and sea surface temperature, vertical wind shear and temperature gradient over the North Indian Ocean and also examined the impact of warming due to GHGs over the changing monsoon climate of India.

2. Data and methodology

For the present study, the authors used monthly wind and air temperature data from the National Centers for Environmental Prediction-National Center for Atmospheric Research (NCEP-NCAR) reanalysis dataset (Kalnay et al. 1996). The source of data of tropical cyclone frequency in the North Indian Ocean for the period 1877-2009 is an India Meteorological Department (1979 and 1996). The data for 1986 – 2009 have been obtained from different volumes of the quarterly journal Mausam. GISS mean monthly surface-air-temperature anomalies data for the study of annual and seasonal variations during 1880-2006, which is obtained from the NASA website (http://www.cdc.noaa.gov/cdc/data.gisstemp.html). Analysis of the NASA GISS Surface Temperature (GISTEMP) provides a measure of the changing global surface temperature with monthly time scale since 1880, when reasonably global distributions of meteorological stations were established. Above dataset is available on an equal area grid (1°x1°). In this study, we used the 250 km smoothing data over whole of India for 280 grids. For the study of climate change over India, the mean monthly Indian rainfall data is used for a period of 1871-2006, which is supplied by the Indian Institute of Tropical Meteorology (IITM), Pune (www.tropmet.res.in). The NCEP/NCAR reanalysis wind data is also obtained for the period 1970-2006 for circulation changes (http://www.cdc.noaa.gov).

In methodology, the homogeneity of the temperature and rainfall datasets have been tested by Swed & Eisenhart's test (WMO 1966). The median and the number of runs above and below the median have been obtained and these are given in table 1. Next, Mann-Kendall test is used for long term trends, while Cramer's running mean test is applied to isolate periods of above and below normal temperature and rainfall. Finally correlation and regression analyses are also used to detect relationship between temperature and rainfall.

Tropical vertical wind shear and temperature gradient over the NIO (5° - 20°N and 40° - 100°E) is given by

Wind Shear (WS) $_{200\text{-}850} = U_{200hPa} - U_{850hPa}$, $WS_{200\text{-}925} = U_{200hPa} - U_{925hPa}$ and

$WS_{150\text{-}850} = U_{150hPa} - U_{850hPa}$,

Temperature gradient = $T_{500hPa} - T_{100hPa}$

Details of the statistical tests used in this study are given below.

Mann-Kendall test:

In climatological time series, the successive values are not likely to be statistically independent of one another, owing to the presence of persistence, cycles, trends or some other non-random component in the series. In view of this, Mann-Kendall test is applied for trends for surface-air temperatures and rainfall as follows:

$$\tau = 4\sum_{i=1}^{n-1} \frac{n_i}{N(n-1)} - 1 \quad (1)$$

Where n_i is the number of values subsequent to i^{th} value in the series exceeding i^{th} value. The value of τ was tested for the significance by the statistic $(\tau)_t$, which is given by

$$(\tau)_t = t_g \sqrt{\frac{4N+10}{9N(N-1)}} \quad (2)$$

Cramer's test:
The aim is to examine the stability of a long term records in terms of comparison between the overall mean of an entire record and the mean of the certain part of the record (WMO 1966).

$$t_k = \sqrt{\left[\frac{n(N-2)}{N-n(1+r_k^2)}\right]} r_k \quad (3)$$

The statistics t_k is distributed as 'Student's t' with N-2 degress of freedom. This test may be repeated for any desired number and choice of sub periods in the whole record. The time plot of the t-value gives the pictorial representation of variability.
Student's t-test: In order to estimate trends in wind shear, temperature gradient and cyclones, simple linear regression technique was used. These trends have been tested by using Student's *t*-test. The statistic, t is given by:

$$t = b\left[\frac{(N-2)\sum(x-\overline{x})^2}{\sum(y-\widehat{y})^2}\right]^{\frac{1}{2}} \quad (4)$$

Where $b, N, x, \overline{x}, y$ and $\widehat{y}$ represent the slope of the regression, the number of years of data, the year, mean of the years, actual shear and estimated shear respectively. This regression analysis gives an indication of the overall tendency of the wind shear and temperature gradient.

3. Results and discussions

3.1 North Indian ocean warming

A number of features of the tropical climate are relevance to tropical cyclone activity appear to be changing in a trend-like fashion. Based on Hadley Center Sea surface temperature, there is an increasing trend in the recent era (1981-2009) compared with the previous eras (1870-1949 & 1950-1980) over the NIO (Fig 1.). Strong positive SST anomalies (0.8°C) are observed during the period of 1981-2009. This is one of reason for the formation of intense tropical cyclones over the NIO. Strong negative (cc = - 0.26, significant at 99.9 %) relationship is observed between SST anomaly and depressions, cyclones (-0.33, significant 99.9%) and positive (0.05) relation with severe cyclones during 1877-2009. But in 20th

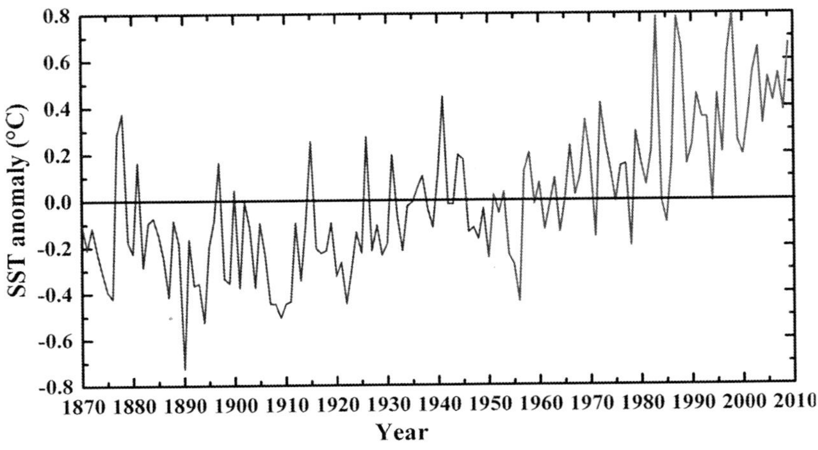

Fig. 1. SST anomaly over the NIO (40°E - 100°E, 5°N - 25°N) during southwest monsoon season. Blue, pink and red colour indicates 1870 - 1949, 1950 - 1980 and 1981 - 2009 respectively.

century 1981-2009) very high negative correlation is noticed between depressions (-0.38, significant at 99%), decrease with cyclone (-0.04) and strong positive correlation with severe cyclones (0.27 significant at 99%). It enlightens that the depressions are decreasing and intense severe cyclones are increasing in the 20th century over the North Indian Ocean. For example, in 2007 category 5 severe cyclone (Gonu) formed in the Arabian Sea after 70 years. The SST increase is a response to the long-term increases in greenhouse gas concentrations. Human induced change in greenhouse gas forcing is the main cause of the rapid increase in SST during 1981-2009 warming. This result is supported by the several other model simulations (Knutson et al., 2006).

3.2 Vertical wind shear

The averaged vertical wind shear for the southwest monsoon season between the upper and lower atmospheric layer over the North Indian ocean is shown in Fig 2. A decrease of 4.2 m/s in easterly shear is observed in a period of 60 years over the NIO. The decrease in the shear is high in 20th century compared with the previous decades. This feature is coinciding with the North Indian Ocean warming. The correlation between shear (in three layers i.e, u(200-925), u(200-850 hPa), u(150-850 hPa) and SST anomaly is -0.5, which is highly significant at 99.9 % level.

The relation between the vertical wind shear over different atmospheric layers and frequency of weather disturbances over the North Indian Ocean is given in Table 1. The relationship between depression and vertical wind shear in all the atmospheric layers shows negative in both before (1950-1980) and in the global warming era (1981-2009). This relation is also statistically significant at 95 % and 99% level. It means that higher easterly wind shear generates more depressions. Fascinatingly the relationship between wind shear and severe cyclonic storm is opposite (positive) i.e. lower shear is more flattering for the formation of more number of severe cyclonic storm. Gray (1968) was suggested that the tropical cyclones of hurricane intensity over several basins including North Indian Ocean basin occur only when the vertical wind shear is small (around 10 m/s between 850-200 hPa). Recently formed sever cyclonic storm (Phet, during 31 May 2010 - 7 June 2010) over the Arabian Sea supporting the impact of global warming on the intensity of severe cyclonic storm. It is the second strongest severe cyclonic storm (first one is Gonu in 2007) to hit the Arabian Peninsula since record keeping began more than 60 years ago.

Layer	Depressions	Cyclones	Severe Cyclones	SST anomaly
200-925 hPa	-0.23 (-0.31)	0.01 (0.13)	0.44*** (0.15)	0.50***
200-850 hPa	-0.28* (-0.27)	-0.05 (0.08)	0.42*** (0.11)	0.50***
150-850 hPa	-0.46** (-0.36)	-0.15 (0.07)	0.33** (0.07)	0.50***

Table 1. Relationship between the weather disturbances and vertical wind shear, SST anomaly over the North Indian Ocean during 1950-2009. Brackets represents for the period of 1981-2009. (*, ** and *** indicate the levels of significance, 95%, 99% and 99.9% respectively).

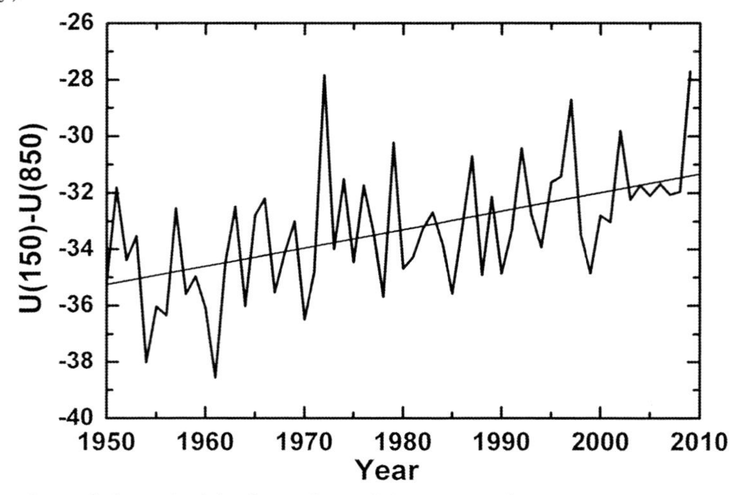

Fig. 2. Vertical wind shear (m/s) of zonal wind during southwest monsoon season over the North Indian Ocean.

3.3 Air temperature gradient

The air temperature difference between lower and upper atmosphere shows an increasing trend (Fig 3) during the southwest monsoon season over the North Indian Ocean. The increasing linear trend is 0.5°C in 60 years which is also significant at 99.9 % level. It shows a strong environmental warming in the layers of the atmosphere over the North Indian Ocean. The increase trend in air temperature and SST anomaly in 20th century over the tropical NIO, capitulates reduced wind shear.

3.4 Mid tropospheric humidity

Mid tropospheric (relative humidity difference between 700 hPa and 500 hPa) is one of favorable condition for the formation and intensification of tropical cyclone in the North Indian Ocean (Fig 4). In 20th century the mid tropospheric humidity (MTH) trend is increasing (~ 6 %). Along with the increase SST anomaly, decrease in wind shear and increasing trend in

MTH is also give an indication of increase intense tropical cyclone over the North Indian Ocean. The correlation between severe cyclones and MTH is 0.23. A strong relation is observed with vertical wind shear (0.24, 0.30 and 0.51 with the layers 200-925, 200-850 and 150-850 respectively). The relation is strong with the upper layers (150-850 hPa) of the atmosphere (which is the most important wind shear for the formation of intense cyclones). SST anomaly is also shows good correlation with MTH (0.61, significant at 99.9 % level).

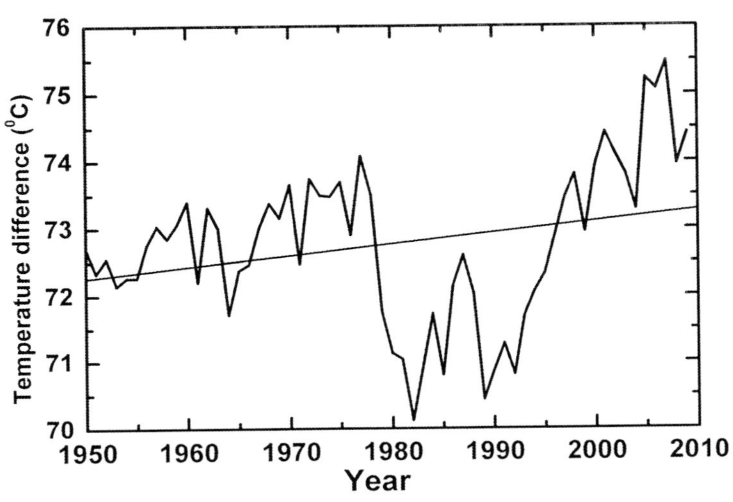

Fig. 3. Air temperature difference between lower (500 hPa) and upper (100 hPa) atmosphere during southwest monsoon over the North Indian Ocean.

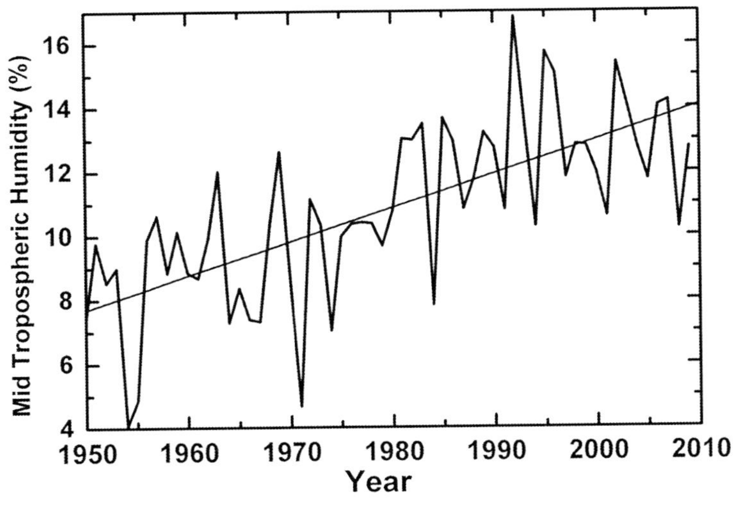

Fig. 4. Mid tropospheric humidity (700-500 hPa) during southwest monsoon over the North Indian Ocean

3.5 Trends of surface-air-temperatures and rainfall over India

The earth's climate is dynamical and always changing. The climate of a place is the average weather that it experiences over a period of time. The factors generally determining the climate of a region are temperature and rain in this study. For temperature and rainfall series, values of statistics for Mann-Kendall rank statistic test have been calculated and the results are given in table 2. The test statistic for N=127 significant at 5% is ±0.1167 and significant at 1% level is ±0.1552. On examination of the table, there is no suggestion of non randomness in the series and that for the purpose of our statistical analysis, these series could be taken as random.

Analysis of mean monthly surface-air-temperature for 280 grids over India is averaged from January through December for annual, SW monsoon (June, July, August and September; JJAS) and NE monsoon (October, November and December; OND) seasons. Fig. 2a indicates year-to-year variations of annual surface-air-temperatures over India for the study period and it clearly indicates that there are 10 hot episodes (based on ±0.5 anomaly; 1910, 1938, 1955, 1984,1985, 1994,1995,2000,2003 and 2006) over the study region. Of them, 7 episodes were recorded during 1970-2006. Trends are also evaluated for the whole of the study period and recent three and half decades separately, which amount to 0.57 and 0.68 (significant at 5% level) respectively. This trend line clearly indicates that global warming is significantly increased during 1970-2006. This is due to a reason that GHGs emissions have grown since pre-industrial time (1970-2006) with an increase up to 70%. Along with the CO_2, the production of CH_4 is also a maximum extent over India and both may lead to climate variability.

Season	Median	No. of runs above and below the median	Mann-Kendall rank statistic test
Annual	-0.097 (1094.3)	46 (60)	6.5 (-0.21)
Southwest	0.001 (860.0)	53 (70)	6.0 (-1.12)
Northeast	0.153 (124.4)	39 (64)	6.5 (0.72)

Table 2. Median, Swed & Heisenhart and Mann-Kendall rank statistic tests for Surface-air temperatures and rainfall (in brackets).

Coming to monsoon season (Fig. 5b), the aberrations of the temperatures are reduced drastically due to the influence of monsoon. There are 8 warm episodes and figure indicates that the many of the warm episodes were noticed during 1970-2006 as similar to annual. The trend values are very close to 0.54, which is significant at 5% level for above specified periods. For the NE monsoon season (Fig. 5c), the surface-air-temperatures are relatively higher for the last three and half decades (1970-2006), but at the beginning i.e. from 1880 onwards up to 1970 the anomalies of the surface-air-temperatures were negative. The trend value for the NE monsoon is same (0.57) for both periods. Similarly an attempt is also made to find out trend values for rainfall series during the study period and recent three and half decades. Those are not at all significant (not shown here).

3.6 Decadal variability of surface-air-temperature and rainfall over India

To have a broader outlook of smoothed temperature and rainfall variations, decadal variability is also evaluated with Cramer's t-statistic test (Fig.6). Fig 6a shows values of Cramer's t-statistic for the 31-year running means of surface-air-temperatures (line format) and all India annual rainfall (bar format). The most striking features are the epochs of above and below normal temperatures and rainfall. It throws light that the temperatures were running above normal during the decades 1930-2006, while there appears to be an inherent internal epochal variability in the rainfall series. The period 1915-50 (1880-1915 and 1950-76) are characterized by above (below) normal rainfall with a very few (frequent) droughts. The turning points are noted around 1915 and 1950. The transition from one state of above (below) normal is an interesting sinusoidal feature. The fall from an extreme state of below normal occurs in a short span of about a decade (1940-1950). However, the rise above normal state is gradual and may take about four decades (1910-50).

The Cramer's t-statistic test for surface-air-temperatures of SW monsoon season shows that there is a turning point around 1900 and the above normal temperatures are continuing till 2006 (Fig.6b). The 31-year sliding Cramer's t-statistic test for all India monsoon rainfall (Fig.6b) shows that the most striking feature is the presence of multi-decadal epochs of

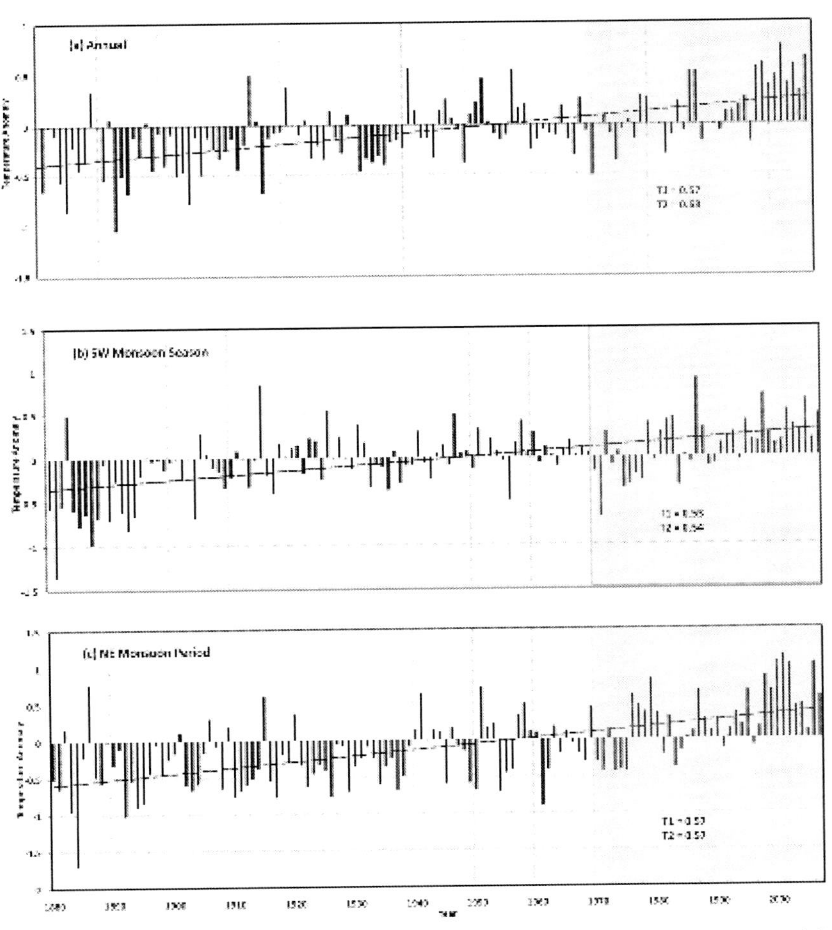

Fig. 5. Variation of all India surface-air-temperature (Dashed line is trend; T1-trend for 1880-2006, T2-trend for 1970-2006).

above and below normal rainfall. The rainfall shows major turning points around 1915 and 1955. The transition from one state of above or below normal monsoon rainfall is an interesting sinusoidal feature like annual rainfall series above. The monsoon rainfall series is free from any sub-period (31-year) trend since nowhere the Cramer's test for 31-year running mean is statistically significant. Thus there is a lot of similarity in the trend and variability of rainfall in both annual and monsoon seasonal rainfall. Similarly for NE monsoon period, temperatures attained increasing tendency since 1960, while rainfall shows major turning points during 1910, 1960 and 1970 (Fig.6c). In general it has been observed that variability is below during the epochs of above normal rainfall.

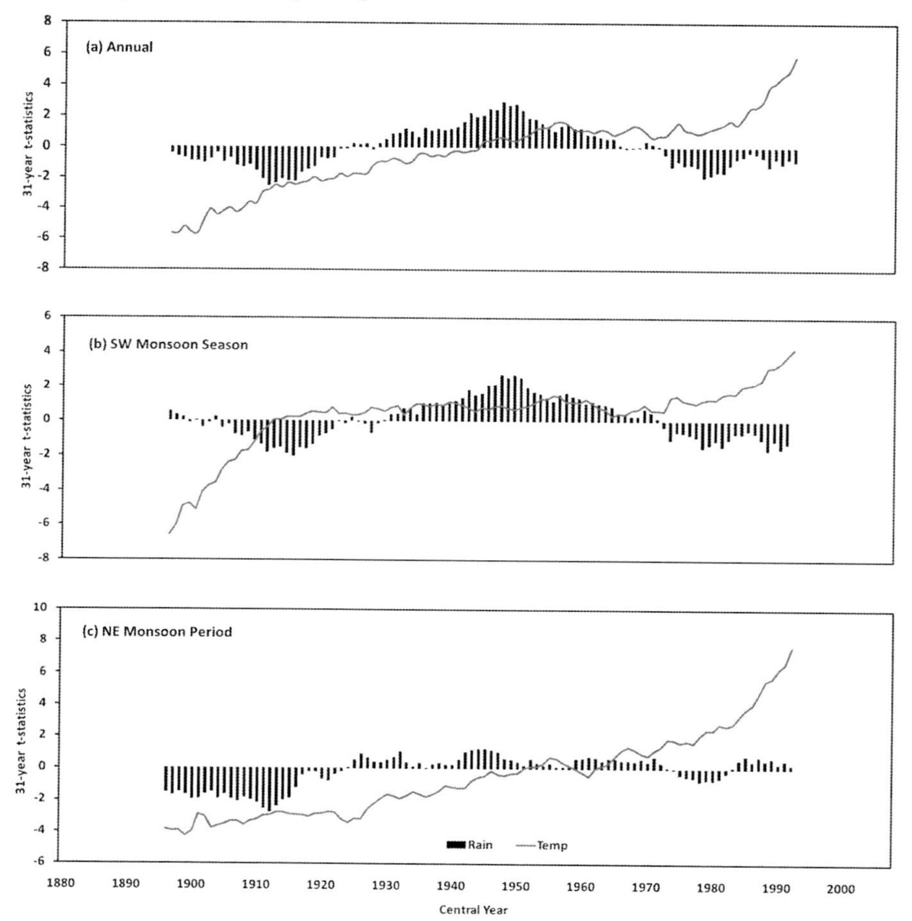

Fig. 6. Values of Cramer's t-statistics for the 31-year running mean depicting climatological variability and epochs of above and below normal rainfall and surface and air temperatures. Values are plotted at the centre of 31-year period.

To further examine the signature of above surface-air-temperatures on rainfall of annual and seasons over India, correlation coefficient is found for 11-year running mean datasets. This study clearly indicates that the impact of temperatures on monsoon rainfall is significant (r = -0.4). Hence the stability of Indian monsoon rainfall is more or less influence to some extent with considerable year-to-year variability in surface-air-temperatures over India.

3.7 Observational evidence of circulation changes during warm/cold temperature episodes

To substantiate above significant inverse relationship between global warming and monsoon rainfall, an attempt is made to investigate contrasting circulation changes in the typical years of clod (1998) and warm (2002) episodes. The chief amounts of monsoon seasonal rainfall were 105% in 1998 and 81% in 2002. Figure 7a shows the anomaly U-wind

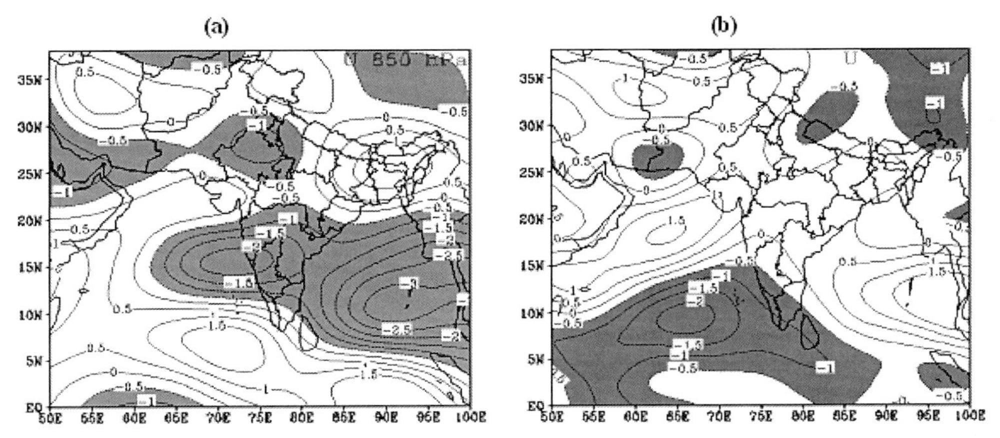

Fig. 7. Anomaly U-wind at 850 hPa level during (a) cold episode-1998 and (b) warm episode-2002.

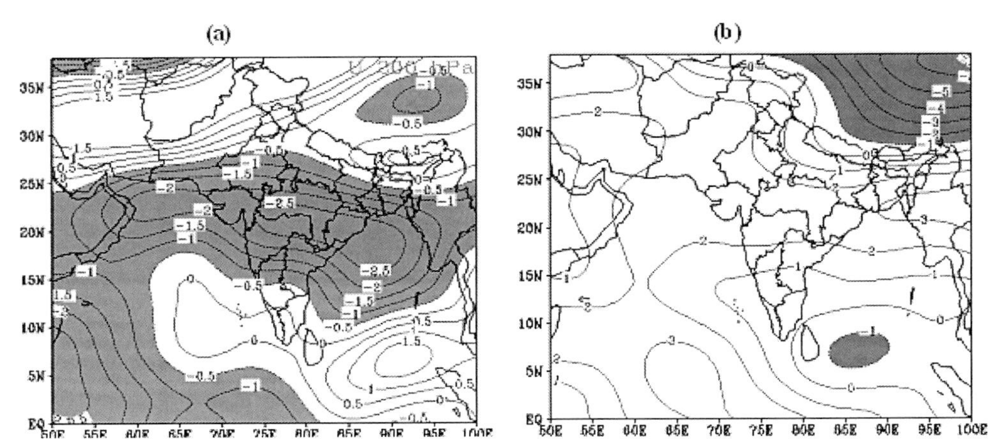

Fig. 8. Same as above except for U-wind at 200 hPa level.

at 850 hPa level for cold episode (1998) and it indicates excess westerly wind (2 m/s) over southern India and parts of Bay of Bengal and Arabian Sea. In warm episode (2002), the anomaly wind pattern is weak (Fig.7b). Similarly the wind filed at 200 hPa during 1998 is negative over parts of Arabian Sea, Bay of Bengal and whole of India except southern tip of India. Anomaly wind speed of -1.5 m/s is observed in the region of tropical easterly jet, while opposite wind appear in 2002 (Fig. 8b). Thus Indian summer is due to a series of feed back mechanisms where in global warming is one important such parameters.

4. Conclusions

The negative anomaly of SSTs over the North Indian ocean is one of the major impacting factor in explaining the lack of major intensification of sever storm during summer monsoon season. The occurrence of intense tropical cyclones in the North Indian Ocean has chronicled increasing trends during southwest monsoon. The increasing trend has been primarily due to increase in SST anomaly, mid tropospheric humidity, temperature difference between lower and upper atmosphere and decrease in the vertical wind shear. In future evolution of North Indian Ocean storm activity will critically depend on the warming of the sea surface waters and also the vertical wind shear. Strong relationship between SST anomaly and vertical wind shear supporting the formation of intense tropical cyclone in the North Indian Ocean. Given the strong correlation between the decreasing easterly wind shear and the increasing number of severe cyclonic storms, decreased TEJ may lead to additional severe tropical storms of hurricane intensity over North Indian Ocean. The catastrophic storms in June 2007 portend disastrous conditions for the large fraction of the global population in the Indian sub-continent and adjacent regions. Other parameters than SST, however, such as the vertical stability of the atmosphere or changes in oceanic mixed layer depth also need to be considered in future projections of cyclonic activity over the North Indian Ocean. There is a growing concern that global warming may be affecting the monsoons and tropical cyclones, their frequency and intensity. The present study shows a good relationship between both ocean and atmospheric variables and severe cyclonic storms. If this trend is continue in future more and more intense cyclonic storms will occur in the North Indian Ocean.

The present study highlights that the increasing trend of temperatures is very similar to that of global warming increasing trend with a little difference of magnitude. The impact of climate change on the Indian monsoons in terms of seasonal rainfall is conspicuous to some extent, but it may be responsible for extreme weather events like Mumbai rainfall on 26th July, 2005 when the warm temperature episode was prevalent. The NCEP circulation changes at 850 and 200 hPa levels in two contrasting episodes show striking contrast in terms of Indian monsoon westerlies and strength of easterly jet stream etc. Climate change may exacerbate water shortage especially during the dry season, as India has 17% of world population with 4% of its water resources. Thus global warming may cut per capita water availability in India in future. This aspect may be further substantiated with global coupled ocean-atmospheric models. Thus more work needed to understand local manifestations of monsoon changes and the possible role of land-surface changes/process.

5. Acknowledgments

The authors are thankful to the Department of Science and Technology, New Delhi for providing financial support through the research projects (SR/FTP/ES-09/2008 and SR/FTP/ES-31/2008) and also thankful to NASA for providing the GISS surface-air-temperature data, NCEP team for wind, temperature data, I.I.T.M., Pune for sub-divisional rainfall data and IMD for cyclone data for this study.

6. References

Ashrit, R. G.; Kitoh, A.; & Yukimoto, S. (2005). Transient response of ENSO-monsoon teleconnection in MRI-CGCM2.2 climate .change simulations, *Journal of the Meteorological Society of Japan*, 83(3), 273-291.

Battacharya, S.; & Narasimha, R. (2005). Possible association between Indian monsoon rainfall and solar activity. *Geophys. Res.Let.* VOL. 32, L05813, doi:10.1029 /2004GL021044, 2005.

Bhanu Kumar, O.S.R.U.; Muni Krishna, K.; & Ramalingeswara Rao, S. (2007). Study of Global Warming effect over Andhra Pradesh. Proceedings of National Conference on Global Temperature Rise: An Indian Effort Towards Mitigation of Carbon dioxide Emissions - A Brain Storming Session, Andhra University, Visakhapatnam during 21-22 September 2007.

Bhanu Kumar, O.S.R.U., Muni Krishna, K.; & Ramalingeswara Rao, S. (2008). Is Global Warming Affecting Rajasthan State? Presented in 95th Indian Science Congress, Andhra University, Visakhapatnam during January 4-7, 2008, pp.31.

Bhaskaran, B.; & Lal, M. (2007). Climatic response of the Indian sub continent to double CO_2 concentrations. *Int. J. Climat.*, 7, 873-892.

Dash, S. K.; Kumar, J.R.; & Shekhar, M.S. (2004). On the decreasing frequency of monsoon depressions over the Indian region, *Cur. Science*, 86, 1404-1411.

Defra. (2005). Investigating the impacts of climate change in India. Key sheets produced as part of Indo-UK collaboration. www.defra.gov.uk/environment/climatechange.html

DeMaria, M. (1996). The effect of vertical wind shear on tropical cyclone intensity change. *J. Atmos. Sci.*, 53, 2076-2087.

Emanuel, K. (1986). The maximum potential intensity of hurricanes. J. Atmos. Sci., 45, 1143-1155.

Emanuel, K. (2005). Increasing destructiveness of tropical cyclones over the past 30 years, *Nature*, 436, 686-688.

Garner, S.T.; Held, I.M.; Knutson, T.M.; & Sirutis, J. (2009). The role of wind shear and thermal stratification in past and projected changes of Atlantic tropical cyclone activity. J. *Climate*, 22, 4723 - 4734

Goswami, B. N.; Venugopal, V.; Sengupta, D.; Madhusoodanan, M.S.; Xavier, P.K. (2006). Increasing trend of extreme rain events over India in a warming environment. *Science*, 314: 1442-1445.

Gray, W.M. (1968). A global view of the origin of tropical disturbances and storms. *Mon. Weather. Rev.*, 96, 669-700.

Hebert, P.J. (1978). Intensification criteria for tropical depressions of the western North Atlantic. *Mon.Wea. Rev.*, 106, 831-840.

Hingane, L. S.; Rupakumar, K.; Ramana Murthy, Bh.V. (1985). Long-term trends of surface air temperature in India. *J. Climatol.* 5: 521-528.

Holland, G. (1997). The maximum potential intensity of tropical cyclones. J. Atmos. Sci., 54, 2519-2541.

IMD (India Meteorological Department). (1979). Tracks of storms and depressions in the Bay of Bengal and the Arabian Sea 1877-1970, report, Govt. of India, New Delhi.

IMD (India Meteorological Department). (1996). Tracks of storms and depressions in the Bay of Bengal and the Arabian Sea 1971-1990, report, Govt. of India, New Delhi.

IPCC (Intergovernmental Panel on Climate Change). (2007). *Climage Change 2007- The Physical science basis.* In: Solomon, Susan, Qin, Dahe, Manning, Martin (Eds), Cambridge University Press.

Joseph, P.V.; & Simon, A. (2005). Weakening trend of monsoon Low Level Jetstream through India 1950 to 2003. *CLIVAR Exchanges*, Vol 10(3), 27-30.

Kalnay, E.; Kanamitsu, M.; Kistler, M.; Collins, R.; Deaven, W.; Gandin, D.; Iredell, L.; Saha, M.; White, S.; Woollen, G.; Zhu, J.; Leetmaa, Y.; Reynolds, A.; Chelliah, R.; Ebisuzaki, M.; Higgins, W.; Janowiak, W.; Mo, J.; Ropelewski, K.C; Wang, C.; Roy Jenne, J.; & Dennis Joseph. (1996). The NCEP/NCAR 40 year re-analysis project, *Bulletin of the American Meteorological Society*, 77, 437-471.

Knutson, T.R.; & Tuleya, R.E. (2004). Impact of CO_2-induced warming on simulated hurricane intensity and precipitation: sensitivity to choice of climate and convective parameterization. *J. Climate*, 17, 3477-3495.

Knutson, T.R.; Delworth, T.L.; Dixon, K.W.; Held, I.M.; Lu, J.; Ramaswamy, V.; Schwarzkopf, D.; Stenchikov, G.; & Stouffer, R.J. (2006). Assessment of twentieth-century regional surface temperature trends using the GFDL CM2 coupled models. *J. Climate*, 19(9), 1624-1651.

Kumar, K.S.K.; & Parikh, J. (2001a). Socio-economic Impacts of Climate Change on Indian Agriculture, *International Review for Environmental Strategies*, 2(2).

Kumar, K.S.K.; & Parikh, J. (2001b). Indian Agriculture and Climate Sensitivity, *Global Environmental Change*, 11: 147-154.

Lal, M.; Nozawa, T.; Emori, S.; Harasawa, H.; Takahashi, K.; Kimoto, M.; Abe-Ouchi, A.; Nakajima, T.; Takemura, T.; & Numaguti, A. (2001). Future climate change: Implications for Indian summer monsoon and its variability. *Current Science*, 81(9), 1196-1207.

Landsea, C.W. (2005). Hurricanes and global warming. *Nature*, 438, 11-13.

Mall, R. K.; Lal, M.; Bhatia, V.S.; Rathore, L.S.; & Singh, R. (2004). Mitigating climate change impact on soybean productivity in India: a simulation study. *Agric. For. Meteorol.*, 121: 113-125.

Mandke, S.K.; & Bhide, U.V. (2003). A study of decreasing storm frequency over Bay of Bengal. *J. Ind. Geophys. Union*, 7, No.2, 53-58.

Mooley, D.A. (1980). Severe Cyclonic Storms in the Bay of Bengal, 1877-1977. *Mon. Wea. Rev.*, 108, 1647-1655.

Muni Krishna, K. (2009). Intensifying tropical cyclones over the North Indian Ocean during summer monsoon - Global warming, *Global and Planetary Change*, 65, 12-16.

Murthy N.S.; Manoj Panda & Kirit Parikh. (2000). CO_2 Emissions Reduction Strategies and Economic Development of India, IGIDR Discussion paper

Sikka, D.R. (2006). A study on the monsoon low pressure systems over the Indian region and their relationship with drought and excess monsoon seasonal rainfall. COLA Report No. 217, Available at: COLA, 4041, Powder Mill Road, Calverton, MD, USA.

Rao, Y.R. (2002). The Bay of Bengal and tropical cyclones. *Current Science*, 82 (4), 379-381.

Rosenzweig, C.; & Parry, M.L. (1994). Potential impact of climate change on world food supply, *Nature*, 367, 133-138.

Rupakumar, K.; & Hingane, L.S. (1988). Long-term variations of surface air temperatures at major industrial cities of India. *Clim. Change*, 13, 287-307.

Xavier, P.K.; & Joseph, P.V. (2000). Vertical wind shear in relation to frequency of monsoon depressions. Tropical cyclones of Indian Seas, in Proceedings of *TROPMET-2000, National Symposium on Ocean and Atmosphere*, 232-245.

Shen, W.; Tuleya, R.E.; & Ginis, I. (2000). A sensitivity study of the thermodynamic environment on GFDL model hurricane intensity: Implications for global warming. *J. Climate*, 13, 109-121.

Yu, J.; & Wang, Y.(2009). Response of tropical cyclone potential intensity over the north Indian Ocean to global warming. *Geophy. Res. Let.*, Vol. 36, L03709, doi: 10.1029/2008GL036742.

2

Greenhouse Gases and Their Importance to Life

Stuart A. Harris
Department of Geography, University of Calgary
Calgary, Alberta, T3A 1E4
Canada

1. Introduction

Greenhouse gases are those in the atmosphere that are essentially opaque to long-wave radiation but virtually transparent to short-wave radiation (Simpson, 1928; Johnson, 1954). They filter out the long-wave component of solar radiation reaching the outer surface of the atmosphere but permit the short-wave radiation to warm the surface of the Earth. Since the re-radiation from that surface is predominantly long-wave, they prevent this energy from escaping. As a result, Arrhenius (1896) proposed that carbon dioxide emitted by combustion in large industrial centres could raise the near-surface atmospheric temperature. In recent years, this warming of the local microenvironment has been found in the heart of many major cities situated away from the Tropics and is called "the urban heat-island effect". During the last decade, this same process has been claimed to be resulting in "global warming", i.e., resulting in rising temperatures across the entire earth. This has set off a frenzy of concern, fed in part by overexposure in the media. In many recent research papers, the data has tended to be interpreted as though atmospheric carbon dioxide concentrations were the only possible cause of climatic change. It is true that carbon dioxide is a greenhouse gas, but even the most extreme estimates of the ability of potential man-made carbon dioxide increases in the next century suggest a warming of mean annual air temperature (MAAT) of under 4˚C, with most recent models suggesting an increase of less than 2˚C. This confirms that the gas is only a minor factor in climatic change (Table 1). In comparison, changes in ocean currents have resulted in a decrease in MAAT over Northern Ellesmere Island of about 30˚C in the last 2.5Ma.

2. Evolution of the atmosphere

The earth is believed to be 4.5 to over 5 billion years old, and its exact means of formation is still being debated. Initially, the bulk of its surface may have been covered by water (Carver and Vardavas, 1994; 1995), indicating that the mean annual air temperature (MAAT) was below 100 ˚C. Assuming that the equatorial regions were under water, the MAAT would have been higher than now since water absorbs about five times as much solar radiation as soil or rock (Pavlov, 1999: Harris, 2002). Gases are believed to have been vented from volcanoes and probably determined the composition of the atmosphere. These gases included large quantities of water vapour that condensed to form lakes and streams on the land areas, but would ultimately join the oceans. The hydrogen sulphide and sulphur dioxide would have dissolved in the water to form sulphuric acid. This would have

Order	Potential change in temperature (°C)	Control.
1st	c. 30°C.	1. Difference in heat absorption by sea and land as controlled by position of continents and oceans.
		2. Changes in the geometry of the solar system.
2nd	c. 15°C.	3. Changes in ocean currents and thermohaline circulations.
3rd.	5-10°C.	4. Milankovich cycles.
4th.	<5°C.	5. Fluctuations in CO_2 and greenhouse gases.
		6. Large-scale volcanic eruptions.
		7. Elevation of large tracts of land, e.g., Tibet.
		8. El Niño and La Niña.
		9. Short-term cycles, e. g., 2 and 7 years.
		10. Variations in solar output.
		11. Agriculture, deforestation and urbanization.

Table 1. The main suggested controls of climatic change arranged into four orders based on the potential temperature change that they can cause (after Harris, 2005).

reacted with the minerals in the rocks to form metal sulphides and sulphates, whereas the carbon dioxide that dissolved in the water was far less potent. Over time, it would have built up in the atmosphere and oceans to levels far exceeding what is found today, though it is believed that this was partially offset by chemical weathering of rocks. The relatively inert nitrogen would also have slowly built up over time. Any hydrogen or helium which may have been present in the primaeval atmosphere would have slowly escaped into space due to their low molecular weights and the correspondingly weak pull on the molecules by the Earth's gravity.

About 3Ma, there appears to have been a phase of expansion of the land areas, especially around the South Pole to form a continent called Pangaea. By that time, the MAAT around the earth appears to have been similar to that today, because we find glacial deposits intercalated in the rocks of that and subsequent ages (Crowell, 1999). Periods of increased solar radiation are postulated to have occurred (Carver and Vardavas, 1995) but there is no evidence that the sea boiled, in spite of the high carbon dioxide levels in the atmosphere. Sedimentary rocks are common in these old rocks, although they have often been metamorphosed into marbles, schists, etc.. In practice, there are numerous natural sinks or storage places for carbon dioxide including the oceans (Roll, 1965), vegetation, soils, etc.. Excess carbon dioxide in warm, shallow seas can result in precipitation of calcium carbonate deposits such as chalk or fine-grained limestones, as is occurring today around the Bahamas. Meanwhile the concentration of nitrogen would be becoming dominant in the atmosphere.

About 3 billion years ago, calcareous algal reefs are found in some sediments together with graphite, believed to be derived from metamorphism of organic matter. It appears that the reef-forming organisms used carbon dioxide dissolved in the water to both build the reefs and to produce energy and oxygen by photosynthesis. They used carbon dioxide and water to produce organic structures while emitting oxygen into sea. There is always a dynamic equilibrium in both carbon dioxide and oxygen levels in water bodies and the adjacent atmosphere, so the atmospheric levels of carbon dioxide would be decreasing while oxygen would be increasing. In the atmosphere, the mixing of gases is fairly efficient, but in the oceans, the dissolved gases may vary considerably from place to place, depending on the topography, gravity, and currents. There can be appreciable latitudinal variations in the oceans (Grasby and Beauchamp, 2008). Even today, parts of the Baltic Sea are anoxic. This implies that indications of carbon dioxide levels in sediments deposited in water bodies cannot safely be used as indicators of contemporary atmospheric concentrations.

At first, the oxygen produced by the reef organisms would have been mainly used in the oxidation of other minerals, but over time, the composition of the atmosphere would have changed. The algal reefs continued to occur as the main evidence of life until there was an explosion of new life forms in the oceans about 600ka, at the beginning of the Cambrian period. Many early Cambrian fossils exhibit large gill-like structures, clearly adapted to making the most of the low oxygen levels in the oceans. By this time, more advanced aquatic plant life was also present.

Shortly afterwards, the first simple plants appeared on land and flourished, soon producing the large and spectacular forests that were to characterize the middle Palaeozoic era, and they paved the way for the survival of the first animals on land. The fern, *Glossopteris,* even colonized the land around the margins of the vast ice cap covering the South Pole at this time.

In photosynthesis, six molecules of carbon dioxide yield six molecules of oxygen plus one molecule of glucose. The organic compounds in the structures of the plant decompose when the plant dies. In the humid tropical climates, this becomes carbon preserved today in the form of coal. For every 12 kilos of carbon being fixed as glucose by photosynthesis, 32 kilos of oxygen would have been released into the atmosphere. The great volumes of the coal deposits of the Palaeozoic era indicate that initially, there was a great concentration of carbon dioxide in the atmosphere. The extensive tropical forests of the latter part of the Palaeozoic era must have resulted in a vast change in the proportions of atmospheric oxygen and carbon dioxide. This would also mean that the weathering of rocks would have changed to processes similar to those occurring today in comparable climatic regimes. Runoff would have been reduced by the accumulation of a litter layer and substantial quantities of water would have been returned to the atmosphere by evapotranspiration.

The GEOCARB III CO_2 model of Berner and Kothavala (2001) indicates the change in relative proportions of carbon dioxide and oxygen, but suggests that the atmospheric carbon dioxide concentration increased from under 1% by volume in early Permian times to a maximum in the early Cretaceous Period. Since then, it has been steadily decreasing. The volcanic outpourings of magma in the Permo-Triassic period would have been accompanied by degassing, with increasing levels of carbon dioxide, but there was significant formation of coal deposits along the Rocky Mountains under a wet tropical climate at this time to drop the levels towards those seen today (0.03% by volume).

Berner (1990) suggests that the atmospheric oxygen levels rose from about 26% by volume in the Lower Mississippian to a maximum around 35% in the Stephanian (Upper

Carboniferous), before declining to around 15% at the Permian-Triassic boundary. Since then, they have increased with minor fluctuations to the present level (21%).
The ice cap persisted over the South Pole, and when Laurasia separated from Gondwanaland, the carbon dioxide content of the atmosphere was still at least eight times the present-day levels (Retalack, 2001). In spite of this, there was no obvious world-wide climatic change, nor was there when the excess carbon dioxide was re-absorbed into the natural sinks or largely used up in forming the Mesozoic coal deposits, e.g., at the Crows Nest Pass and Banff. If carbon dioxide levels controlled the MAAT the way that has been claimed by proponents of global warming, why were there not considerable changes in the MAAT as the concentration of carbon dioxide dwindled at various times in the past?

3. Greenhouse gases in the atmosphere

The dominant greenhouse gas in the atmosphere is water vapour. For every three parts of carbon dioxide, there are 500-750 parts of water vapour. The latter fluctuates continuously within these limits, depending on the ambient temperature and the relative humidity. Any excess is condensed into water droplets or ice crystals (clouds) which can lead to precipitation (rain, hail, sleet or snow). The clouds also act as layers of greenhouse substances in the atmosphere, modifying diurnal heating and cooling. Given the vast discrepancy in abundance between the concentrations of water vapour and carbon dioxide, it is surprising that so little attention has been paid to the role of water vapour. It would seem likely that the continuous and substantial changes in water vapour would dominate the relatively puny changes in carbon dioxide. Both gases are key elements for the survival of life as we know it since both are required for photosynthesis in green plants. The latter represents the starting point in our food chain, and the carbon dioxide emitted by living animals helps provide the continuous supply needed by plants. The latter return the favour by emitting oxygen which is needed by the animals. This relationship is especially well demonstrated on coral reefs. Note that water is essential for all forms of life as we know it, and as the human population grows, it will become increasingly a limiting factor in our lives.

4. Sources of atmospheric moisture

There are three main sources of atmospheric moisture. The first is evaporation from water bodies, primarily oceans. This source represents about 70% of the surface of the Earth and does not change significantly on the time scale affecting Mankind. A second source is evaporation from bare ground, but this is a relatively minor component, even though it tends to be increasing due to desertification. However the third main source is evapo-transpiration from vegetation. This returns enormous quantities of water to the atmosphere in the case of mature forests, which have tremendous areas of leaf surfaces with stomata. This represents a very important element in the hydrologic cycle, while the trees act as an umbrella, keeping the bulk of the solar radiation from heating the ground. Instead, there is a relatively cool, humid micro-environment beneath a closed forest canopy. Furthermore, the organic surface layer on top of the mineral soil acts like a sponge, storing water and reducing runoff. This water ensures that the humidity remains high beneath the upper canopy between precipitation events and provides the moisture for the evapo-transpiration to continue long after the precipitation has ceased. This ensures that this water is recycled back into the atmosphere instead of mainly draining away as runoff down stream courses to the oceans.

5. Humans, population growth and the loss of mature forests

As human populations increase, they need more places to live, with more food, and there is increasing commercial use of the forests. This results in clearing more land to try to grow more food and provide employment opportunities. The logging industry, together with the slash and burn expansion of agriculture in forested regions greatly affects the global availability of atmospheric moisture due to its reduction in evapo-transpiration in the deforested areas. Since the middle of the last century, 85% of the forests have been logged to provide arable land in Costa Rica, resulting in a decrease in precipitation by 30% in some areas of the country. Since the country is in the subtropical trade wind belt with no obvious barriers to the rain-bearing winds, it would appear that deforestation is the culprit. Replanting of forests is a step in the right direction, but it takes decades to replace the original forest with an equivalent leaf-cover to the original forest in the tropics, and several centuries in the case of forests at higher latitudes. Since fresh water is becoming a scarce commodity in many parts of the world while populations and demands are increasing, this problem needs serious consideration.

Desertification is an ever increasing problem. When Sir Henry Rawlins sailed his gunboat up the Tigris River in 1845, the villagers around the Arch of Ctesiphon requested him to send a shore party to kill a man-eating lion that was dwelling in the jungle around the Arch. He obliged, and then continued up the river, dodging plane tree logs that were floating down to the sea. The last lion is reputed to have been killed at Balad Ruz in 1890. By 1957, the Arch stood alone in the desert (Figure 1). Gone were both the jungle and the villagers. The only remaining forest in Iraq at that time was some scrub Oak in the mountains along the Iranian and Turkish borders.

Fig. 1. The Arch of the Ctesiphon in 1956.

There is a similar trend all along the Silk Road to China. Towns that once saw the hustle and bustle of the trade route in the Middle Ages now lie abandoned to the wind and sand dunes. The same thing has happened in North Africa. In Roman Times, this was the granary of the south, while France and England represented the granary of the north.

6. Contemporary desiccation in western Canada

In western Canada, the glaciers along the crest of the Eastern Cordillera in the National Parks are retreating at an alarming rate (Bolch et al., 2010), though there is far less loss of mass in the case of the glaciers along the Coastal Ranges. The values of mean annual air temperature at stations along the Eastern Cordillera are not currently changing (Harris, 2009), but the glaciers are retreating at an ever-increasing rate. There is no farming or agriculture in the vicinity, nor any large towns. The problem is that the precipitation has decreased markedly since 1910 A.D. There were terrible accidents along the Canadian Pacific Railroad through the Rogers Pass about that time due to enormous avalanches which killed many workers. Today, many of the former deadly avalanche slopes have become reforested with young trees, though many problematic slopes remain. In a study of the increase in size of the avalanche tracks in the Vermilion Pass, Winterbottom (1974) demonstrated that the avalanches were more than double the size of the present-day tracks in the recent past, based on the evidence of the mixed up materials (soil layers, logs, roots, rock, organic matter) in the soil profile. Again, the high precipitation area around the Sunshine Ski Area in Banff National Park has double the number of vascular plant species of the lower precipitation areas to the north and south (Harris, 2007; 2008). It is therefore acting as a refugium for species that require higher precipitation and had migrated into the region after deglaciation but before the present reduction in precipitation. What could be the cause?

To the west is the Central Plateau, some 300km wide, and west of there are the Coast Ranges. These receive their precipitation from the westerly maritime Pacific air masses that have crossed the Pacific Ocean. The precipitation on the Coast Ranges has not changed much. However extensive clear-cut logging along the coast is starting to change that. In the interior on the Plateau, clear-cut logging has essentially decimated large areas of virgin forest, and even relatively puny second and third growth forest is being harvested. The result is that there is now relatively little evapo-transpiration putting moisture back into the air before it continues eastwards to climb over the Eastern Cordillera. This seems to be the cause of the problem of declining precipitation in the source areas of the glaciers along the Eastern Cordillera. Obviously, too many jobs in British Columbia depend on the forests, but the amount of precipitation falling in the mountains determines the flow of the rivers (Fraser, Columbia, Bow, and Athabasca) which originate from the Divide near the Columbia Icefields. The dying glaciers result in lower base river levels as well as reduced precipitation on the mountain slopes, which, in turn, mean that less water is available for use in the dry Prairie Provinces. The current B.C. policy of sacking up to 200 Government forestry officers who are supposed to ensure that there is no over-cutting and that there is replanting of the cut-over areas will not help the problem. Franklin (2001) is among those who have documented the fact that the rate of cutting cannot be sustained at present levels if the forest industry is to survive in B.C. The recent devastation by the pine bark beetle has been used as an excuse for increasing the clear-cutting during the last two years. It would therefore appear that there is a major problem developing as a result of the major exploitation of the forests in the central interior of B.C. What can be done to reduce this?

7. Carbon dioxide levels and tree growth

The striking feature of the Palaeozoic forests was the enormous size of the Horsetails and Coniferous trees. The present-day species of *Equisetum* do not grow taller than about 2 m while most of the modern species of *Pinus* are also nowhere near as big. The Palaeozoic forests were growing in an atmosphere containing far more carbon dioxide than the trees have available today. Is this an indicator of a partial solution to speeding up re-forestation?

Recently, it has been demonstrated in experiments that increasing the carbon dioxide content around poplar seedlings doubles their rate of growth. Assuming that there is replanting of the forests (which is not always being carried out in western Canada), allowing carbon dioxide levels in the atmosphere to rise could have considerable economic importance in increasing the rate of re-growth of forests after harvesting of timber. The latter is not going to stop, but any means of increasing the rate of regrowth of the forests will have an important economic effect, both to the lumber industry and in minimizing the reduction in water supplies to the east.

This leads to the question, is it appropriate to sequester carbon dioxide underground when it could aid reforestation and potentially significantly increase precipitation down-wind of these forests? Increasing the amount of carbon dioxide has a relatively minor effect on atmospheric temperatures compared with other known factors (see table 1, after Harris, 2005), indeed, any such minor increase in temperature would also increase the rate of reforestation. On the other hand, it could make a significant difference to the hydrologic cycle, provide more water in the rivers for use by on the dry Prairie Provinces to the east, and reduce the rate of desertification. The Forest Products industry would gain through more rapid regeneration of the forests, the Prairie Provinces would face a reduced water shortage in the future, and the Energy Industries would not have to spend large sums of money on a very expensive new technology. Thus there appears to be a trade-off between stopping a very minor influence on the temperature of the Earth and taking measures to curb desertification in drought-prone areas such as the Prairies. At the moment, this trade-off is being ignored.

As usual in nature, there is a downside. Research carried out at UC Davis indicates that the increased plant growth is followed by decreased ability of the plants to assimilate nitrogen from the soil. This element is needed to manufacture proteins, so there is a reduction in nutritional value of crops unless a suitable fertilizer is applied to the soil. However, the increase in the number of animals including humans is resulting in greater amounts of nitrogen being returned to the environment, thus potentially mitigating this problem.

8. References

Arrhenius, S., 1896. On the influence of carbonic acid in the air upon the temperature of the ground. *Philosophical Magazine,* 41 (5), 237-276.

Berner, R. A., 1990. Atmospheric carbon dioxide levels over Phanerozoic time. *Science,* vol. 249, pp. 1382-1386.

Berner, R. A. and Kothavala, Z., 2001. GEOCARB III: A revised model of atmospheric CO_2 over Phanerozoic time. *American Journal of Science,* vol. 301, pp. 182-204.

Bolch, T., Menounos, B. and Wheate, R., 2010. Landsat-based inventory of glaciers in western Canada, 1985-2005. *Remote Sensing of Environment,* vol. 114, p. 127-137.

Carver, J. H. and Vardavas, I. M., 1994. Precambrian glaciations and the evolution of the atmosphere. *Annales Geophysicae,* vol. 12, pp. 674-682.

Carver, J. H. and Vardavas, I. M., 1995. Atmospheric carbon dioxide and the long-term control of the Earth's climate. *Annales Geophysicae,* vol. 13, pp. 782-790.

Crowell, J. C., 1999. *Pre-Mesozoic Ice Ages: Their bearing on understanding the climate system.* Boulder, Colorado. Geological Society of America, Bulletin 192, 106pp.

Franklin, S. E., 2001. *Remote Sensing for Sustainable Forest Management.* Boca Raton. Lewis Publishers. 407p.

Glasby, S. E., and Beauchamp, B., 2008. Intrabasin variability of the carbon isotope record across the Permian-Triassic transition, Sverdrup Basin, Arctic Canada. *Chemical Geology,* vol. 263, pp. 141-150.

Harris, S. A., 2002. Global heat budget, plate tectonics and climatic change. *Geografiska Annaler,* vol. 84A, pp. 1-10.

Harris, S. A., 2005. Thermal history of the Arctic Ocean environs adjacent to North America during the last 3.5 Ma and a possible mechanism for the cause of the cold events (major glaciations and permafrost events). *Progress in Physical Geography,* 29(2), 218-237.

Harris, S. A., 2007. Biodiversity of the alpine vascular flora of the N. W. North American Cordillera: The evidence from Phytogeography. *Erdkunde,* vol. 61 (4), pp. 344-357.

Harris, S. A., 2008. Diversity of vascular plant species in the Montane Boreal Forest of Western North America in response to climatic changes during the last 25 ka, fire and land use. *Erdkunde,* vol. 62 (1), pp. 59-73.

Harris, S. A., 2009. Climatic change and permafrost stability in the Eastern Canadian Cordillera: The results of 33 years of measurements. *Sciences in Cold and Arid Regions,* vol. 1(5), pp. 381-403.

Johnson, J. C., 1954. *Physical Meteorology.* New York, Technical Press, M.I.T. and J. Wiley and Sons. 393p.

Pavlov, A. V., 1999. The thermal regime of lakes in Northern Plains Region. *Earth Cryoshere,* vol. 3(3), pp.59-70. [In Russian}.

Retallack, G. J., 2001. A 300-million-year record of atmospheric carbon dioxide from fossil plant cuticles. *Nature,* 411, 287-290.

Roll, H. U., 1965. *Physics of the Marine Atmosphere.* New York and London. Academic Press. 426p.

Simpson, T. C., 1928. Further studies in terrestrial radiation. *Memoires of the Royal Meteorological Society,* #21.

Winterbottom, K. M., 1974. *The effects of slope angle, aspect and fire on snow avalanching in the Field, Lake Louise, and Marble Canyon region of the Rocky Mountains.* Unpublished M.Sc. thesis, Department of Geography, University of Calgary. 149p.

3

Global Warming: CO_2 vs Sun

Georgios A. Florides, Paul Christodoulides and Vassilios Messaritis
Faculty of Engineering and Technology, Cyprus University of Technology, Limassol
Cyprus

1. Introduction

It is an undoubted fact, within the scientific community, that the global temperature has increased by about 0.7°C over the last century, a figure that is considered disproportionally large. With the whole world being alarmed, the scientific community has assumed the task to explain this warming phenomenon. This has resulted in the formation of basically two schools of thought with two opposing theories.

The first theory (the most popular one) claims that the prime guilty for the recent temperature increase is the release of greenhouse gases - mainly Carbon Dioxide (CO_2) - coming mostly from the burning of fossil fuels, the clearing of land and the manufacture of cement. Due to these anthropogenic activities the concentration of CO_2 has increased by about 35% from its 'pre-industrial' values, with all the resulting consequences.

There are though other factors - besides the greenhouse gases - that affect the global temperature, like changes in solar activity, cloud cover, ocean circulation and others. Therefore, the 'second' theory claims that it is the Sun's activity that has caused the recent warming that, *incidentally* in this theory, is considered to be in the generally expected limits of the physical temperature variation throughout the aeons. One assumption on how the Sun is affecting the climate is that the magnetic field and the solar wind modulate the amount of high energy cosmic radiation that the earth receives. This in turn affects the low altitude cloud cover and the amount of water vapor in the atmosphere and thus regulates the climate. (It must be noted that water vapor is considered as the main greenhouse gas.)

In the sequel both of the above-mentioned theories are examined and conclusions about their soundness are drawn. Firstly (section 2), an analysis in past time of the temperature of the Earth is presented, showing that today's temperatures are not in any way extraordinary, unnatural or exceptional, contrary to what many scientists claim. In fact, on large time scale today's temperatures agree with what was expected for this geologic period, while on a small time scale the higher temperatures observed are the result of a natural recovery of the planet from the global coldness of the Little Ice age.

In section 3, the view of the Intergovernmental Panel on Climate Change (IPCC) on the effect of the accumulation of the CO_2 in the atmosphere is presented. Then, in section 4 the CO_2 accumulation effect during the past is examined on large and small time scales. Note that Florides & Christodoulides (2009) using three independent sets of data (collected from ice-cores and Chemistry) presented a specific regression analysis and concluded that forecasts about the correlation between CO_2-concentration and temperature rely heavily on the choice of data used, making it very doubtful if such a correlation exists or even, if

existing, whether it leads to a gentle or any global warming at all. A further analysis of existing data suggested that CO2-change is not a negative factor for the environment. In fact it was shown that biological activity has generally benefited from the CO2-increase, while the CO2-change history has affected the physiology of plants. Moreover, here, an extensive analysis based on data from physical observations is presented.
Section 5 is concerned with the Sun. The Sun radiation varies in amount with respect to time and affects accordingly the environment of the earth. The factors that affect the Sun radiation reaching the earth on long time scales are physical and are mainly due to the so-called Milankovitch cycles. It is also shown that the Sun's activity is another major factor. At the present time during the course of the solar cycle, the total energy output of the sun changes by only 0.1%. On the other hand, the ultraviolet radiation from the sun can change by several percent but the largest changes, however, occur in the intensity of the solar wind and interplanetary magnetic field. Finally, it is claimed that cosmic rays could provide the mechanism by which changes in solar activity affect the climate. We close with our concluding statements in section 6.

2. A matter of time scale

In this section time series of mean global temperatures are presented to examine the presence of any unusual ongoing global warming.
In Fig. 1 we present the observation datasets (variance adjusted, version CRUTE3Vgl) for the last decade's combined land and marine temperature anomalies on a 5°×5° grid-box basis, as published by the Met Office Hadley Centre (2010). A slight downward trend can be observed with a mean anomaly of about 0.45°C.

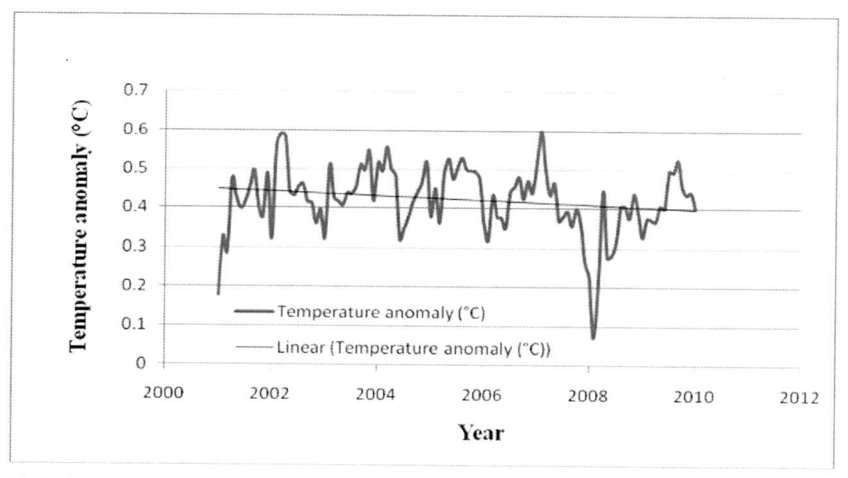

Fig. 1. Global temperature anomaly from 2001 to 2010 showing a slight downward trend.

For a longer period in time the Met Office Hadley Centre data-set shows a slightly negative temperature slope between 1850 and 1915, a positive slope between 1916 and 1943, a slightly negative slope between 1944 and 1968 and a positive slope between 1969 and 2000 (see Fig. 2). In all there is an increase in the global temperature between 1850 and today of about 0.8°C. Note that year 1850 signals the beginning of systematic recordings of temperatures.

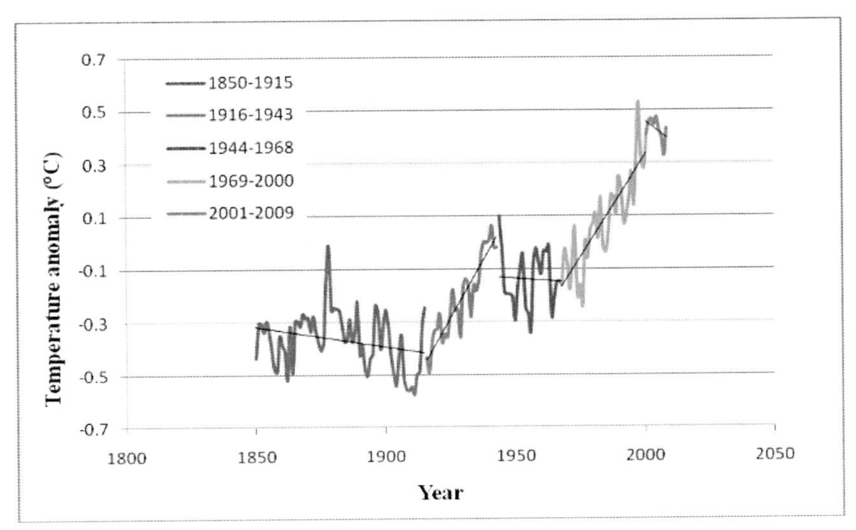

Fig. 2. Global temperature anomaly since 1850 exhibiting differing trends.

Still, a 150 year-span is not sufficient to convincingly answer the question of how temperature varies with time. It is unavoidable that only indirect methods can give insight to this matter. Such methods include extracting temperature information from tree rings, measuring borehole temperatures, the use of pollen and/or diatoms, measuring cave layer thickness, obtaining speleotherm data, collecting stalagmite oxygen isotope data, obtaining $^{\delta 18}O/Mg/Ca$ data and so forth. As expected, the methods are not perfect and assessment of the data requires great knowledge of all possible affecting factors.

One such series of data, studying pre-1850 temperatures, has been used by the Intergovernmental Panel on Climate Change (IPCC) of the United Nations in their Fourth Assessment Report (2007a, p. 467). Their graph, reproduced here in Fig. 3, shows the various

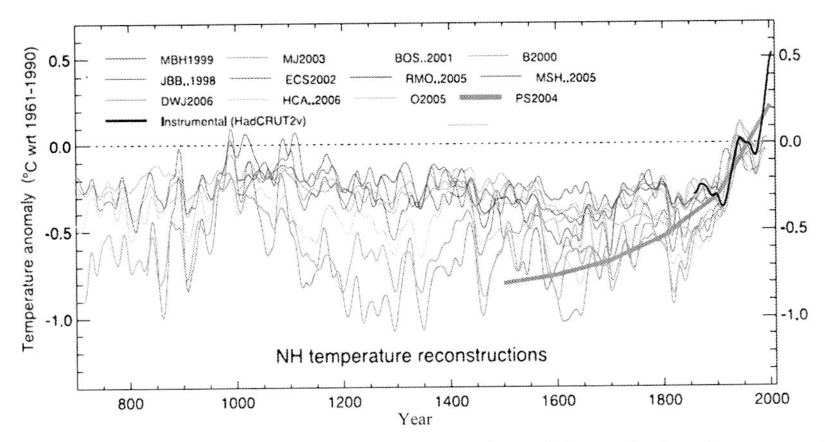

Fig. 3. IPCC's (2007) instrumental and proxy climate data of the variations in average large-scale surface temperatures over the last 1300 years for the North hemisphere.

instrumental and proxy climate data of the variations in average large-scale surface temperatures over the last 1300 years for the North hemisphere. Fig. 3 clearly shows that the warming experienced in the last decades of the 20th century is prominent and it had never been surpassed during the studied time period.
The temperature graphs presented by the IPCC (especially the "hockey stick" reconstruction of Mann et al., 1999) have been severely criticized (see for example Soon & Baliunas, 2003; McIntyre & McKitrick, 2003; the National Academy of Sciences (NAS) Report, 2006; the Wegman Report, 2006; McIntyre, 2008). It has been independently ascertained that Mann's PC method produces unreal "hockey-stick" shapes. It is worth inspecting McIntyre & McKitrick's (2010) reconstruction of the MBH98 data series of Fig. 3. After a critical analysis of the data the reconstruction presents high early 15th century values, as shown in Fig. 4. In recent years there have been numerous publications examining and updating old proxy data and improving the statistical methods for extracting more reliable results for the past temperature variation. Two such studies are presented below.

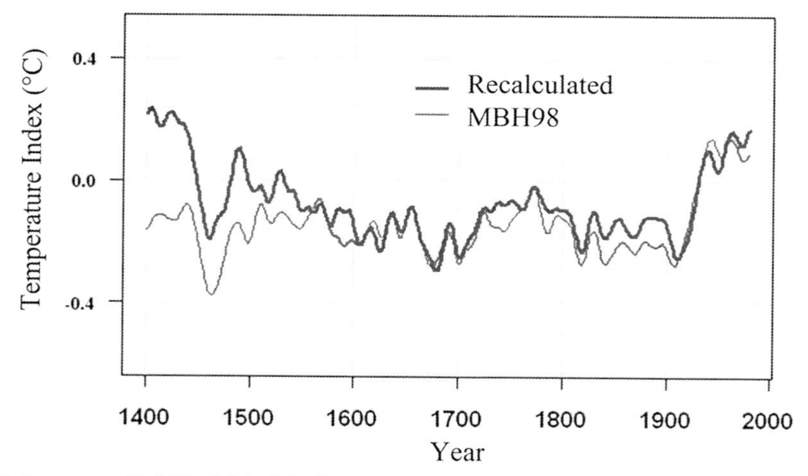

Fig. 4. McIntyre and McKitrick's (2010) reconstruction of the MBH98 data series.

(a) Grudd (2008) presented updated tree-ring width (TRW) and maximum density (MXD) from Tornetrask in northern Sweden, covering the period AD 500-2004. This reconstruction is compared with two previously published temperature reconstructions based on tree-ring data from Tornetrask (Fig. 5). The red curve is from Briffa et al. (1992) based on TRW and MXD. These data subsequently found their way in various other multiproxy reconstructions. The hatched curve is from Grudd et al. (2002) and is based on TRW. The three reconstructions were equally smoothed with a 100-year spline filter and have AD 1951-1970 as a common base period. Fig. 5 also shows the Grudd (2008) revised Tornetrask MXD low frequency reconstruction of April-August temperatures (blue curve), with a 95% confidence interval (grey shading). Grudd mentions that the updated data enable a much improved reconstruction of summer temperature for the last 1500 years in northern Fennoscandia and concludes that the late-twentieth century is not exceptionally warm in the updated Tornetrask record. On decadal-to-century timescales, periods around AD 750, 1000, 1400 and 1750 were all equally warm, or warmer. The warmest summers in this reconstruction occur in a 200-year period centered on AD 1000. A "Medieval Warm Period" is supported by other paleoclimate evidence from northern Fennoscandia, although the new tree-ring evidence from Tornetrask suggests that this period was much warmer than previously recognized.

(b) - Loehle & McCulloch (2008) presented a reconstruction using data that largely excluded tree-ring records to investigate the possible effect of proxy type on reconstruction outcome (Fig. 6). Again confidence intervals were computed for more robust evaluation of the results.

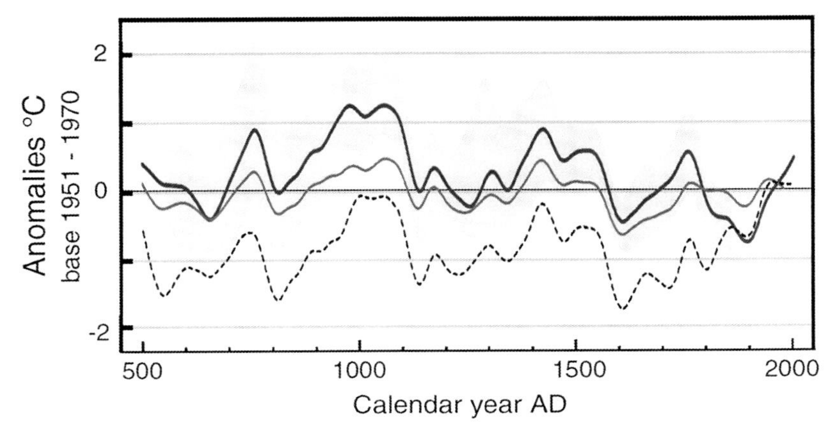

Fig. 5. Reconstructions by Grudd (2008) (blue curve) with a 95% confidence interval (grey shading) compared to Briffa et al. (1992) (red curve) and Grudd et al. (2002) (hatched curve).

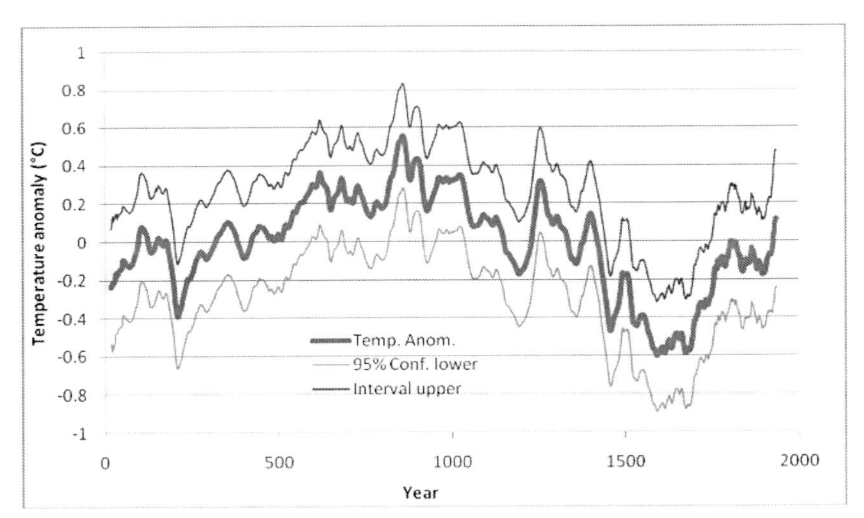

Fig. 6. Loehle & McCulloch (2008) temperature reconstruction.

The obtained data are for long series that had been previously calibrated and converted to temperature by their respective authors. All data that were used had at least 20 dates over

the 2000-year period. The final results were smoothed (29-year running mean), therefore peaks and troughs are damped compared to annual data. Thus it is not possible to compare recent annual data to this figure and ask about anomalous years or decades. The results continue to show the Medieval Warm Period (MWP) and the Little Ice Age (LIA) quite clearly. The 95% confidence intervals indicate that the MWP was significantly warmer than the bimillennial average during most of approximately 820-1040 AD. Likewise, the LIA was significantly cooler than the bimillennial average during most of approximately 1440-1740 AD. The peak value of the MWP is 0.526°C above the mean over the period (again as a 29-year mean, not annual, value). This is 0.412°C above the last reported value in 1935 (which includes data through 1949) of 0.114°C. The main significance of these results is the overall picture of the 2000 year pattern showing the MWP and LIA timing and curve shapes.
Finally, observing the temperature fluctuation on an even larger scale, we present the results derived from the European Project for Ice Coring in Antarctica (EPICA, 2010). EPICA is a multinational European project for deep ice core drilling. Its main objective is to obtain full documentation of the climatic and atmospheric record archived in Antarctic ice by drilling and analyzing two ice cores and comparing these with their Greenland counterparts. The site of Concordia Station, Dome C, was chosen to obtain the longest undisturbed chronicle of environmental change, in order to characterize climate variability over several glacial cycles. Drilling, reaching a depth of 3270.2 m, 5 m above bedrock, was completed in December 2004. Presenting the results of the project above, Jouzel et al. (2007) mention that a high-resolution deuterium profile is now available along the entire European Project for Ice Coring in Antarctica Dome C ice core, extending this climate record back to marine isotope stage 20.2, about 800000 years ago.
The general correspondence between Dansgaard-Oeschger events and their smoothed Antarctic counterparts for this Dome C record were assessed and revealed the presence of such features with similar amplitudes during previous glacial periods. It was suggested that the interplay between obliquity and precession accounts for the variable intensity of interglacial periods in ice core records. Fig. 7, presents the temperature variation ΔT for the

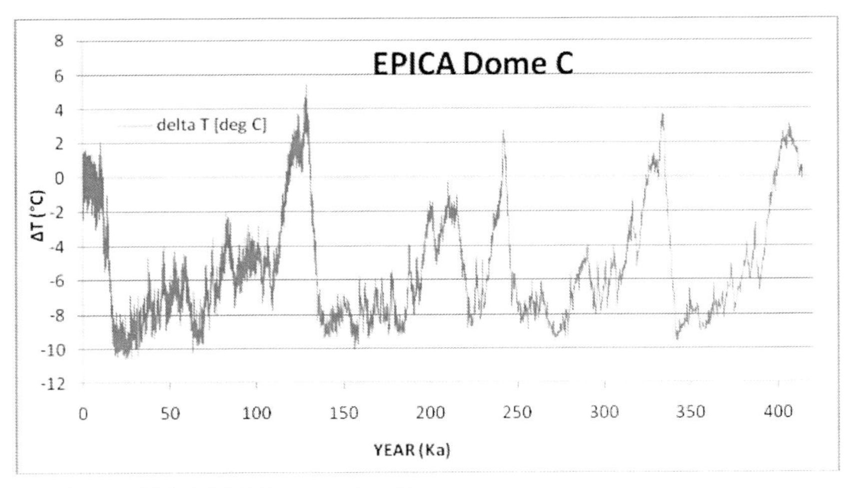

Fig. 7. Jouzel et al. (2007) EPICA reconstruction.

past 420000 years indicating that nowadays we are at a pick point (at about 1.5-2°C) and that the temperature was higher during all four previous recorded picks (128 ka at about 5°C, 242 ka at about 2.4°C, 333 ka at about 3.7°C and 405 ka at about 2.6°C).
We believe that the analysis presented above clearly shows that the mean environmental temperature of the planet has varied continuously through time and there is probably nothing unusual to last century's 0.7°C increase of temperature.

3. The greenhouse effect

According to IPCC (2007b, p. 144) the Sun powers the climate of the Earth, radiating energy at very short wavelengths. Roughly one-third of the solar energy that reaches the top of Earth's atmosphere is reflected directly back to space. The remaining two-thirds are absorbed by the surface and, to a lesser extent, by the atmosphere. The Earth balances the absorbed incoming energy, by radiating on average the same amount of energy back to space at much longer wavelengths primarily in the infrared part of the spectrum. Much of this emitted thermal radiation is absorbed by the atmosphere and clouds, and is reradiated back to Earth warming the surface of the planet. This is what is called the greenhouse effect.
The natural greenhouse effect makes life as we know it possible because without it the average temperature at the Earth's surface would be below the freezing point of water. However, as IPCC maintains, human activities, primarily the burning of fossil fuels and clearing of forests, have greatly intensified the natural greenhouse effect, causing global warming. The greenhouse effect comes from molecules that are complex and much less common, with water vapor being the most important greenhouse gas (GHG), and carbon dioxide (CO2) being the second-most important one.
In regions where there is water vapor in large amounts, adding a small additional amount of CO2 or water vapor has only a small direct impact on downward infrared radiation. However, in the cold and dry Polar Regions the effect of a small increase in CO2 or water vapor is much greater. The same is true for the cold and dry upper atmosphere where a small increase in water vapor has a greater influence on the greenhouse effect than the same change in water vapor would have near the surface. Adding more of a greenhouse gas, such as CO2, to the atmosphere intensifies the greenhouse effect, thus warming the Earth's climate.
The amount of warming depends on various feedback mechanisms. For example, as the atmosphere warms up due to rising levels of greenhouse gases, its concentration of water vapor increases, further intensifying the greenhouse effect. This in turn causes more warming, which causes an additional increase in water vapour in a self-reinforcing cycle. This water vapor feedback may be strong enough to approximately double the increase of the greenhouse effect due to the added CO2 alone.
IPCC (2007b, p. 121) reports that climate has changed in some defined statistical sense and assumes that the reason for that is anthropogenic forcing. As it states, traditional approaches with controlled experimentation with the Earth's climate system is not possible. Therefore, in order to establish the most likely causes for the detected change with some defined level of confidence, IPCC uses computer model simulations that demonstrate that the detected change is not consistent with alternative physically plausible explanations of recent climate change that exclude important anthropogenic forcing. The results of the computer simulations are that anthropogenic CO2 emissions to the atmosphere are the main reason for the observed warming and that doubling the amount of CO2 in the atmosphere will increase the temperature by about 1.5°C to 4.5°C. A similar result is mentioned in IPCC

(2007c, p. 749), where the equilibrium global mean warming for a doubling of atmospheric CO2, is likely to lie in the range 2°C to 4.5°C, with a most likely value of about 3°C.

In IPCC (1997, p. 11) the formula for calculating the radiative forcing for a CO2 doubling gives 4.0–4.5 W×m^{-2} before adjustment of stratospheric temperatures. Allowing for stratospheric adjustment reduces the forcing by about 0.5 W×m^{-2}, to 3.5–4.0 W×m^{-2}. If temperature were the only climatic variable to change in response to this radiative forcing, then the climate would have to warm by 1.2°C in order to restore radiative balance. The new formula for radiative forcing in W×m^{-2} is given as:

$$\Delta Q = 4.37 \frac{\ln C(t) - \ln C_0}{\ln 2} \tag{1}$$

where C(t) is today's CO2 concentration and C_0 the preindustrial level of 285 ppmv. As seen in Fig. 8, for the present CO2 concentration (385 ppmv) the warming calculated by the above-mentioned formula is 0.6°C, i.e. all the warming occurring from preindustrial era is allocated to the CO2 increase. Also, note that formula (1) above would give an increase of 1.2°C for the doubling of the CO2 concentration to 570 ppmv. In addition, the IPCC models consider a positive feedback because of this increase and depending on the model the final result is between 2°C and 4.5°C for a doubling of atmospheric CO2.

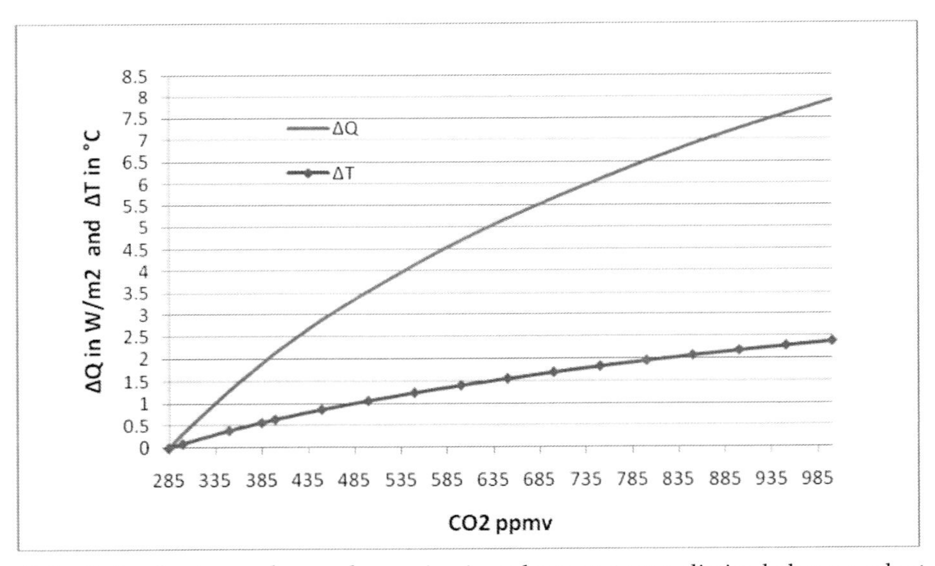

Fig. 8. Radiative forcing and caused warming in order to restore radiative balance evaluated by IPCC with no amplification considered.

As mentioned by the Committee on the Science of Climate Change-National Research Council (2001, p. 5), "the central value of 3°C is an amplification by a factor of 2.5 over the direct effect of 1.2°C. Well-documented climate changes during the history of Earth, especially the changes between the last major ice age (20000 years ago) and the current warm period, imply that the climate sensitivity is near the 3°C value. However, the true climate sensitivity remains uncertain, in part because it is difficult to model the effect of feedback. In particular, the magnitude and even the sign of the feedback can differ according to the composition, thickness, and altitude of the clouds, and some studies have suggested a lesser climate sensitivity." Also on p. 15 of the same book, it is stated that "climate models calculate outcomes after taking into account the great number of climate variables and the complex interactions inherent in the climate system. Their purpose is the creation of a synthetic reality but although they are the appropriate high-end tool for forecasting hypothetical climates in the years and centuries ahead, climate models are imperfect. Their simulation skill is limited by uncertainties in their formulation, the limited size of their calculations, and the difficulty of interpreting their answers that exhibit almost as much complexity as in nature."

4. How much of the global warming is caused by CO2?

4.1. Physical observations

Assuming that the above-mentioned theory of IPCC on CO2 concentration is correct, then one should expect a strong relation between CO2 concentration in the atmosphere and global temperature increase.

Plotting the CO2 concentration and temperature anomaly over the last 40 years (Fig. 9) one can observe that as from 2001 the relation that existed since 1969 has now deviated and although the CO2 concentration is still increasing as before, the temperature has slightly decreased.

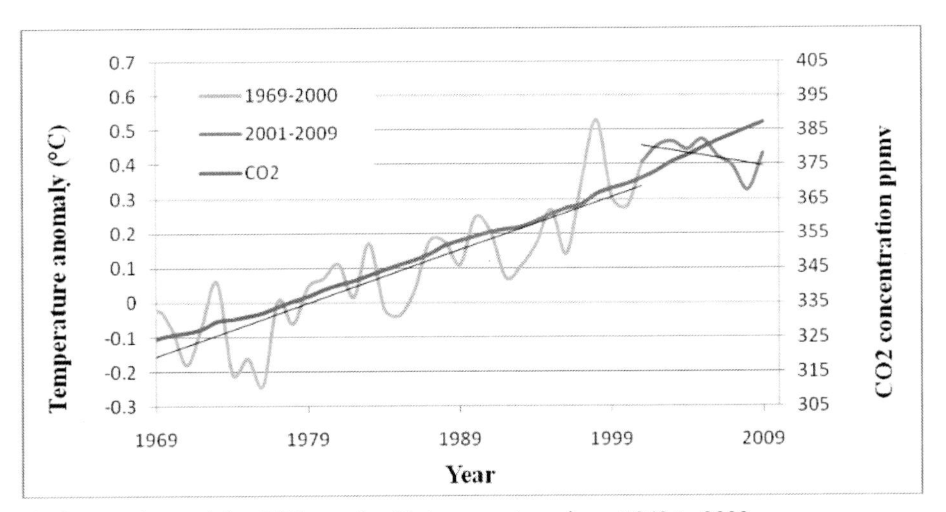

Fig. 9. Comparison of the CO2 trend with temperature from 1969 to 2009. (Temperature data from: Met Office Hadley Centre, 2010. CO2 data from: Mauna Loa CO2 annual mean data, 2010).

As before (recall Fig. 2), plotting the temperature anomaly along with CO2 concentration since year 1850, one cannot avoid observing that from 1850 to 1915 there was an opposite trend with the temperature cooling, from 1916 to 1943 the trend reversed as the temperature was increasing at a high rate and again from 1944 to 1968 when the CO2 was accumulating at an increased rate the temperature was not increasing (see Fig. 10).

Let us now compare the temperature data and the CO2 variation during greater time spans in order to obtain a deeper insight on how the CO2-concentration change affects the temperature.

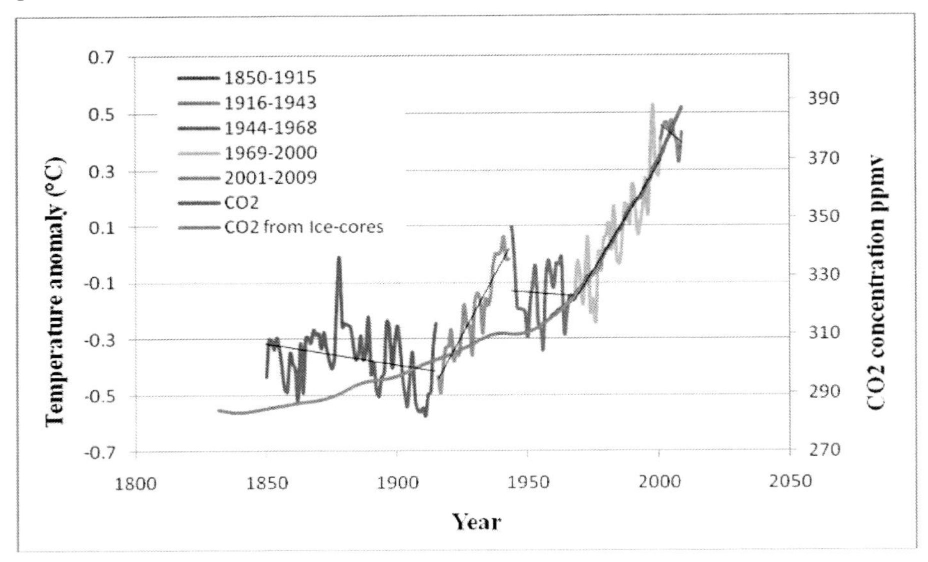

Fig. 10. Comparison of the CO2-concentration trend with temperature anomaly since 1850. (Temperature data from: Met Office Hadley Centre, 2010. CO2 data from: Mauna Loa CO2 annual mean data, 2010 and Historical CO2 record, 1998).

In Fig. 11 the temperature difference in Antarctica (as measured in ice cores by Jouzel et al., 2007) is compared to various CO2 concentrations: Petit et al. (1999) for the past 420000 years from the Vostok ice cores, Monnin et al. (2004) for the High resolution records of atmospheric CO2 concentration during the Holocene as obtained from the Dome Concordia and Dronning Maud Land ice cores, and others. It is clear (see circled points) that the temperature increase by natural causes precedes the CO2-concentration increase.

In fact, concentrating on the period between 400-650 thousand years before present (see Fig. 12) it is even clearer that very frequently CO2-concentration increase (blue circles) follows the temperature-increase that takes place many thousand years in advance (orange circles). Actually only in the era between 580-600 thousand years before now (red circles), did the CO2-rise precede the temperature-rise. The behavior described by Figs. 11 and 12 could possibly show that physical phenomena like the degassing/dilution of CO2 in the oceans, biological effects (plan growth and microbial activity) and so forth, may be the reason for the CO2 change following the temperature fluctuation and not the other way round.

Let us finally check how CO2-concentration has fluctuated throughout the Earth's history and draw conclusions about its correlation with the temperature over the geologic aeons. Palaeo-climatologists calculated palaeolevels of atmospheric CO2 using the GEOCARB III model (Berner & Kothavala, 2001). GEOCARB III models the carbon cycle on long time-scales (million years resolution) considering a variety of factors that are thought to affect the CO2 levels. The results are in general agreement with independent values calculated from the abundance of terrigenous sediments expressed as a mean value in 10 million year time-steps (Royer, 2004).

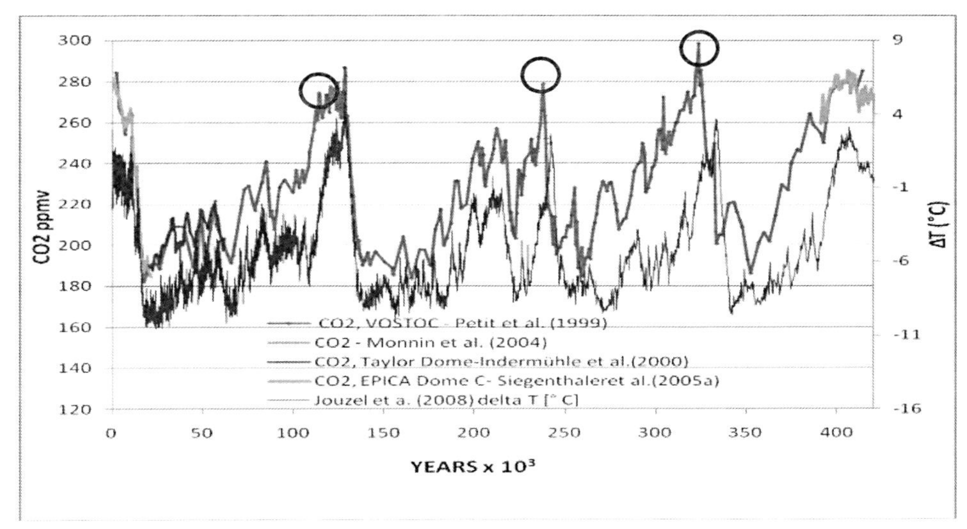

Fig. 11. Circled points indicate CO2-concentration increase following the temperature increase by natural causes.

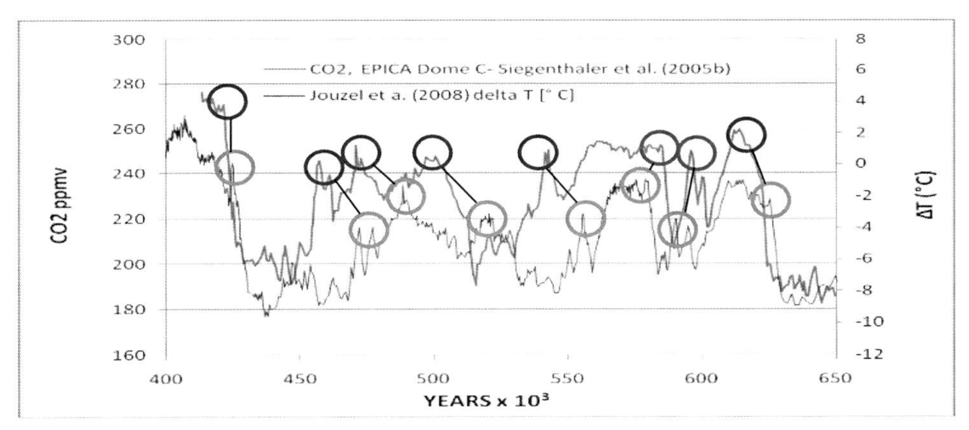

Fig. 12. CO2-concentration increase sometimes follows temperature-increase (blue circles) and sometimes precedes the temperature-rise (red circles).

As shown in Fig. 13A, CO2 levels were very high, about 20-26 times higher than at present, during the early Palaeozoic - about 550 million years ago (Ma). Then a large drop occurred during the Devonian (417–354 Ma) and Carboniferous (354–290 Ma), followed by a considerable increase during the early Mesozoic (248–170 Ma). Finally, a gradual decrease occurred during the late Mesozoic (170–65 Ma) and the Cainozoic (65 Ma to present). In Fig. 13B, C and D the range of global temperature through the last 500 million years is reconstructed. Figure 13B presents the intervals of glacial (dark color) and cool climates (dashed lines). Figure 13C shows the estimated temperatures, drawn to time-scale, from mapped data that can determine the past climate of the Earth (Scotese, 2008). These data include the distribution of ancient coals, desert deposits, tropical soils, salt deposits, glacial material, as well as the distribution of plants and animals that are sensitive to climate, such as alligators, palm trees and mangrove swamps. Figure 13D presents the temperature deviations relative to today from $\delta^{18}O$ records (solid line) and the temperature deviations corrected for *p*H (dashed line).

As indicated in Figure 13B, one of the highest levels of CO2-concentration (about 16 times higher than at present) occurred during a major ice-age about 450 Ma, indicating that it is not the CO2-concentration in the atmosphere that drives the temperature. The logical conclusion drawn from Fig. 13 is that the temperature of the Earth fluctuates continuously and the CO2-concentration is not a driving factor.

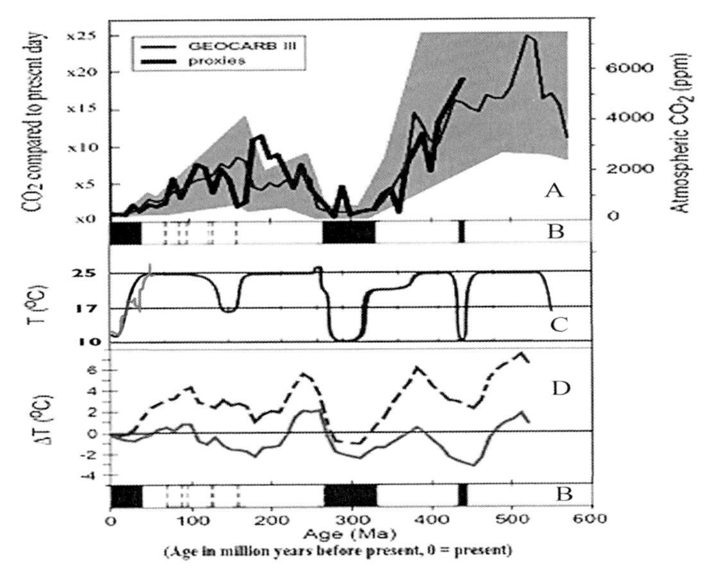

Fig. 13. (A) GEOCARB III model results with range in error shown for comparison with combined atmospheric CO2-concentration record as determined from multiple proxies in average values in 10 Ma time-steps, (redrawn from Royer, 2004). (B) Intervals of glacial (dark color) and cool climates (lighter color) (redrawn from Royer, 2004). (C) Estimated temperature drawn to time scale (Scotese, 2008). (D) Temperature deviations relative to today (solid line - Shaviv and Veizer, 2003) from the "10/50" $\delta^{18}O$ compilation presented in Veizer et al. (2000) and temperature deviations corrected for *p*H (dashed line) reconstructed in Royer (2004) and redrawn from Veizer et al. (2000).

4.2 Physical observations of glacier melting

Huss et al. (2008) determined the seasonal mass balance of four Alpine glaciers in the Swiss Alps (Grosser Aletschgletscher, Rhonegletscher, Griesgletscher and Silvrettagletscher) for the 142-year period 1865–2006. They report that the cumulative mass balance curves show similar behavior during the entire study period, the mass balances in the 1940s were more negative than those of 1998–2006 and the most negative mass balance year since the end of the Little Ice Age was 1947 and not the year 2003 despite its exceptional European summer heat wave. As correctly argued in the NIPCC (2009, p. 145) and shown on the redrawn Fig. 14, the most important observation is the fact that the rate of shrinkage has not accelerated over time, as evidenced by the long-term trend lines they have fit to the data. There is no compelling evidence that this 14-decade-long glacial decline has had anything to do with the air's CO2 concentration.

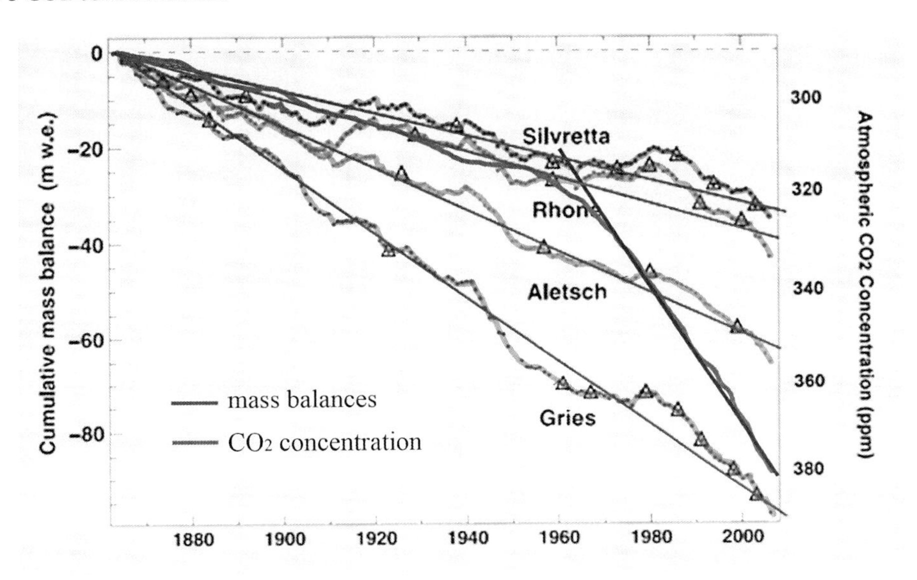

Fig. 14. Time series of the cumulative mass balances of the four Swiss Alps glaciers of Huss et al. (2008). CO2 increase in air concentration shows no inverse effect on their melting rate (redrawn from NIPCC, 2009).

As it is stressed, from 1950 to 1970 the rate-of-rise of the atmosphere's CO2 concentration increased by more than five-fold, yet there were no related increases in the long-term mass balance trends of the four glaciers. It is clear that the ice loss history of the glaciers was not unduly influenced by the increase in the rate-of-rise of the air's CO2 concentration that occurred between 1950 and 1970, and that their rate of shrinkage was also not materially altered by what the IPCC calls the unprecedented warming of the past few decades.

A similar argument to the one above can be applied to a worldwide study of 169 receding glaciers of Oerlemans (2005). Fig. 15 shows the composite average of up to 169 glaciers (the number varies in different time periods) indicating the pattern that is consistent for most glaciers. Exactly as in Fig. 14 for the four Alpine glaciers in the Swiss Alps, the recession of glaciers started long before anthropogenic CO2 levels rose, and naturally there is no indication that since 1970, when the anthropogenic CO2 began increasing at a higher rate, the recession rate of the glaciers has increased.

4.3. A chemist view: vibrational modes and emission spectra

According to Barrett (2005) greenhouse molecules absorb terrestrial radiation, which is emitted by the Earth's surface as a result of the warming effect of incoming solar radiation. Their absorption characteristics allow them to act in the retention of heat in the atmosphere

increasing the global mean temperature. The absorption characteristics of CO_2 depend on the form of the molecule, which is linear and symmetrical about the central carbon atom.
The three vibrational modes of the molecule and their fundamental wave numbers are the symmetric stretch at 1388 cm^{-1}, the antisymmetric stretch at 2349 cm^{-1}, and the bend at 677 cm^{-1}. The CO_2 spectrum is dominated by the bending vibration, centered at 667 cm^{-1}.
As calculated in Barrett (2005), the contributions to the absorption of the Earth's radiance by the first 100 meters of the atmosphere for the pre-industrial CO_2 concentration (285 ppmv) are 68.2% for the water vapor, 17% for CO_2, 1.2% for CH_4 and 0.5% for N_2O. These absorption values add up to 86.9%, which is significantly higher than the actual resulting combined value of 72.9%. This discrepancy occurs because there is considerable overlap between the spectral bands of water vapor and those of the other GHGs. If the concentration of CO_2 were to be doubled in the absence of the other GHGs, the increase in absorption would be 1.5%. But in the presence of the other GHGs the same doubling of CO_2's concentration would yield an increase in absorption of only 0.5%.

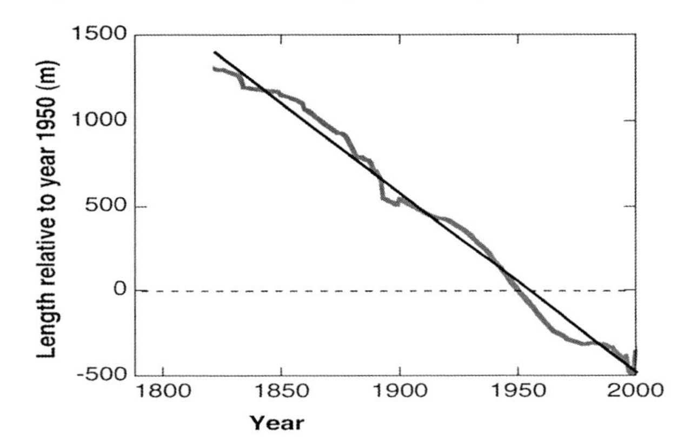

Fig. 15. Curve for the change in mean global glacier length. CO_2-increase in air concentration especially after 1970 shows no inverse effect on their melting rate (modified from Oerlemans, 2005).

As far as temperature-rise is concerned, Barrett & Bellamy (2010) explain that by using the MODTRAN program (which is a state of the art atmospheric transfer code used as a basic tool of research) they compute an increase of 1.5 K, resulting from doubling the pre-industrial CO_2 concentration.
The GHGs absorb 72.9% of the available radiance, leaving 27.1% of it to be transmitted, of which, 22.5% passes through the window, leaving a small amount of 4.6% to be transmitted by the other parts of the spectral range. For the doubled CO_2 case this small percentage decreases slightly to 4.1%. The above-mentioned small percentage transmissions (4.6 and 4.1%) are further reduced by 72.9% and 73.4% respectively, by the second layer of 100 m of the atmosphere so that only about 1%, in both cases, is transmitted to the region higher than 200 m.
Moreover Barrett (2005), states that the 19.6% of pre-industrial CO_2 contribution to the greenhouse effect are responsible for a 6.7 K temperature rise. Doubling of the CO_2 concentration will increase its contribution to 20.9% (which will, by simple analogy, correspond to a 0.44 K additional temperature rise), but at the same time the water vapor contribution will diminish from 78.5 to 77.1%.

To support his results Barrett presents an analysis of three emission spectra of the Earth as recorded by satellites (Fig. 16). As explained from 400–600 cm^{-1} the spectra consist of rotational transitions of water molecules, while the regions from 600–800 cm^{-1} are dominated by the main bending mode of CO2 and some combination bands that are of smaller intensity, together with some of the water rotation bands which overlap with the CO2 bands. Between 800 and 1300 cm^{-1} the spectra correspond to IR window regions with some ozone absorption and emission spectra at around 1043 cm^{-1}, which essentially demonstrate the temperature of the surface when compared to the Planck emission spectra that are incorporated into each spectrum. The Saharan surface temperature is around 320 K (spectrum a), that of the Mediterranean is around 285 K (spectrum b) and that of the Antarctica is around 210 K (spectrum c). Above 1300 cm^{-1}, the spectra consist of vibration-rotation bands of water molecules and the bending vibrations of methane and dinitrogen monoxide molecules.

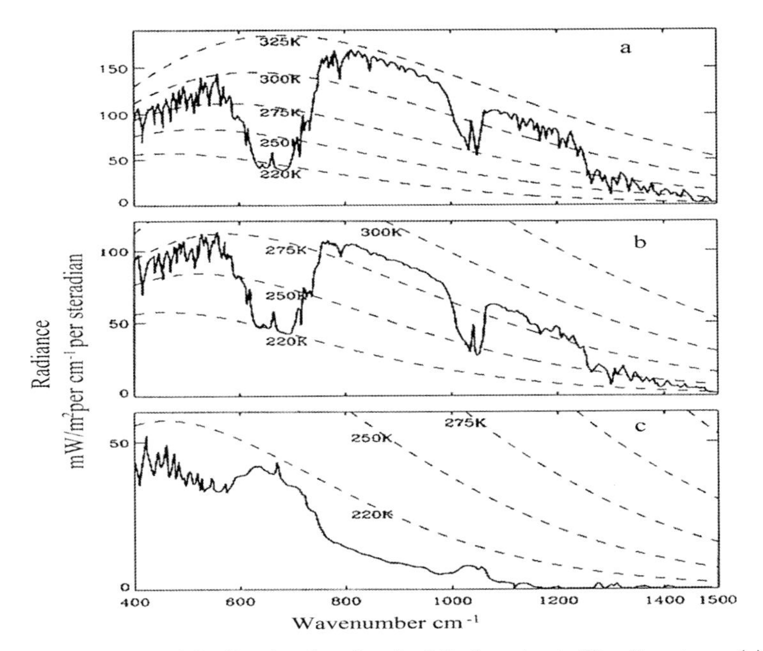

Fig. 16. Emission spectra of the Earth taken by the Nimbus 4 satellite. Spectrum (a) is measured over the Sahara Desert, spectrum (b) over the Mediterranean, and spectrum (c) over the Antarctic

The influence of CO2 can be seen from the Saharan and Mediterranean spectra as absorbing all of the radiation in the 600–800 cm^{-1} regions and emitting about 25% of it. The large absorption areas in the two spectra show how CO2 is an important GHG. Spectrum (c) shows that emission by CO2 in the Polar Regions is relatively more important and arises because of the low water vapor pressure. A detailed estimate of the effects of the GHGs from the analysis of spectra such as those shown in Fig. 16 indicates that CO2 provides about 7–8 K of global warming, in agreement with the conclusion yielding from a study of its absorption characteristics.

Fig. 16 shows that much of the CO_2 emission originates from the atmosphere at a temperature of about 218 K. This part of the atmosphere is at an altitude of about 15 km (tropopause) and is the dividing layer between the troposphere and the stratosphere. Emission from the CO_2 occurs at that level because the air is 'thin', as around 90% of the atmosphere is below that level. The doubling of CO_2 concentration will have as a result the widening of the wings of the saturated band where the additional absorption will occur. It is stressed though (Barrett & Bellamy, 2010) that this increase represents the instantaneous effects of CO_2 changes and it does not include the ameliorating effects of clouds, nor does it include the eventual global consequences of the instantaneous changes.

4.4 A Geophysicist view

The 1970 radiative spectrum from the Earth, measured by Nimbus 4 satellite through clear skies at several locations, clearly shows a deep 'notch' at the 4.77 micron wavelength band caused by the 325 ppmv atmospheric CO_2 concentration of the time (Fig. 16 & Fig. 17). The depth and width of this 'notch' demonstrate that over 90% of the Earth's thermal radiation from this wavelength band that could possibly be affected by CO_2, had already been affected at that concentration (of just 325ppmv). As Karmanovitch & Geoph (2009) mention, we now know that about three quarters of the Earth's 34 K total greenhouse effect is from clouds and only 10% of the effect is from CO_2. Ten per cent of 34 K is 3.4 K and this is exactly the total effect that has resulted from the observed notch in the spectrum from CO_2 as measured by the Nimbus 4 satellite. Since this 3.4 K effect results from 90% of the available energy within this wavelength band, the energy remaining in this band is only capable of adding another 10% to the 3.4 K greenhouse effect already in place. Regardless of how great the concentration of CO_2 in the atmosphere becomes, there is only 10% of the available energy left to capture and the possible additional effect from CO_2 increases is therefore limited to something in the order of just 0.34 K, which is nowhere near the 5–6 K predicted by Arrhenius (the first to provide a simplified expression, still used today, linking the CO_2 concentration to the temperature increase). This result alone clearly falsifies the equation and the numerical values used to determine the forcing parameter of the climate models that support the anthropogenic global warming (AGW) hypothesis of IPCC.
Recent measurements (Karmanovitch & Geoph, 2009) of the thermal radiation spectrum from Mars reveal a spectral notch that is virtually identical on both the 1970 Earth spectra with a 325ppmv (see Fig. 17). Mars has a very thin atmosphere consisting of 95% CO_2, virtually zero water vapor and 5% O_2, N_2 and Ar. So, CO_2 is essentially the only "GHG' with a concentration of 950000 ppmv, being 9 times that of the Earth's in absolute terms. Again, using the formula of the IPCC one should expect, for Mars, a spectral notch from CO_2 that represents an increase in forcing analogous to 9 times difference in CO_2, i.e. a value of 11.755 $W \times m^{-2}$. This is not at all the case, as shown if Fig. 17, indicating that there is virtually no effect increases in CO_2 beyond 325ppmv.

4.5 A planetary climatologist view: a comparison of the CO2 greenhouse effect on Mars, Earth and Venus

As Idso (1988) mentions, in two separate assessments of the magnitude of the CO_2 greenhouse effect, the U.S. National Research Council concluded that the likely consequence of a 300–600 ppmv doubling of the Earth's atmospheric CO_2 concentration would be a 3 ± 1.5 K increase in the planet's mean surface air temperature.

Idso (1988), presented a comparison for the CO2 greenhouse effect on Mars, Earth and Venus by plotting the CO2 warming and the CO2 atmospheric partial pressure on a log-log scale (Fig. 18). He concludes that considering the consistency of all empirical data, atmospheric CO2 fluctuations influence surface air temperature largely, independently of atmospheric moisture conditions, because water vapor is practically non-existent on Mars, is intermediate on Earth and large on Venus (in an absolute sense). Hence, the long-espoused claim of a many-fold amplification of direct CO2 effects by a positive water vapor feedback mechanism would appear to be rebuffed by the analysis. As a result, the final conclusion is that the scientific consensus on the strength of the CO2 greenhouse effect, as expressed in past reports of the U.S. National Research Council, is likely to be in error by nearly a full order of magnitude. Based on the comparative planetary climatology relationship of Fig. 18, a 300–600 ppmv doubling of Earth's atmospheric CO2 concentration should only warm the planet by about 0.4 K.

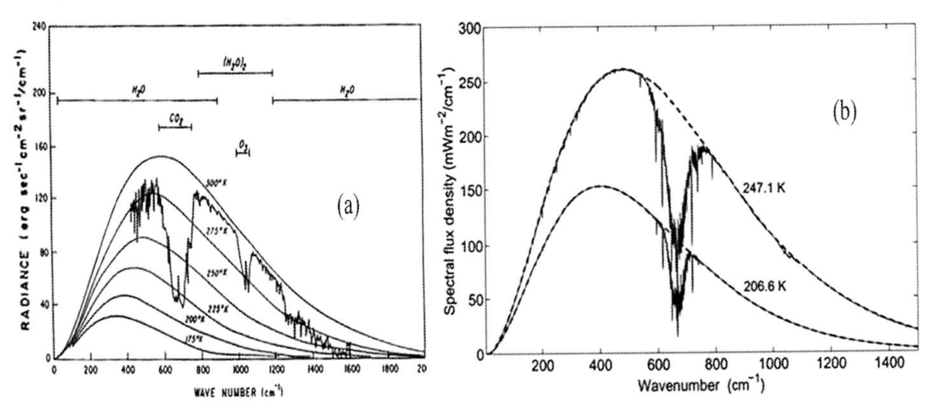

Fig. 17. Earth Thermal Radiative Spectrum (a) as measured by Nimbus 4 satellite in 1970, when the CO2 concentration was 325ppmv and Mars Thermal Radiative Spectrum (b) redrawn from Karmanovitch & Geoph, 2009.

Additionally Idso, (1998) in a worth noting review enumerates a number of analyses he performed on natural phenomena that reveal how the Earth's near-surface air temperature responds to surface radiative perturbations. All of these studies confirm that a 300–600 ppmv doubling of the atmospheric CO2 concentration could raise the planet's mean surface air temperature by only about 0.4 K. He even expresses doubts that even this modicum of warming will ever be realized, for it could be negated by a number of planetary cooling forces that are intensified by warmer temperatures and by the strengthening of biological processes that are enhanced by the same rise in atmospheric CO2 concentration that drives the warming. At the same time he is skeptical of the predictions of significant CO2-induced global warming that are being made by state-of-the-art climate models and believes that much more work on a wide variety of research fronts will be required to properly resolve the issue.

4.6 A view from a general science graduate

Archibald (2009), using the MODTRAN facility maintained by the University of Chicago, estimated the relationship between atmospheric CO2 concentration and increase in average global atmospheric temperature, concluding that anthropogenic warming is real but at the same time minute. Archibald used the temperature response, demonstrated by Idso (1998), of 0.1 K per W×m^{-2} as a base for his calculations. The effect of CO2 on temperature is logarithmic and thus climate sensitivity decreases with increasing concentration. The first 20 ppmv of CO2 has a greater temperature effect than the next 400 ppmv.

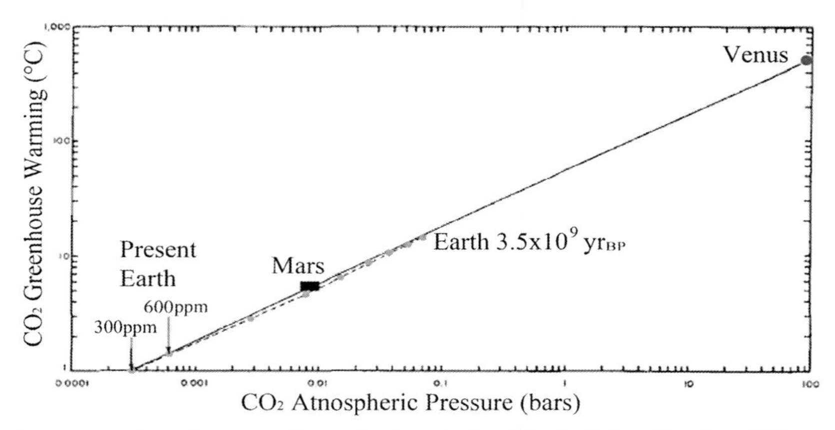

Fig. 18. A comparative planetary climatologist relationship for Mars, Earth and Venus based on the greenhouse warming of Mars and Venus, which are produced by their atmospheric partial pressures of CO2 (solid line). Also shown is the almost identical relationship derived from standard considerations related to the Earth's paleoclimatic record and the first early Sun paradox (dashed line) (redrawn from Idso, 1988).

The increase in atmospheric CO2 concentration from the pre-industrial level of 285 ppmv to the current level of 384 ppmv is calculated to have resulted in a 0.1 K rise in the atmospheric temperature. If the atmospheric CO2 level would increase to 600 ppmv, a further 0.3 K increase in temperature would be projected due to this factor (see Fig. 19).

In addition Archibald (2008), presents an interesting comparison of estimates of the effect that CO2 would have if its concentration in the atmosphere doubled from its pre-industrial level, and he concludes that the models of the IPCC apply an enormous amount of compounding water vapor feedback and, at their worst, the IPCC models take 1 K of heating and turn it into 6.4 K (see Fig. 20).

For Fig. 20 Archibald explains that:

a. The Stefan-Boltzmann figure of 1 K is based on the Stefan-Boltzmann equation without the application of feedbacks and as he comments everybody agrees with this figure when no feedbacks are involved.
b. Kininmonth estimates a 0.6 K and this is based on water vapor amplification but also includes the strong damping effect of surface evaporation.
c. Lindzen's estimation is based on water vapor and negative cloud feedback.
d. Idso derived an estimate of climate sensitivity from nature observations and Spencer used data from the Aqua satellite for his estimates and proved that these are very close to what happens in Nature.

In particular, what Spencer (2008) has done was to examine the satellite data in great detail, and then built the simplest model that can explain the observed behavior of the climate system whilst, as he explains, the currently popular practice is to build immensely complex and expensive climate models and then make only simple comparisons to satellite data. His main conclusion is that the net feedbacks in the real climate system are probably negative. A misinterpretation of cloud behavior has led climate modelers (of the IPCC) to build models in which cloud feedbacks are instead positive, which has led the models to predict too much global warming in response to anthropogenic greenhouse gas emissions.

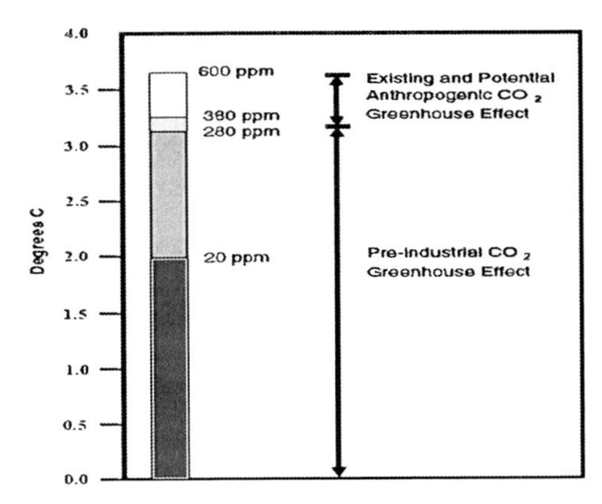

Fig. 19. Relative Contributions of Pre-Industrial and Anthropogenic CO2 (redrawn from Archibald, 2009).

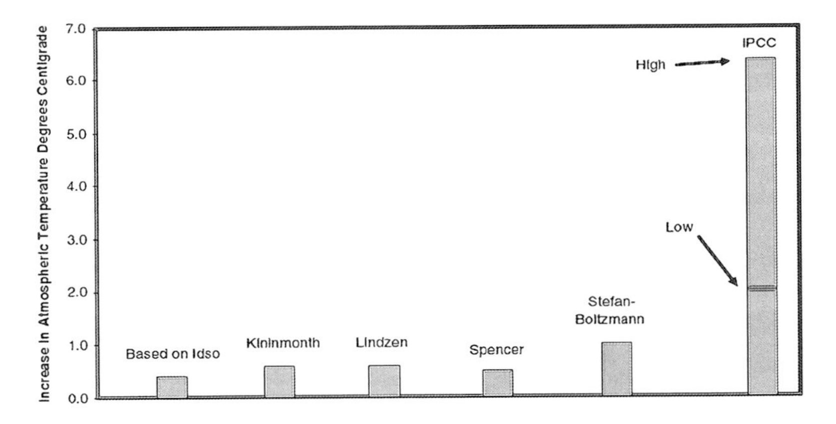

Fig. 20. Archibald's (2008) comparison of estimates for doubling the atmospheric CO2 from its pre-industrial level.

4.7 Unique fingerprint for anthropogenic warming

The concentration of water reduces very rapidly with altitude and therefore, the water contribution to the warming of the atmosphere is accordingly diminishing in comparison to that of CO2 as altitude increases. This is because the water vapor concentration is affected by temperature but CO2 concentration decreases only with the decreasing pressure. At sea

level the mean molecular ratio of water vapor to CO2 is around 23, but at an altitude of 10 km the value is as low as 0.2. It would be expected that more CO2 would have a greater effect on atmospheric warming at higher altitudes (Barret, 2005), but this seems not to be occurring in spite of the predictions of most greenhouse computer models (GCMs).
All Climate models predict that, if GHG is driving climate change, i.e. if the current global warming is anthropogenic, there will be a unique fingerprint in the form of a warming trend increasing with altitude in the tropical troposphere (the region of the atmosphere up to about 15 km). Climate changes due to solar variability or other known natural factors do not produce this pattern and only sustained greenhouse warming does. While all greenhouse models show an increasing warming trend with altitude, especially at around 10 km at roughly two times the surface value, the temperature data from balloons give the opposite result showing no increasing warming, but rather a slight cooling with altitude in the tropical zone (compare Figs. 21 and 22). The Climate Change Science Program executive summary (CCSP, 2006) tried inexplicably to reconcile this difference, claiming agreement between observed and calculated patterns, the opposite of what the report itself documents. The obvious disagreement shown in the body of the report is dismissed by suggesting that there might be something wrong with both balloon and satellite data (NIPCC, 2009, p. 106).
The main conclusion of the CCSP Report (CCSP, 2006, p. 118) is that many factors - both natural and human-related - have probably contributed to the climate changes. Analyses of observations alone cannot provide definitive answers because of important uncertainties in

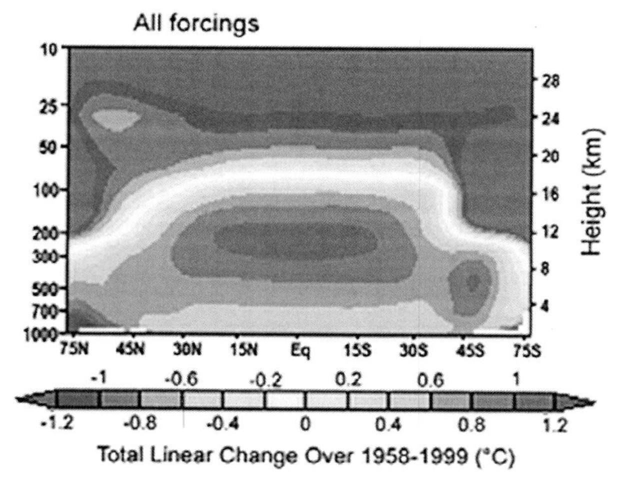

Fig. 21. Greenhouse-model-predicted mean atmospheric temperature change versus latitude and altitude for all forcings (CCSP, 2006, p. 25). The figure clearly shows the increased temperature trends in the tropical mid-troposphere (8–12 Km).

the observations and in the climate forcings that have affected them. Although computer models of the climate system are useful in studying cause-effect relationships, they, too, have limitations. Finally the Report notes that advancing our understanding of the causes of recent lapse-rate changes will best be achieved by comprehensive comparisons of observations, models, and theory - it is unlikely to arise from analysis of a single model or observational data set.

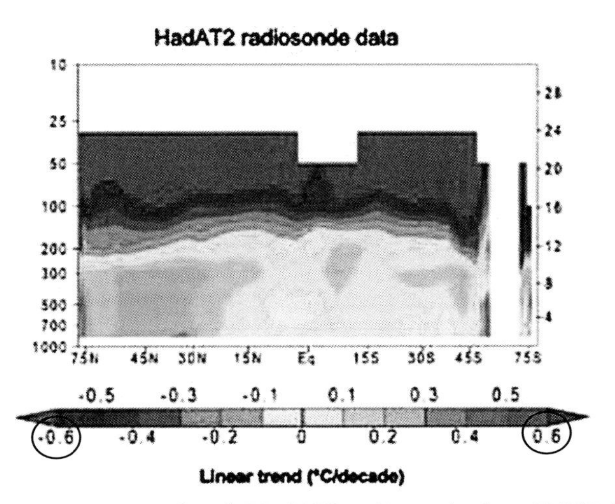

Fig. 22. Observed changes estimated with HadAT2 radiosonde data (CCSP, 2006, pp. 116). All temperature changes were calculated from monthly-mean data and are expressed as linear trends (in K/decade) over 1979 to 1999. Note that the colours used in this figure represent lower temperature increases when compared to figure 21 above.

4.8 A view from physics: the adiabatic theory of the greenhouse effect

As stressed by Sorokhtin et al. (2007), until recently a sound theory, using laws of Physics, for the greenhouse effect was lacking and all predictions were based on intuitive models using numerous poorly defined parameters. Their examination showed that at least 30 such parameters were contained in the models making the numerical solution of the problem incorrect. For this reason they devised a model based on well-established relationships among physical fields describing the mass and heat transfer in the atmosphere. Basic formulas describe, among others, the heat transfer in the atmosphere by radiation, the atmospheric pressure and air density change with elevation, the effect of the angle of the Earth's precession and the adiabatic process. For the adiabatic process the formula considers the partial pressures and specific heats of the gases forming the atmosphere, an adiabatic constant and corrective coefficients for the heating caused by water condensation in the wet atmosphere and for the absorption of infrared radiation by the atmosphere. The adiabatic constant and the heat coefficients are estimated using actual experimental data.

The convective heat transfer dominates in the troposphere (the lower and denser layer of the atmosphere, with pressures greater than 0.2 atm). When infrared radiation is absorbed by the GHGs, the radiation energy is transformed into oscillations of the gas molecules, heating the exposed volume of gaseous mixture. Then, further heat transfer can occur either due to diffusion or by convective transfer of expanded volumes of gas. Since the specific heat of air is very small, the rates of heat transfer by diffusion do not exceed several $cm \times s^{-1}$, whereas the rates of heat transfer by convection can reach many $m \times s^{-1}$. Analogous situation occurs upon heating of air as a result of water vapor condensation: the rates of convective transfer of heated volumes of air in the troposphere are many orders of magnitude higher than the rates of heat transfer by diffusion.

The adiabatic model was verified, with a precision of 0.1%, by comparing the results obtained for the temperature distribution in the troposphere of the Earth with the standard

model used worldwide for the calibration of the aircraft gauges and which is based on experimental data. The model was additionally verified with a precision of 0.5–1.0% for elevations of up to 40 km, by comparing the results with the measured temperature distribution in the dense troposphere of Venus consisting mainly of CO2.
The results of their analysis are the following:

a. Convection accounts for approximately 67% of the total amount of heat transfer from the Earth's surface to the troposphere, the condensation of water vapor for 25% and radiation accounts for only 8%. As the heat transfer in the troposphere occurs mostly by convection, accumulation of CO2 in the troposphere intensifies the convective processes of heat and mass transfer, because of the intense absorption of infrared radiation, and leads to subsequent cooling and not warming as believed.
b. If the nitrogen–oxygen atmosphere of the Earth was to be replaced by a CO2 atmosphere with the same pressure of 1 atm, then the average near-surface temperature would decrease by approximately 2.5 K and not increase as commonly assumed.
c. The opposite will happen by analogy if the CO2 atmosphere of Venus was to be replaced by a nitrogen–oxygen atmosphere at a pressure of 90.9 atm. The average near-surface temperature would increase from 462°C to 657°C. This is explained easily by observing how the results of the derived formulas are affected, considering that the molecular weight of CO2 is about 1.5 times greater and its specific heat 1.2 times smaller than those of the Earth's air.
d. If the CO2 concentration in the atmosphere increases from 0.035% to its double value of 0.070%, the atmospheric pressure will increase slightly (by 0.00015 atm). Consequently the temperature at sea level will increase by about 0.01 K and the increase in temperature at an altitude of 10 km will be less than 0.03 K. These amounts are negligible compared to the natural temporal fluctuations of the global temperature.
e. In evaluating the above consequences of the doubling of the CO2, one has to consider the dissolution of CO2 in oceanic water and also that, together with carbon, a part of atmospheric oxygen is also transferred into carbonates. Therefore instead of a slight increase in the atmospheric pressure one should expect a slight decrease with a corresponding insignificant climate cooling.

4.9 The verdict on the effect of CO2 on climate change

To examine the increase in the greenhouse effect in recent years corresponding to the CO2 concentration increase in the atmosphere, Harries et al. (2001), analyzed the difference between the spectra of outgoing longwave radiation (OLR) obtained in 1970 by the IRIS satellite and in 1997 by the IMG satellite of the Japanese Space Agency. The data utilized over a 26-year period showed a number of differences in the land-masked and cloud-cleared data, which the authors attributed to changes in atmospheric concentrations of CH4, CO2, O3, CFC-11 and CFC-12. Hence, they concluded that their results provided direct experimental evidence for a significant increase in the earth's greenhouse effect over the examined period.
In a related article Griggs & Harries (2007), observe the changes in the earth's spectrally resolved outgoing longwave radiation (OLR), comparing the outgoing longwave spectra from the 1970 (IRIS) the 1997 (IMG) and the 2003 (AIRS) instruments. As they mention, in all three difference spectra, the brightness temperature difference in the atmospheric window between 800–1000 cm^{-1} is zero or slightly positive, indicating a warming of surface temperatures between the three time periods. Although these observations are within the limit of uncertainty due to noise, the positive anomaly in this region seems to increase. This

is in agreement with the trend in sea surface temperatures. A negative brightness temperature difference is observed in the CO2 band at 720 cm^{-1} in the IMG–IRIS (1997–1970) and the AIRS–IRIS (2003–1970) difference spectra, indicating increasing CO2 concentrations, consistent with the Mauna Loa record. However, they note that this channel in the difference is also sensitive to temperature, and in the 2003–1997 difference, despite a growth in CO2 between these years, there is no signal at 720 cm^{-1}. As a key result of their analysis they state that "the CO2 band at 720 cm^{-1}, though asymmetric, nevertheless shows some interesting behavior, with strong negative brightness temperature difference features for 1997–1970 and 2003–1970: whereas, the 2003–1997 (a much shorter period, of course) shows a zero signature. Since we know independently that the CO2 concentration globally continued to rise between 1997 and 2003, we must conclude that the 2003–1997 result must be due to changes in temperature that compensate for the increase in CO2. This would mean a warming of the atmosphere at those heights that are the source of the emission in the center of this band. This is somewhat contrary to the general (small) cooling of the stratosphere at tropical latitudes."

Such a conclusion, however, does not provide direct experimental evidence for a significant increase in the Earth's total greenhouse effect. What the above-mentioned studies show is that for the cloud-free part of the atmosphere, there was a drop in outgoing radiation at the wavelength bands that greenhouse gases such as CO2 and methane (CH4) absorb energy. These results do not show the earth's climatic response to the inferred increase in radiative forcing and do not consider negative or positive feedback cycles.

Lindzen & Choi (2009), estimated climate feedbacks from fluctuations in the outgoing radiation budget from the Earth Radiation Budget Experiment (ERBE) nonscanner data in general and with all OLR wavelengths. For the entire tropics, the observed outgoing radiation fluxes increase with the increase in sea surface temperatures (SSTs). The observed behavior of radiation fluxes implies negative feedback processes associated with relatively low climate sensitivity. This is the opposite of the behavior of 11 atmospheric models forced by the same SSTs. Their conclusion is that the models display much higher climate sensitivity than is inferred from ERBE, though it is difficult to pin down such high sensitivities with any precision. Results also show that the feedback in ERBE is mostly from shortwave radiation while the feedback in the models is mostly from longwave radiation. Although such a test does not distinguish the mechanisms, this is important since the inconsistency of climate feedbacks constitutes a very fundamental problem in climate prediction.

Having in mind all the above it is doubtful if further increase in CO2 concentration will have any significant effect on the global temperature. IPCC is not sure at what extent the doubling of CO2 will affect the temperature (its guess is 2–4.5 K) and this is because the initial equation was derived empirically without any theoretical background. For this reason computer modelling, as used by IPCC, is merely a form of data curve fitting.

Our focus must now be turned to the main source of energy that drives the climate system, the radiation from the Sun. All meteorological phenomena are the results of this energy which is absorbed, transformed and redistributed on Earth. Therefore any phenomena observed in the climate change should be in a way related to this source.

IPCC (2007d), states that continuous monitoring of total solar irradiance now covers the last 28 years. The data show a well-established 11-year cycle in irradiance that varies by 0.08% from solar cycle minima to maxima, with no significant long-term trend. The primary

known cause of contemporary irradiance variability is the presence on the Sun's disk of sunspots and faculae. The estimated direct radiative forcing due to changes in the solar output since 1750 is +0.12 W×m^{-2}, which is less than half of the estimate given in the Third Assessment Report (TAR), with a low level of scientific understanding. While this leads to an elevation in the level of scientific understanding from very low in the TAR to low in this assessment, uncertainties remain large because of the lack of direct observations and incomplete understanding of solar variability mechanisms over long time scales.
The IPPC report then mentions that empirical associations have been reported between solar-modulated cosmic ray ionization of the atmosphere and global average low-level cloud cover, but evidence for a systematic indirect solar effect remains ambiguous. It has been suggested that galactic cosmic rays with sufficient energy to reach the troposphere could alter the population of cloud condensation nuclei and hence microphysical cloud properties (droplet number and concentration), inducing changes in cloud processes analogous to the indirect cloud albedo effect of tropospheric aerosols and thus causing an indirect solar forcing of climate. Studies have probed various correlations with clouds in particular regions or using limited cloud types or limited time periods; however, the cosmic ray time series does not appear to correspond to global total cloud cover after 1991 or to global low-level cloud cover after 1994. Together with the lack of a proven physical mechanism and the plausibility of other causal factors affecting changes in cloud cover, this makes the association between galactic cosmic ray-induced changes in aerosol and cloud formation controversial.

5. The Sun

The Sun is one of over 100 billion stars in the Milky Way Galaxy. It is by far the largest object in the solar system and it contains 99.8% of the total mass of the Solar System (Fig. 23).

5.1 Solar activity and energy output

Before studying the effect of the Sun on global warming we give a brief description of Sun's activity.
The Sun's magnetic field changes from a simple overall shape to an extremely distorted one in cycles (NASA - The Sun, 2007). The sun's magnetic fields rise through the convection zone and erupt through the photosphere into the chromosphere and corona. The eruptions lead to solar activity, which includes such phenomena as sunspots, flares, and coronal mass ejections.
Sunspots are dark, often roughly circular features on the solar surface. They form where denser bundles of magnetic field lines from the solar interior break through the surface. These are "cool" regions, only 3800 K (they look dark only by comparison to the surrounding regions). Sunspots can be very large, as much as 50000 km in diameter. The number of sunspots on the sun depends on the amount of distortion in the magnetic field.
The change in the number of sunspots, from a minimum to a maximum and back to a minimum, is known as the sunspot cycle. The average period of the sunspot cycle is about 11 years during which the activity varies from a solar minimum at the beginning of a sunspot cycle to a solar maximum about 5 years later. The number of sunspots that exist at a given time varies from none to approximately 250 individual sunspots and clusters of sunspots. At the end of a sunspot cycle, the magnetic field quickly reverses its polarity, from north to south, and loses most of its distortion. A change of polarity from one orientation to

the other and back again covers two successive sunspot cycles and is therefore about 22 years.

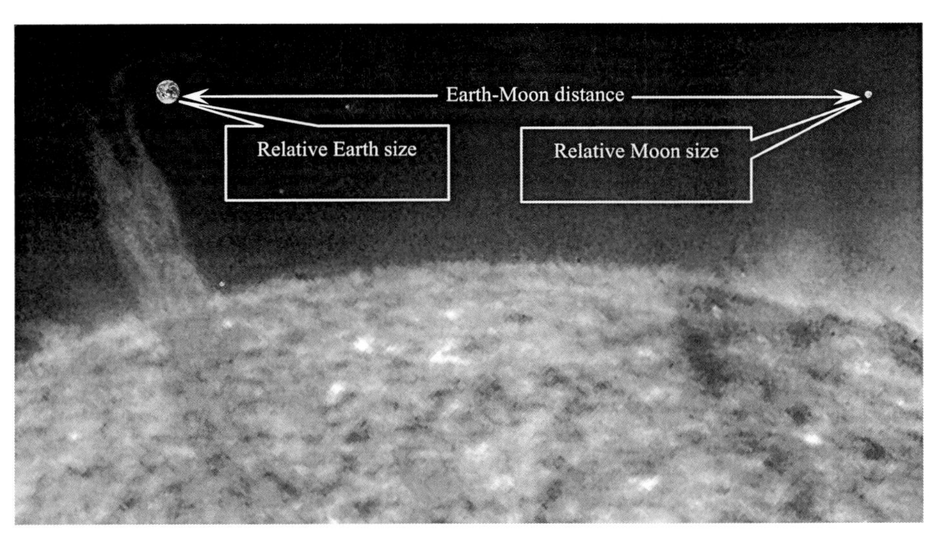

Fig. 23. Comparative sizes of the Sun, Earth, Moon and Earth-Moon distance. (Sun diameter: 1390000 km, Earth diameter: 12700 km, Moon diameter: 3474 km, Earth-Moon distance: 382500 km).

Flares are the most violent eruptions in the solar system. Coronal mass ejections, though less violent than flares, involve a tremendous mass (amount of matter). A single ejection can spew approximately 18 billion metric tons of matter into space.

The Sun emits electromagnetic radiation in the form of visible light, infrared radiation that we feel as heat, ultraviolet rays, microwaves, X-rays, gamma rays and so forth. This radiation can be thought of as waves of energy or as particle-like "packets" of energy, the so-called photons. The energy of an individual photon is related to its frequency. All forms of electromagnetic radiation travel through space at the speed of light (299792 km$\times$s^{-1}). At this rate, a photon emitted by the sun takes 8 minutes to reach the Earth that travels around the sun at an average distance of about 149600000 km.

The amount of electromagnetic radiation from the sun that reaches the top of the Earth's atmosphere is known as the solar constant. This amount is about 1370 W$\times$m^{-2}. Of this energy only about 40% reaches the Earth's surface. The atmosphere blocks some of the visible and infrared radiation, almost all the ultraviolet rays, and all the X-rays and gamma rays. Nearly all the radio energy reaches the Earth's surface.

The Sun also emits particle radiation, consisting mostly of protons and electrons comprising the solar wind. These particles come close to the Earth, but the Earth's magnetic field (Fig. 24) prevents them from reaching the surface. More intense ejections, known as solar cosmic rays, reach the Earth's atmosphere. Solar cosmic rays cannot reach the Earth's surface but they are extremely energetic, they collide with atoms at the top of the atmosphere and may cause major disturbances in the Earth's magnetic field disrupting electrical equipment and overloading power lines.

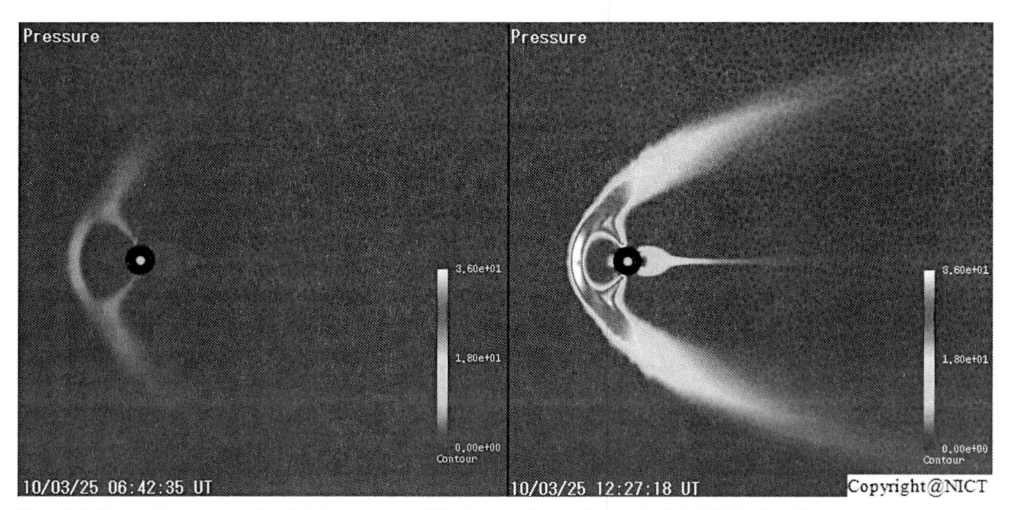

Fig. 24. Earth's magnetic field protects life from the solar wind (NICT, 2010).

5.2 Earth's orbit and variations in climate

Astronomers have linked the earth climate to various changes related with the earth orbit around the sun and the amount of energy that it receives. These orbital processes are thought to be the most significant drivers of ice ages according to the theory of Milankovitch, and such changes are (NASA-Milankovitch, 2007):

(a) The shape of the orbit (eccentricity) of the Earth around the Sun, which changes from a nearly perfect circle to an oval shape on a 90000–100000-year cycle. At present there is a 3% (5000000 km) difference between the closest approach (perihelion) in January and the furthest departure (aphelion) in July. This difference in distance causes about a 6% increase in the incoming solar radiation from July to January.

(b) The angle of the Earth's axis with respect to its orbital plane (axial obliquity). Today, the Earth's axis is tilted at 23.5°. This tilt varies between 22.1–24.5° during a 40000 years cycle causing warmer summers and colder winters when the tilt is greater.

(c) The orientation of the Earth's axis is slowly but continuously changing (precession), like a wobbling top, tracing out a conical shape in a cycle of approximately 26000 years. Changes in axial precession alter the dates of perihelion and aphelion, and therefore increase the seasonal differences in one hemisphere and decrease the seasonal differences in the other hemisphere.

The Milankovitch cycles were recently observed and confirmed in the Antarctica Dome C ice-core samples recording the climate variability over the past 800000 years (Jouzel et al., 2007).

Precession is caused by the deviation of the Earth's mass distribution from the spherical symmetry and is mainly due to the non-uniformity of the Earth's crust in its continental and oceanic regions and also in the possible heterogeneity in the mantle density. According to Sorokhtin et al. (2007), the attraction of the Earth by the Moon and the Sun plays a leading role in the reduction of the precession angle. In their work to estimate the climatic temperature deviations due to the above-mentioned attraction effect they also considered the main harmonic components of the Milankovitch cycles giving rise to a temperature deviation of about ± 3%. Their work could be used for forecasting the climatic changes in the future considering a best fit of theoretical to experimental data. Such a fit is presented in Fig. 25 where one observes that there were slow periods of climatic cooling of about 8–10°C which lasted approximately 100000–120000 years. After the formation of thick ice covers, a rapid warming – by the same 8–10°C – occurred degrading the glaciers completely in a few thousand years. Their forecast is that in the future we should expect a significant cooling.

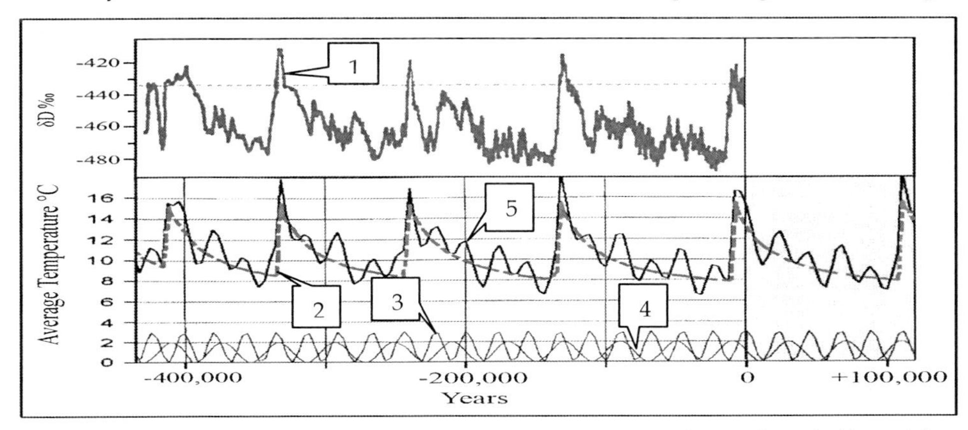

Fig. 25. Temperature deviations with respect to time presenting the combined effect of the attraction of the Earth by the Moon and the Sun and the main harmonic components of the Milankovitch cycles compared to the Vostok isotope temperature measurements (redrawn from Sorokhtin et al., 2007). (1) Vostok isotope temperature measurements, (2) Earth temperature change due to the attraction of the Earth by the Moon and the Sun, (3)–(4) Temperature change due to the main harmonic components of the Milankovitch cycles, (5) Resulting temperature change.

Undoubtedly, as argued above, solar variability plays an important role in global climate change. The total average solar energy flux currently reaching the Earth's surface is S_0 = 1.75×1014 kW and is determined by the so-called solar constant which is 1.37 kW×m^{-2}. The total heat flux through the Earth's surface due to energy generated in the mantle and the crust is approximately 0.0257% of the total Earth's solar irradiation. Additionally, the world total energy production is estimated to be about 0.0077% of the total solar irradiation reaching the Earth's body. Therefore it can easily be estimated that the solar radiation supplies more than 99.95% of total energy driving the world climate (Khilyuk & Chilingar, 2006).

The effect of solar irradiation on global atmospheric temperature can be evaluated using the adiabatic model of the heat transfer in the Earth's atmosphere (Sorokhtin et al., 2007). For a rough estimate of the global atmospheric temperature change ΔT at sea level, attributed to the natural variations in insolation S, the formula is:

$$\Delta T = 288[(S/S_0)^{1/4} - 1] \quad (2)$$

and as they have calculated a 1% increase in the current solar radiation reaching the Earth's body translates directly into approximately 0.86 K increase in the Earth's global temperature.

To examine how the total solar irradiance (TSI) variation affects the climate, we use the reconstruction of Steinhilber et al. (2009), based on the relationship between TSI and the open solar magnetic field obtained from the cosmogenic radionuclide 10Be measured in ice cores. As shown in Fig. 26 a relation may exist but it is obvious that other factors are affecting the climate as well. One can observe that TSI reached an absolute maximum in about 1990, whereas the absolute maximum for temperature occurred with a time-lag of about +10 years. Over the 10 next years, while the temperature has remained essentially constant, the TSI has continued dropping. It would, therefore, be physically highly unlikely that there would be no temperature drop in the succeeding years.

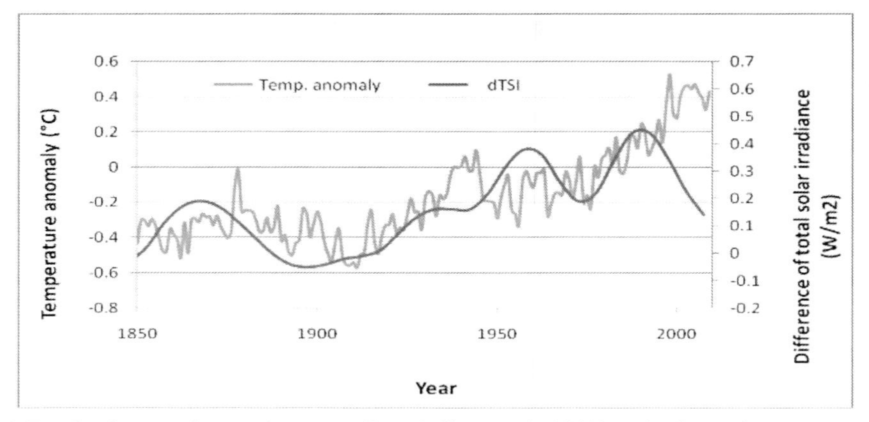

Fig. 26. Total solar irradiance variation (Steinhilber et al., 2009) and adjusted variance (HadCRUT3gv) taken from the Met Office Hadley Centre (2010).

A different solar parameter showing long-term changes is the length of the cycle. This parameter is known to vary with solar activity so that high activity implies short solar cycles whereas long solar cycles imply low activity levels of the Sun. Gleissberg demonstrated that the variation occurs in a systematic manner with a long-term periodicity of 80 to 90 years, known after his name as the Gleissberg period (Friis-Christensen & Lassen, 1991). Friis-Christensen and Lassen, (1991) showed that there is a close inverse relationship between sunspot cycle length and Northern Hemisphere land temperatures over the 1860–1985 period (Fig. 27).

5.3 Simultaneous warming of other planets in our solar system

Of course if the cause for global warming was solar irradiation then it would be expected that this cause should affect other planets of the solar system in the same manner.

It is accepted by NASA that Mars has warmed up by about 0.5 K since the 1970s. The climate change is so rapid that the red planet could lose its southern ice cap in the coming years. This is similar to the warming experienced on Earth over approximately the same period. If there is no life on Mars, as it is presently assumed, it means that rapid change in climate could be a natural phenomenon and essentially not anthropogenic.

A theory trying to explain the warming of Mars claims that variations in radiation and temperature across the surface of the red planet are generating strong winds caused by widespread changes in some areas which have become darker since 1970 (Fenton et al., 2007).

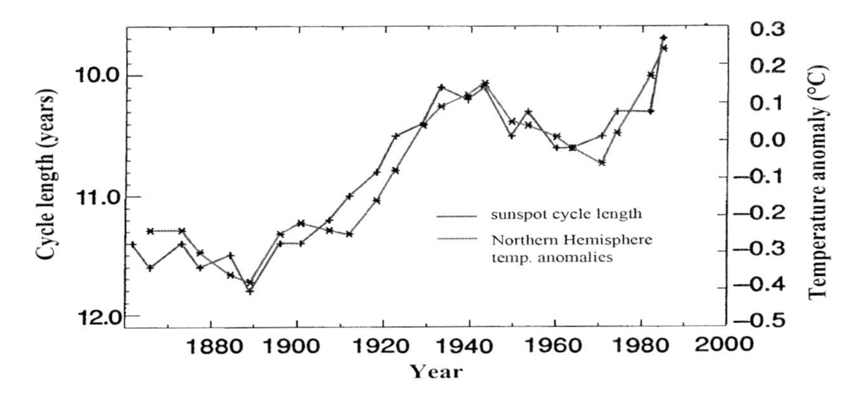

Fig. 27. Variation of the sunspot cycle length compared with the Northern Hemisphere temperature anomalies (modified from Friis-Christensen & Lassen, 1991).

The main gas in Pluto's atmosphere is nitrogen, and Pluto has nitrogen ice on its surface that can evaporate into the atmosphere when it gets warmer, causing an increase in surface pressure (MIT, 2002). Observations using eight telescopes made by Elliot et al. (2003), proved that the average surface temperature of the nitrogen ice on Pluto has increased slightly less than 2°C over a period of 14 years (1988–2002). The results have surprised the observers, who thought that Pluto's atmosphere may be cooling because now Pluto is orbiting away from the sun, being at its closest in 1989. Pluto's atmospheric temperature varies between around -235°C and -170°C, depending on the altitude above the surface.
Long-term, over half a century, photometric measurements of Neptune show variations of brightness. The detailed variations may partially be explain by seasonal change in Neptune's atmosphere but also the possibility of solar-driven changes, i.e. changes incurred by innate solar variability perhaps coupled with changing seasonal insolation may be considered as well. As Hammel & Lockwood (2007) point out the striking similarity of the temporal patterns of variation of Neptune's brightness and the Earth's temperature anomaly should not be ignored simply because of low formal statistical significance. If changing brightnesses and temperatures of two different planets are correlated, then some planetary climate changes may be due to variations in the solar system environment.
Based on the observed changes of temperature on other planets, one can conclude that it is possible that the main reason for these changes are variations in the electromagnetic solar environment that may trigger secondary processes affecting the temperature. Seasonal changes due to changes in orbit, internally generated heat or unknown mechanisms cannot be excluded neither.

5.4 A solar forcing scenario

As Solanki et al. (2004) mention, the level of solar activity during the past 70 years is exceptional, and the previous period of equally high activity occurred more than 8000 years ago (Fig. 28). The sunspot number covering the past 11400 years was reconstructed based on dendrochronologically dated radiocarbon concentrations averaged in 10-year intervals. Solanki et al. (2004) find that during the past 11400 years the Sun spent only about 10% of time at a similarly high level of magnetic activity and almost all of the earlier high-activity

periods were shorter than the present episode. They conclude that the Sun may have contributed to the unusual climate change during the 20th century although it is unlikely to have been the dominant cause. The big question until recently of a solar forcing scenario for climate change has been that the Sun's energy output through the sunspot cycle varies only by about 0.1%. This energy output variability is insufficient on its own, to cause the 0.6 K increase in global temperature observed through the 20th century.
Considering that the level of solar activity now is exceptional the response of the water vapor amplification must be re-examined especially since water vapor is the most dominant GHG with uncertain contribution in the range of 55–95% and since before present there was no anthropogenic CO2 to upset the balance.

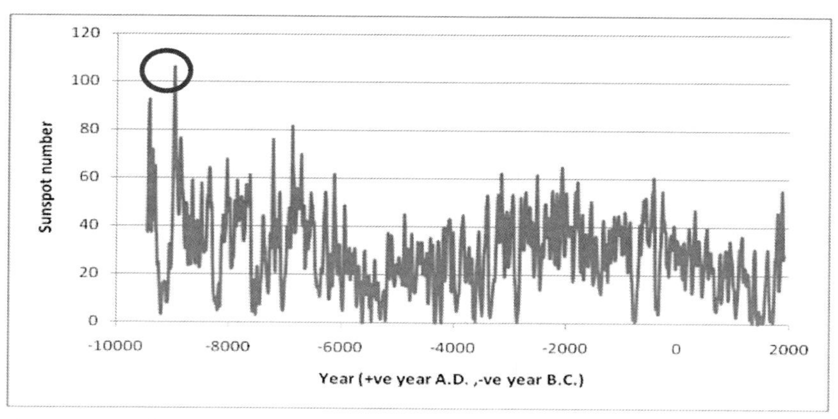

Fig. 28. Sunspot number (Solanki et al. 2004).

Time dependent experiments produce a global mean warming of 0.2–0.5 K in response to the estimated 0.7 $W \times m^{-2}$ change of solar radiative forcing from the Maunder Minimum to the present. However, the spatial response pattern of surface air temperature to an increase in solar forcing was found to be quite similar to that in response to increases in GHG forcing (IPCC, 2001).
Most of the Sun's heat is deposited into the tropics of the Earth with small variation of solar heating throughout the year. The amount of solar heating of the polar latitudes varies greatly, with the polar latitudes receiving much solar energy in summer, whilst in winter they receive no solar heat at all. As a result, in the winter hemisphere, the difference in solar heating between the equator and the pole is very large. This causes the large-scale circulation patterns observed in the atmosphere. The difference in solar heating between day and night also drives the strong diurnal cycle of surface temperature over land.
On the other hand, clouds block much of the solar radiation and reflect it back to space before it reaches the Earth. The more plentiful and thicker the clouds are, the cooler the Earth is. At the same time, clouds also block the emission of heat to space from the Earth acting like GHGs. The altitude of clouds changes the amount of thermal infrared blocking. Because temperature decreases with altitude, high clouds are colder and more effective in absorbing the surface emitted heat in the atmosphere, whilst they emit very little to space. Therefore clouds can either cool or warm the planet depending on the area of the Earth they cover, their thickness, and their altitude. Also the effectiveness of clouds depends on their

structure. The climate is so sensitive to clouds that present models of global climate can vary in their global warming predictions by more than a factor of 3, depending on how clouds are modeled (Goddard Space Flight Center, 1999).

5.5 Cosmic rays

During the 1990s Svensmark H. and Friis-Christensen E., presented a new astronomical cause for climate change, that of the cosmic ray hypothesis. Cosmic radiation originates from all luminous objects in the universe and it comprises primary particles with very high energy (mainly protons, 92%, and alpha particles, 6%). When the ray particles reach the Earth they cause ionization in the upper layers of the atmosphere. The particles loose energy colliding with other particles in the atmosphere and many of the lower energy particles are absorbed by the atmosphere on their way down to the surface. Magnetic fields deflect these rays and since the solar wind expands the magnetic field of the Sun, the Earth is shielded more from the incoming cosmic rays. Solar wind increases in strength with sunspot activity. According to the cosmic ray hypothesis, periods with low solar activity would allow more cosmic radiation to reach the earth, more clouds (low clouds) would be formed and finally a lower global mean temperature would result, and vice versa.
In examining the above-mentioned hypothesis the Danish National Space Center (DNSC, 2007), identified five external forcing parameters that are modulated by solar variability and have the potential to influence the Earth's lower atmosphere below 50 km. These are (a) the Total Solar Irradiance (TSI), (b) the Ultra-Violet (UV) component of solar radiation, (c) the direct input from the Solar Wind (SW), (d) the total Hemispheric Power Input (HPI) reflecting properties of precipitating particles within the magnetosphere, and (e) the Galactic Cosmic Rays (GCR). Their conclusion is that UV and GCR present a striking correlation with the global coverage of low clouds, over nearly two and a half solar cycles as shown in Fig. 29.
Currently, the National Space Institute of Denmark (DNSI, 2007) has been investigating the hypothesis that solar variability is linked to climate variability by a chain that involves the solar wind, cosmic rays and clouds. The reported variation of cloud cover was approximately 2% over the course of a sunspot cycle but simple estimates indicate that the resultant global warming could be comparable to that presently attributed to GHGs from the burning of fossil fuels.
Recent work has directed attention to a mechanism involving aerosol production and the effects on low clouds. This idea suggests that ions and radicals produced in the atmosphere by cosmic rays could influence aerosol production and thereby cloud properties. Cosmic rays ionize the atmosphere, and an experiment performed at DNSI has found that the production of aerosols in a sample atmosphere with condensable gases (such as sulphuric acid and water vapor) depends on the amount of ionization. Since aerosols work as precursors for the formation of cloud droplets, this is an indication that cosmic rays influence cloud formation.
As the National Space Institute of Denmark (2009) informs us, the European Organization for Nuclear Research, CERN, has currently been creating an atmospheric research facility at its Particle Physics laboratory. Called CLOUD, it will consist of a special cloud chamber exposed to pulses of high-energy particles from one of CERN's particle accelerators. Conditions prevailing the Earth's atmosphere will be recreated in CLOUD, and the incoming particles will simulate the action of cosmic rays. The main cloud chamber for the CLOUD facility is expected to begin operating in 2010.

Recently the Sun's magnetic activity unexpectedly declined and its surface is almost free of sunspots. Because of this, there is a concern that the Sun might now fall asleep in a deep minimum that may continue through the next years. During the period from 1650-1715 almost no sunspots were observed on the sun's surface. This extended absence of solar activity may have been partly responsible for the Little Ice Age in Europe and may reflect cyclic or irregular changes in the sun's output over hundreds of years. During that period, winters in Europe were longer and colder by about 1 K than they are today. A consequence of this phenomenon occurring will be the clarification of global warming causes.

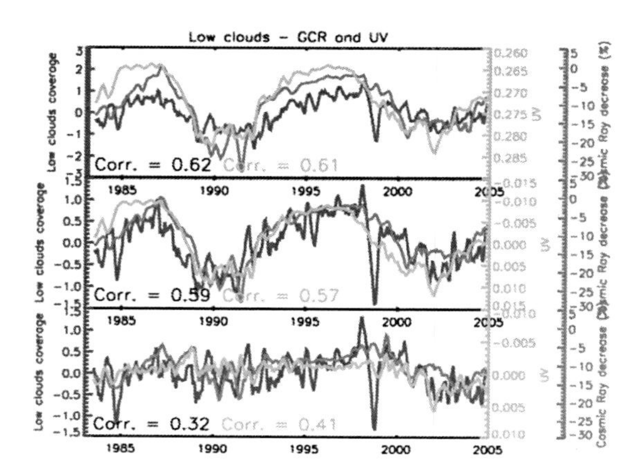

Fig. 29. Correlation between GCR (red) and UV (green) and coverage of low clouds (blue). ISCCP data for coverage of low altitude clouds after adjustment for their offset when compared with the independent data set of low cloud provided by the SSM/I microwave instrument aboard the DMSP satellites. From top: annual cycle removed, trend and internal modes removed, solar cycle removed (modified from DNSC, 2007).

Recently UK's National Centre for Atmospheric Science (NCAS) and the Science and Technology Facilities Council (STFC) (Osprey et al. 2009), showed that the number of high-energy cosmic-rays reaching a detector deep underground strongly matches temperature measurements in the upper atmosphere during short-term atmospheric (10-day) events. The effects were seen by correlating data from the underground detector used in a U.S.-led particle physics experiment called MINOS (managed by the U.S. Department of Energy's Fermi National Accelerator Laboratory) and temperatures from the European Center for Medium Range Weather Forecasts during the winter periods from 2003–2007.

As it was shown the relationship can be used to identify weather events that occur very abruptly in the stratosphere during the Northern Hemisphere winter. These events can have a significant effect on the severity of winters we experience, and also on the amount of ozone over the poles.

The cosmic-rays, known as muons are produced following the decay of other cosmic rays, known as mesons. Increasing the temperature of the atmosphere expands the atmosphere so that fewer mesons are destroyed on impact with air, leaving more to decay naturally to muons. Consequently, if temperature increases so does the number of muons detected. The relation for the winter period from 2006-2007 is shown in Fig. 30.

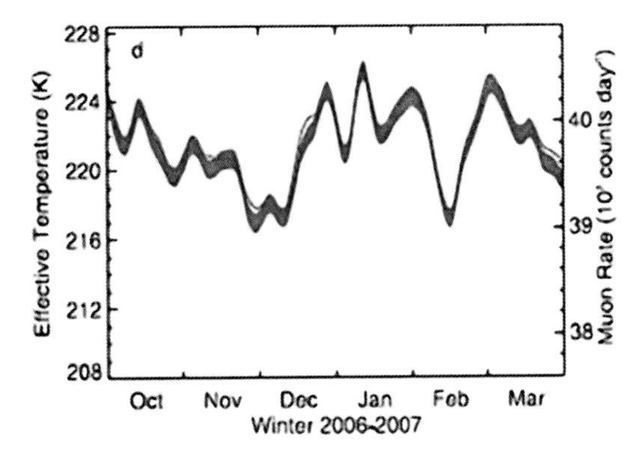

Fig. 30. Relationship between the cosmic-rays and stratospheric temperature for the winter of 2006–2007 (Osprey et al., 2009).

5.6 Palaeoclimate and cosmic-rays

The geological record of the past 550 million years shows variations between ice-free and glaciated climates. Since there were four alternations between "hothouse" and "icehouse" conditions during the Phanerozoic the greenhouse-warming theory could not account for these changes (Svensmark, 2007). Reconstructions of atmospheric CO2 show just two major peaks as shown in Fig. 13.

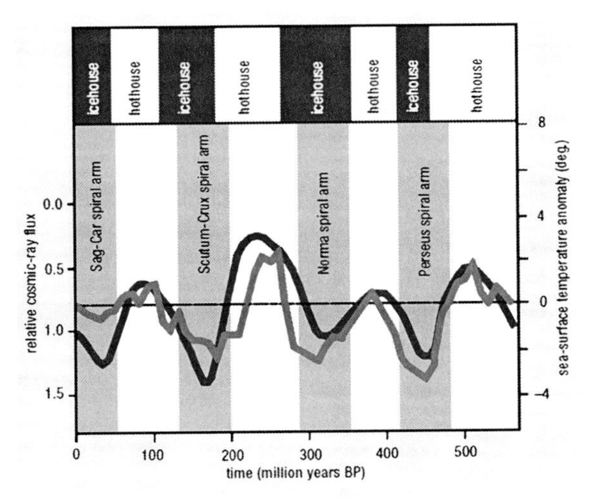

Fig. 31. Variations in tropical sea-surface temperatures corresponding to four encounters with spiral arms of the Milky Way and the resulting increases in the cosmic-ray flux (Redrawn from Svensmark, 2007 – after Shaviv and Veizer, 2003).

A more persuasive explanation comes from cosmoclimatology, which attributes the icehouse episodes to four encounters with spiral arms of the Milky Way, where explosive blue stars and cosmic rays are more concentrated. As Shaviv & Veizer (2003) showed, the relative motion of the spiral arm pattern of our galaxy with respect to the solar orbit around the galactic centre gave a good fit with the climatic record, in cycles of about 140 million years. The matches between spiral-arm encounters and icehouse episodes occurred during the Ordovician to Silurian Periods with the Perseus Arm, during the Carboniferous with the Norma Arm, the Jurassic to Early Cretaceous Periods with the Scutum-Crux Arm, the Miocene Epoch with the Sagittarius-Carina Arm leading almost immediately to the Orion Spur during the Pliocene to Pleistocene Epochs. This is demonstrated in Fig 31, where four switches from warm "hothouse" to cold "icehouse" conditions during the Phanerozoic are shown with respect to variations in tropical sea-surface temperatures (several degrees K).

5.7 A bold prediction

Archibald (2006), predicted weak solar maxima for solar cycles 24 and 25 and correlated the terrestrial climate response to solar cycles over the last 300 years. He also predicted a temperature decline of 1.5 K by 2020, equating to the experience of the Dalton Minimum from 1796–1820. In his 2009 paper he compares solar Cycles 4 and 23 aligned on the month of minimum. From this comparison it is apparent that solar cycles 22 and 23 are very similar to solar cycles 3 and 4, which preceded the Dalton Minimum, and assumes that the coming solar cycles will be similar to cycles 5 and 6 of the Dalton Minimum (see Fig. 32).

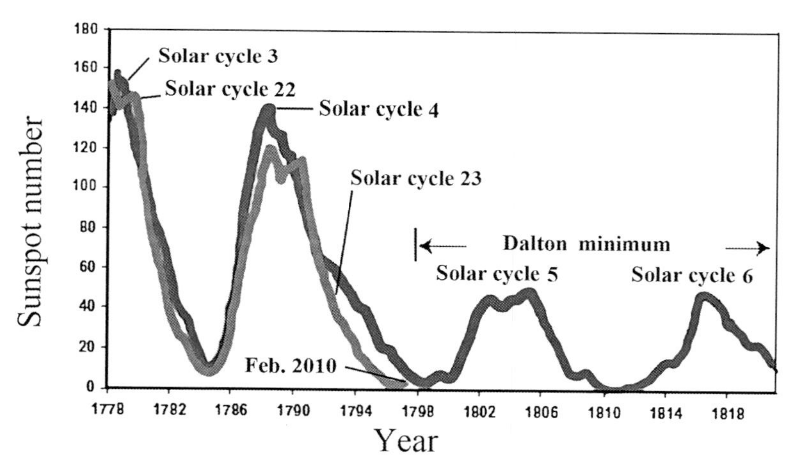

Fig. 32. Solar cycles 22 and 23 compared to solar cycles 3 and 4 (revised from Archibald, 2009).

Furthermore Archibald applied the methodology of Friis-Christensen and Lassen (1991), that demonstrated a relationship between solar cycle length (in one cycle) and annual-average temperature over the following solar cycle, to predict the annual average temperature of Hanover New Hampshire to be 2.2 K cooler during solar cycle 24 than it had been on average over solar cycle 23. His prediction in 2009, assumed that the solar minimum would be in July making solar cycle 23 over 13 years long, which in turn would mean that solar cycle 23 would be 3.2 years longer than solar cycle 22. His plot is presented in Fig. 33.

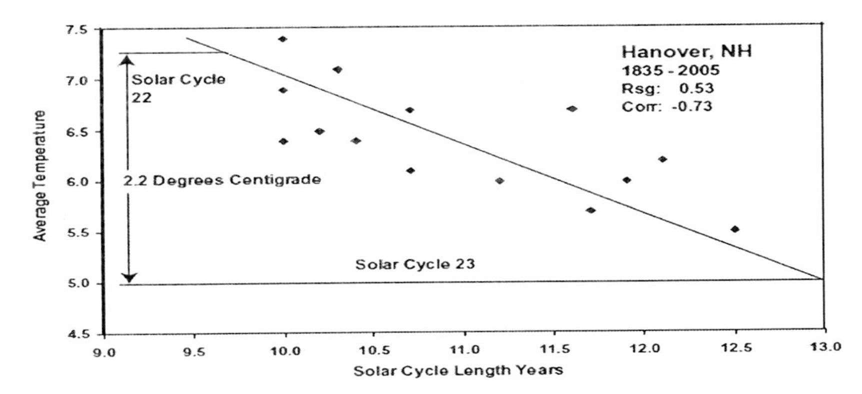

Fig. 33. Average temperature over the following solar cycle 24, for Hanover NH (modified from Archibald 2009).

6. Epilogue

The main conclusion drawn from this study is that contrary to what has become common place, there are mechanisms other than atmospheric CO2 concentration like the solar radiation and cosmic rays that may be largely responsible for the observed change in temperature. Besides, the phenomenon of global temperature rise is not only observed on this planet but it is also detected on other planets as well.

Anthropogenic CO2-concentration increase has contributed to global warming in a small part during the 20th century, but CO2 increase is in part a result of the temperature rise and various natural processes, like ocean changes in CO2 solubility. The procedure used by the IPCC may have mistaken a natural signal for an anthropogenic change because the spatial response pattern of surface air temperature to an increase in solar forcing is quite similar to that of the increase in GHG forcing.

Computer models have failed up to now to reproduce convincingly the observed climate changes. This is greatly due to the lack of implementation of any appropriate laws from Physics. Science does not really have a complete and total understanding of the functioning of the Earth's complex climate system, especially how any effects by or on water can be modeled. In our view, scientists involved in climate research will most likely not agree on a universal conclusion in the near future.

7. References

Archibald, D. (2006). Solar Cycles 24 and 25 and Predicted Climate Response, *Energy and Environment*, Vol. 17, pp.29-38.

Archibald, D. (2008). *Solar Cycle 24*. Success Print, ISBN 978-0-646-50269-4. http://www.davidarchibald.info/

Archibald, D. (2009). Solar cycles 24 and 25 and predicted climate Response, *Energy & Environment*, Vol. 20, No. 1&2.

Barrett, J. (2005). Greenhouse Molecules, their Spectra and Function in the Atmosphere, *Energy & Environment, Vol.* 16:6.

Barrett- Bellamy Climate (2010). http://www.barrettbellamyclimate.com/index.htm

Berner, R.A. & Kothavala, Z. (2001). GEOCARB III: A Revised Model of Atmospheric CO2 over Phanerozoic Time, IGBP PAGES/World Data Center for Paleoclimatology, Data Contribution Series # 2002-051. NOAA/NGDC Paleoclimatology Program, Boulder CO, USA.http://www1.ncdc.noaa.gov/pub/data/paleo/climate_forcing/trace_gases/phanerozoic_co2.txt

Briffa, K.R., Jones, P.D., Bartholin, T.S., Eckstein, D., Schweingruber, F.H., Karlen, W., Zetterberg, P. & Eronen, M. (1992). Fennoscandian summer from AD 500: temperature changes on short and long timescales, *Clim. Dyn.*, Vol. 7, pp. 111–119.

CCSP (2006). Climate Change Science Program Report. U. S. Climate Change Science Program. Temperature Trends in the Lower Atmosphere-Steps for Understanding and Reconciling Differences. http://www.climatescience.gov/Library/sap/sap1-1/finalreport/sap1-1-final-all.pdf

Committee on the Science of Climate Change- National Research Council (2001). Climate Change Science: An Analysis of Some Key Questions. National Academy Press. DNSC (2007). Christiansen, F., Haigh, J. & Lundstedt, H. Influence of Solar Activity Cycles on Earth's Climate, Second Scientific Report (2/2007), Danish National Space Center,
http://www.space.dtu.dk/upload/institutter/space/forskning/07_reports/scientific_reports/isac_final_report.pdf.

DNSI (2007). The National Space Institute in Denmark (http://www.space.dtu.dk/English/Research/Research_divisions/Sun_Climate.aspx) Source: Svensmark H. (2007). Cosmoclimatology: a new theory emerges, *Astronomy & Geophysics*, Vol. 48, No. 1, pp. 1.18-1.24.

Elliot, J.L., Ates, A., Babcock, B.A., Bosh, A.S., Buie, M.W., Clancy, K.B., Dunham, E.W., Eikenberry, S.S., Hall, D.T., Kern, S.K., Leggett, S.E., Levine, D.-S., Moon, C.B., Olkin, D.J., Osip, J.M., Pasachoff, B.E., Penprase, M.J., Person, S.Q., Rayner, J.T., Roberts, L.C.Jr, Salyk, C.V., Souza, S.P., Stone, R.C., Taylor, B.W., Tholen , J.E., Thomas-Osip, Ticehurst D.R. & Wasserman, L.H. (2003). The recent expansion of Pluto's atmosphere. *Letters to Nature, Nature*, Vol. 424, pp. 165-168, doi:10.1038/nature01762.

EPICA (2010). http://www.esf.org/activities/research-networking-programmes/life-earth-and-environmental-sciences-lesc/completed-esf-research-networking-programmes-in-life-earth-and-environmental-sciences/european-project-for-ice-coring-in-antarctica-epica-page-1/more-information.html.

Fenton, L. K., Geissler, P. E. & Haberle, R. M. (2007). Global warming and climate forcing by recent albedo changes on Mars. *Letter, Nature*, Vol. 446, pp. 646-649, doi:10.1038/nature05718

Florides, G.A. & Christodoulides, P. (2009). Global Warming and Carbon Dioxide through Sciences, *Environmental International*, Vol. 35, pp. 390–401.

Friis-Christensen, E. & Lassen K. (1991). Length of the solar cycle: an indicator of solar activity closely associated with climate, *Science*, Vol. 254, No. 5032, pp.698-700.

Goddard Space Flight Center (1999). Earth's EnergyBalance http://earthobservatory.nasa.gov/Newsroom/MediaResources/Energy_Balance.pdf.

Griggs, J.A. & Harries, J.E. (2007). Comparison of Spectrally Resolved Outgoing Longwave Radiation over the Tropical Pacific between 1970 and 2003 Using IRIS, IMG and AIRS, *Journal of Climate*, American Meteorological Society, Vol. 20 pp. 3982-4001, DOI: 10.1175/JCLI4204.1.

Grudd, H. (2008). Tornetrask tree-ring width and density AD 500-2004: a test of climatic sensitivity and a new 1500-year reconstruction of north Fennoscandian summers, *Clim. Dyn.*, Vol. 31, pp. 843–857.

Grudd, H., Briffa, K.R., Karlen, W., Bartholin, T.S., Jones, P.D. & Kromer, B (2002). A 7400-year tree-ring chronology in northern Swedish Lapland: natural climatic variability expressed on annual to millennial timescales, *Holocene*, Vol. 12, pp. 643–656.

Hammel, H. B. & G. W. Lockwood (2007). Suggestive correlations between the brightness of Neptune, solar variability, and Earth's temperature, *Geophys. Res. Lett.*, Vol. 34, L08203, doi:10.1029/2006GL028764.

Harries, E. J., Brindley, E.H., Sagoo, J.P. & Bantges, J.R. (2001). Increases in greenhouse forcing inferred from the outgoing longwave radiation spectra of the Earth in 1970 and 1997, *Letters to Nature, Nature,* Vol. 410, pp. 355-357, doi:10.1038/35066553.

Historical CO2 record (1998). The Law Dome DE08, DE08-2, and DSS ice cores, Carbon Dioxide Information Analysis Center (CDIAC). Source: Etheridge, D.M., Steele, L.P., Langenfelds, R.L., Francey, R.J., Barnola, J.-M. & Morgan, V.I. http://cdiac.ornl.gov/ftp/trends/co2/lawdome.combined.dat.

Huss, M., Bauder, A., Funk, M. & Hock, R. (2008). Determination of the seasonal mass balance of four Alpine glaciers since 1865, *Journal of Geophysical Research*, 113:10.1029/2007JF000803.

Idso, S. (1988). The CO2 greenhouse effect on Mars, Earth and Venus, *The Science of the Total Environment*, Vol. 77, pp. 291-294.

Idso, B.S. (1998). Review: CO2-induced global warming: a skeptic's view of potential climate change. *Climate Research*, Vol. 10, pp. 69–82.

Indermühle, A., Monnin, E., Stauffer, B., Stocker, T.F. & Wahlen, M. (2000). A high-resolution record of the atmospheric CO2 concentration from 60-20 kyr BP from the Taylor Dome ice core, Antarctica. doi:10.1594/PANGAEA.710905 http://doi.pangaea.de/10.1594/PANGAEA.710905?format=html.

IPCC (1997). Technical Paper II, http://www.ipcc.ch/pdf/technical-papers/paper-II-en.pdf

IPCC (2001). The Intergovernmental Panel on Climate Change. Detection of climate change and attribution of causes p 709. CLIMATE CHANGE 2001: THE SCIENTIFIC BASIS. http://www.grida.no/climate/ipcc_tar/wg1/449.htm

IPCC Fourth Assessment Report, Climate Change (2007a). Chapter 6: Palaeoclimate, http://www.ipcc.ch/publications_and_data/ar4/wg1/en/contents.html.

IPCC Fourth Assessment Report, Climate Change IPCC (2007b). Chapter 1: Historical Overview of Climate Change Science. http://www.ipcc.ch/publications_and_data/ar4/wg1/en/contents.html.

IPCC Fourth Assessment Report, Climate Change IPCC (2007c). Chapter 10: Global Climate Projections. http://www.ipcc.ch/publications_and_data/ar4/wg1/en/contents.html.

IPCC Fourth Assessment Report: Climate Change (2007d). Working Group I: The Physical Science Basis, TS.2.4 Radiative Forcing Due to Solar Activity and Volcanic Eruptions (p. 30).

Jouzel, J., et al. (2007). Orbital and millennial Antarctic climate variability over the past 800,000 years. *Science*, Vol. 317, No.5839, pp. 793-797. doi:10.1126/science.1141038 http://doi.pangaea.de/10.1594/PANGAEA.683655?format=html

Karmanovitch, N. & Geoph, P. (2009). Hansen Mars Challenge - A Challenge to Hansen et al. 1988. *Australian Institute of Geoscientists (AIG), NEWS* No 96, May 2009. http://aig.org.au/assets/194/AIGnews_May09.pdf

Khilyuk, L.F. & Chilingar, G. V. (2006). On global forces of nature driving the Earth's climate. Are humans involved?, *Environ Geol*, Vol 50, pp. 899–910, DOI 10.1007/s00254-006-0261-x.

Lindzen, R.S. & Choi Y.S. (2009). On the determination of climate feedbacks from ERBE data. *Geophys. Res. Lett.*, Vol.36, L16705, doi:10.1029/2009GL039628.

Loehle, C. & McCulloch, J.H. (2008). Correction to: A 2000-Year Global Temperature Reconstruction Based on Non-Tree Ring Proxies, Energy and Environment, 19(1), 93–100. http://www.econ.ohio-state.edu/jhm/AGW/Loehle/

Mann, M. E., Bradley, R. S. and Hughes, M. K. (1999). Northern Hemisphere temperatures during the past millennium: Inferences, uncertainties and limitations, *Geophys. Res. Lett.*, Vol. 26, pp.759–762.

Mauna Loa CO2 annual mean data (2010).Trends in Atmospheric Carbon Dioxide, NOAA Earth System Research Laboratory, http://www.esrl.noaa.gov/gmd/ccgg/trends/#mlo_full.

McIntyre S. (2008). How do we "know" that 1998 was the warmest year of the millennium?, presentation at Ohio State University, May 16, 2008, http://climateaudit.files.wordpress.com/2005/09/ohioshort.pdf and http://climateaudit.org/2008/05/22/ohio-state-presentation/.

McIntyre, S. & McKitrick, R. (2003). Corrections to the Mann et. al. (1998) Proxy Data Base and Northern Hemisphere Average Temperature Series, *Environment and Energy*, Vol. 14, No.6, pp. 751–771.

McIntyre, S. & McKitrick, R. (2010). http://www.uoguelph.ca/~rmckitri/research/trc.html

Met Office Hadley Centre (2010). Temperature record file CRUTE3Vgl, variance adjusted, http://www.cru.uea.ac.uk/cru/data/temperature/#datdow.

MIT (Massachusetts Institute of Technology) (2002). News office. Pluto is undergoing global warming, researchers find. October 9, 2002. http://web.mit.edu/newsoffice/2002/pluto.html.

Monnin, E. et al. (2004). Evidence for substantial accumulation rate variability in Antarctica during the Holocene, through synchronization of CO2 in the Taylor Dome, Dome C and DML ice cores. *Earth and Planetary Science Letters*, Vol. 224, No.1-2, pp.45-54. doi:10.1016/j.epsl.2004.05.007 http://doi.pangaea.de/10.1594/PANGAEA.472488?format=html.

National Space Institute of Denmark (2009). The CLOUD experiment. http://www.space.dtu.dk/English/Research/Research_divisions/Sun_Climate/Experiments_SC.aspx

NAS (2006). Surface Temperature Reconstructions for the Last 2,000 Years. Committee on Surface Temperature Reconstructions for the Last 2,000 Years, United States National Research Council, Washington. http://www.nap.edu/catalog.php?record_id=11676#toc.

NASA - Milankovitch (2007).

http://earthobservatory.nasa.gov/Library/Giants/Milankovitch/
NASA - The Sun (2007). http://www.nasa.gov/worldbook/sun_worldbook.html.
NICT - National Institute of Information and Communications Technology, Japan (2010). http://www2.nict.go.jp/y/y223/simulation/realtime/home.html
NIPCC (Nongovernmental International Panel on Climate Change), (2009). Climate Change Reconsidered, http://www.nipccreport.org/reports/2009/2009report.html.
Oerlemans, J. (2005). Extracting a Climate Signal from 169 Glacier Records. *Science,* Vol. 308, No. 5722, pp. 675–677. DOI: 10.1126 / science. 1107046.
Osprey, S., et al. (2009). Sudden stratospheric warmings seen in MINOS deep underground muon data, *Geophys. Res. Lett.*, 36, L05809, doi:10.1029/2008GL036359.
Petit, J.-R., et al. (1999). Climate and atmospheric history of the past 420,000 years from the Vostok ice core, Antarctica. *Nature,* Vol. 399, No. 6735, pp.429-436. doi:10.1038/20859.http://doi.pangaea.de/10.1594/PANGAEA.55501?format=html
Royer, D.L., Berner, R.A., Montañez, I.P., Tabor, N.J. & Beerling, D.J. (2004). CO2 as a primary driver of Phanerozoic climate. *GSA Today,* 14:3. doi: 10.1130/1052-5173. http://www.soest.hawaii.edu/GG/FACULTY/POPP/Royer%20et%20al.%202004%20GSA%20Today.pdf
Scotese, C.R. (2008). PALEOMAP Project, Arlington, Texas; http://www.scotese.com
Shaviv, N.J. & Veizer, J. (2003). Celestial driver of Phanerozoic climate? *GSA Today,* Vol. 13, No.7, pp 4–10.
Siegenthaler, U et al. (2005a). EPICA Dome C carbon dioxide concentrations from 423 to 391 kyr BP. Laboratoire de Glaciologie et Géophysique de l`Environnement, Saint Martin, doi:10.1594/PANGAEA.472482, http://doi.pangaea.de/10.1594/PANGAEA.472482?format=html
Siegenthaler, U et al. (2005b). EPICA Dome C carbon dioxide concentrations from 650 to 413 Kyr BP. *Physikalisches Institut, Universität Bern,* doi:10.1594/PANGAEA.472481, http://doi.pangaea.de/10.1594/PANGAEA.472481?format=html
Solanki, S.K, Usoskin, I. G., Kromer, B., Schüssler, M. & Beer, J. (2004). Unusual activity of the Sun during recent decades compared to the previous 11,000 years, *Nature,* Vol. 431, No. 7012, pp. 1084 - 1087. Data: NOAA Paleoclimatology Program and World Data Center for Paleoclimatology, Boulder. ftp://ftp.ncdc.noaa.gov/pub/data/paleo/climate_forcing/solar_variability/solanki2004-ssn.txt.
Soon, W. & Baliunas, S. (2003). Proxy climatic and environmental changes of the past 1000 years, *Climate Research,* Vol. 23, pp. 89–110.
Sorokhtin, O.G., Chilingar, G. V., Khilyuk, L.F. (2007). *Global Warming and GlobalCooling. Evolution of Climate on Earth,* Developments in Earth & Environmental Sciences 5, Elsevier, ISBN 978-0-444-52815-5.
Spencer, W. R. (2008). Satellite and climate model evidence against substantial manmade climate change. http://www.drroyspencer.com/research-articles/satellite-and-climate-model-evidence/
Steinhilber, F., Beer, J.& Frohlich, C. (2009).Total solar irradiance during the Holocene, *Geophys. Res. Lett.,* Vol. 36, L19704, doi:10.1029/2009GL040142. http://www.ncdc.noaa.gov/paleo/forcing.html

Reconstructed Solar Irradiance, 9,300 Years, Steinhilber et al., 2009, ftp://ftp.ncdc.noaa.gov/pub/data/paleo/climate_forcing/solar_variability/steinhilber2009tsi.txt

Sunspot number data: NOOA U.S. Government. File yearrg.dat. ftp://ftp.ngdc.noaa.gov/STP/SOLAR_DATA/SUNSPOT_NUMBERS/GROUP_SUNSPOT_NUMBERS/.

Svensmark, H. (2007). Cosmoclimatology: a new theory emerges, *Astronomy and Geophysics*, Vol. 48. Link: Danish National Space Center. http://www.spacecenter.dk/research/sun climate/Scientific%20work%20and%20publications

SWPC (2010). U.S. Dept. of Commerce, NOAA, Space Weather Prediction Center. http://www.swpc.noaa.gov/ftpdir/weekly/RecentIndices.txt

Veizer, J., Godderis, Y. & François, L.M. (2000). Evidence for decoupling of atmospheric CO2 and global climate during the Phanerozoic eon. *Nature,* Vol. 408, pp. 698-701.

Wegman, E. et al. (2006). Ad hoc Committee Report on the "Hockeystick" Global Climate Reconstruction, U.S. House of Representatives Committee on Energy & Commerce. http://www.climateaudit.org/pdf/others/07142006_Wegman_Report.pdf.

Section 2

4

Global Warming, Glacier Melt & Sea Level Rise: New Perspectives

Madhav L Khandekar
Climate Scientist, Markham Ontario
Canada

1. Introduction

There is a heightened interest at present on the possibility of rapid melting of world-wide glaciers and ice caps (e.g., Greenland & Antarctic Ice Caps) as a result of ongoing global warming, which could lead to escalated sea level rise in future. Several stories and news items in the print & TV media and also in scientific magazines seem to strongly suggest that glaciers and ice caps are melting faster than ever and this could lead to significant rise in global and regional sea level. Recent popular Hollywood movies like *An Inconvenient Truth* showing big ice shelves breaking off and sliding down into cold Arctic Ocean seem to reinforce this perception that the ice caps and glaciers are indeed melting rapidly and causing sea level to rise dramatically. From a scientific perspective, satellite altimeter data (Topex/Poseidon) have estimated recent sea level rise of about 3mm/yr and possibly higher, this value being significantly higher than the traditionally held value of about 1.5 to 2 mm per yr (for most of the twentieth century) has generated sufficient interest among the scientific community about 'escalating sea level rise' in near future. In a recent (Journal of Climate 2007) paper, the lead author Prof (Emeritus) Carl Wunsch states that " *Modern sea level is a matter of urgent concern from a variety of points of view, but especially because of possibility of its acceleration and consequent threats to many low-lying parts of the inhabited world* ".
The Intergovernmental Panel on Climate Change (IPCC), a UN Body of scientists, has been making periodic review of the earth's climate change since 1990. In its most recent climate change documents (Meehl 2007), the IPCC projects global sea level rise (SLR) for the next 100 years to be between 14 and 43 cm (with a mean value of 29 cm) under an emission scenario A1B (these emission scenarios have been developed using economic development indices for world countries) in which the earth's mean temperature is projected to rise between 2.3C and 4.1C by 2100. The IPCC projects largest increase in SLR (about 230 mm by 2100) as due to thermal contribution, resulting from warming of the earth's surface and associated expansion of water in future. The contribution due to melting of glaciers and ice caps is estimated by the IPCC to be about 60 mm over next 100 years. These two components namely thermal (or *steric)* and mass balance (or *eustatic)* are being closely examined at present in the context of several recent studies on climate sensitivity and observational evidence of increased glacier melting.
In this chapter, a brief background of the global warming science is presented in section 2. This is followed by a discussion on sea level rise over the past several thousand years and

the present state of SLR as determined by world-wide sea level records like tide gauge and modern satellite altimetry. The impact of recent warming on global & regional SLR by way of glacier melt (eustaic rise) and thermal expansion (steric rise) is discussed in the context of observational and other theoretical studies as well as several uncertainties associated with some of the estimates. Also discussed is the present state of sea level change at some of the most vulnerable areas like The Maldives in the equatorial Indian Ocean and the Tuvalu Island in the south Pacific. An estimate of present and future SLR is obtained using most recent studies and the possibility of escalated sea level rise at selected locations is analyzed.

2. A brief overview of the global warming science

The present debate on the global warming science may have begun with a landmark paper (Revelle & Suess 1958) in which the lead author, (late) Roger Revelle, an eminent geophysicist, suggested that *humans are carrying out a large-scale geophysical experiment through world-wide industrial activity that could lead to the build-up of CO_2 greater than the rate of CO_2 production by volcanoes.* Revelle & Suess estimated human-added CO_2 in the earth's atmosphere and expressed concern that this continued build-up of carbon dioxide by human activity could impact earth's climate in future through absorption of long-wave radiation emanating from the earth's surface. Revelle was instrumental in establishing the first carbon-dioxide measuring station at Mauna Loa, Hawaii in 1956. Another carbon-dioxide measuring station was later established in the Antarctic and at present there are several dozen locations providing carbon-dioxide measurements world-wide. Based on these measurements it is now well established that the concentration of atmospheric carbon dioxide, the principal greenhouse gas, has increased from about 350 ppmv (parts per million by volume) in 1956 to about 385 ppmv today. Besides carbon dioxide, there are two other atmospheric gases namely methane (CH_4) and nitrous oxide (N_2O) which are also considered as 'greenhouse gases' whose concentration in the earth's atmosphere is much smaller than carbon dioxide. It should be noted here that the earth's atmosphere consists of nitrogen (~78%), oxygen (~21%) and argon (~0.93%), while carbon dioxide makes up for just about 0.03% of total atmospheric gases. It should be also noted that atmosphere-ocean system is continuously exchanging carbon dioxide which is estimated at about 150 billion tons (Giga-tons OR Gt) annually, while the human-added CO_2 is about 20-22 billion tons annually, just about 15% of total carbon dioxide exchange between atmosphere and ocean.
The publication of Revelle/Suess paper sparked rapid development of a number of computer-based climate models which attempted to simulate the impact of increasing future carbon-dioxide concentration on the earth's mean temperature. Using steadily increasing concentration of carbon dioxide based on some assumed emission scenarios, some of the climate models developed in the 1980s and 1990s have projected mean temperature increase of 3C to 6C or more, by the end of 2100. A recent study (Knutti et al 2008) examines a suite of state-of-the-art climate models and associated uncertainties. This study (by Knutti et al) obtains a best guess value of 2.8C with a range of 1.7C to 4.4C for mean temperature increase by 2100. Many climate models now estimate the mean temperature increase for a doubling of the present value of carbon dioxide concentration (e.g, from a present value of 350 to value say 700 or about) over next fifty to one hundred years and this value is often referred to as climate sensitivity. Several recent studies (e.g., Lindzen 2007; Schwartz 2007;

Chylek & Lohmann 2008) now suggest that the climate sensitivity obtained by most climate models is too large and a more realistic value seems to be just about 1C to 1.2C. Whether the earth's mean temperature warms by as much as 3C or more by 2100 is one of the most contentious issues in the global warming science today.

The evolution of the earth's mean surface temperature over the last 150 years is shown in **Figure 1**. This mean temperature graph is obtained by calculating for each year, a mean value using all available land station data which are suitably combined with Sea Surface Temperature (SST) data to come up with one single value for each year. Such calculations are subject to contamination due to urbanization impact, large-scale circulation changes etc, as discussed in Khandekar et al (2005) and many other recent papers. Despite such uncertainties in mean temperature calculation, Figure 1 does provide a general representation of the earth's mean temperature evolution over last 150 years. The temperature curve shows two distinct periods of warming, one from 1910-1945 and the recent warming from about 1977 till present. It may also be noted that from 1945 till about 1977, the earth's mean temperature declined by about 0.25C. The year 1998 has been designated as the 'hottest year' according to the IPCC benchmarks. In **Figure 2** global mean temperature trends for the recent years (since 2002) are shown together with atmospheric carbon dioxide trend. Note that the carbon dioxide concentration is increasing while global mean temperature is steadily declining in the last seven years or about. This has become another contentious issue in the present global warming debate.

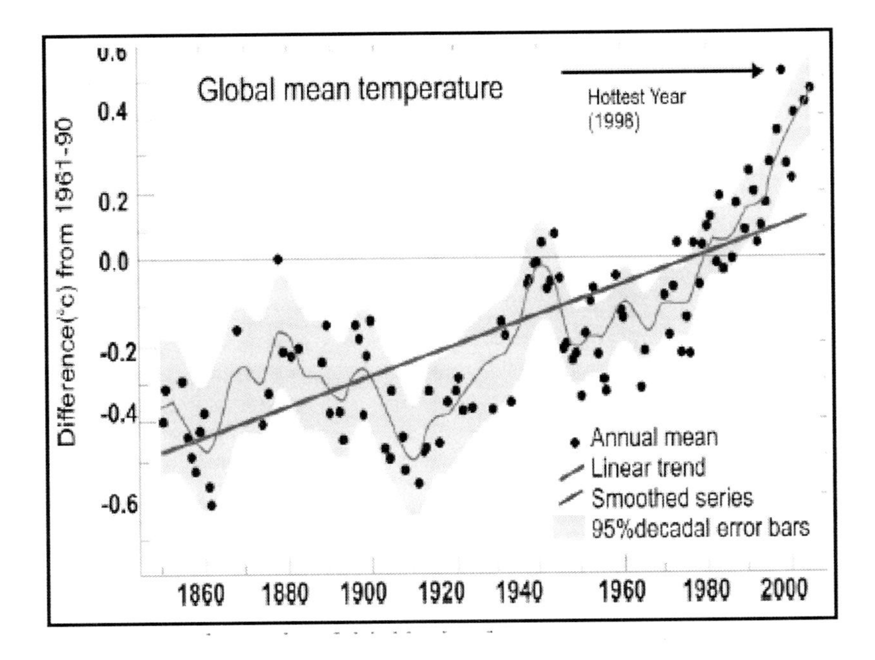

Fig. 1. Global mean temperature anomalies, 1860-2005 (IPCC 2007)
The temperature anomalies are calculated with respect to base period 1960-1990.

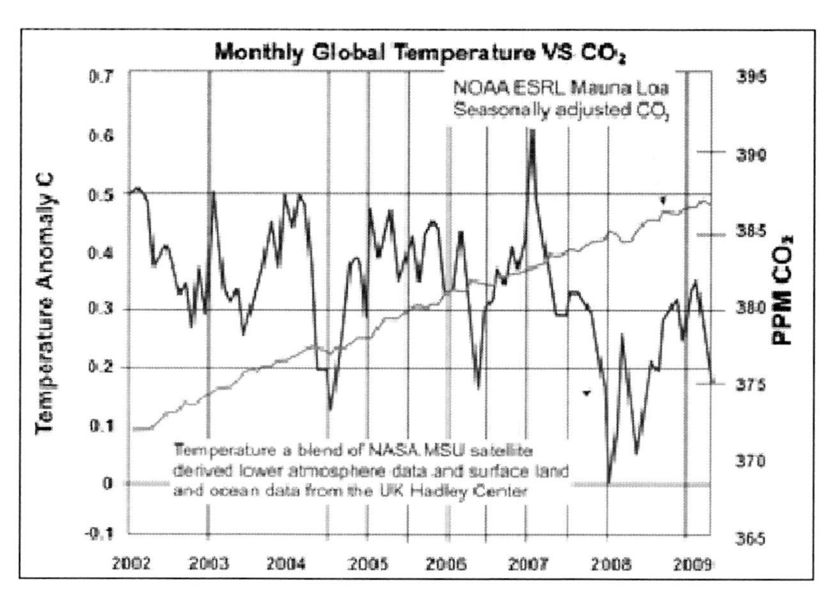

Fig. 2. Global mean temperature variation since 2002, together with variation of atmospheric carbon dioxide (parts per million, by volume).

3. Sea level rise, past & present

It is now generally accepted that since the Last Glacial Maximum (LGM) 21000 yr BP (Before Present), the sea level has risen by about 120 m. Most of this rise occurred as the vast amounts of ice sheets and ice caps in northern and Arctic latitudes started to melt as the earth's mean temperature steadily warmed. The melting of the high-latitude ice mass was essentially completed by about 5000-6000 yr BP (Douglas & Peltier 2002). Thereafter, global sea level rise was very small and appears to have ceased by 3 to 4 thousand yr BP. The rate of SLR over the last 1000 years and prior to the twentieth century is estimated to be just about 0.2 mm/yr (Fleming et al 1998; Lambeck 2002). The sea level rise in the late twentieth century is most intriguing and has sparked a large number of studies in the last decade (Douglas & Peltier 2002; Munk 2002; Church et al 2004; Holgate & Woodworth 2004; Jevrejeva et al 2006; Holgate 2007). These and many other studies provide following general assessment for the twentieth century sea level rise: **a.** The rate of sea level rise was larger in the first half of the 20th century (2.03 mm/yr) than in the second half (1.45 mm/yr). **b.** There is evidence of 2-13 yr variability in sea level records increasing during the last fifty years over most ocean basins. **c.** Among major sources of uncertainty are; inadequate distribution of tide gauges particularly in the southern hemisphere, inadequate information on geophysical signatures (e.g. glacial isostatic adjustment & tectonic activity) and relatively short duration of satellite altimetric data for the latter half of 20th century.

The GIA (Glacial Isostatic Adjustment) refers to the gradual springing back of the earth's surface, especially in the higher latitudes, in response to the removal of ice loads of the LGM which were at their maximum extent around 21000 yr BP. The GIA is still significant in the region around the Gulf of Bothnia (often referred to as Fennoscandia) which was covered with ice to a depth of several kilometers during the LGM and where relative sea level is currently falling at the rate of 5-10 mm/yr as the land in that region continues to rebound. The issue of GIA is well articulated by Prof R Peltier (University of Toronto Canada) and his associates in a series of papers (Peltier 1996, 1998, 2001). A recent paper (Peltier 2009) obtains revised estimates of sea level rise due to glacier melts using the GRACE (Gravity Recovery & Climate Experiment) data from a special satellite launched in March 2002. The GRACE Data provide finer estimates of glacier melt contribution against the backdrop of post-glacial rebound of the earth's surface; these estimates will be discussed later.

4. How fast are glaciers melting?

This has become an intensely debated topic in the global warming science today. Several media reports of rapid melting of polar glaciers and news items about ice shelves of various sizes and magnitude breaking away and sliding into polar oceans have crated a heightened interest about escalated sea level rise in the next decade or more. This issue also appears to have generated considerable interest in the scientific community with a number of papers, articles & scientific commentaries appearing in recent literature (e.g., Munk 2003; Cazenave 2006; Meier et al 2007; Mitrovica et al 2009; Bahr et al 2009). These and many other recent papers now strongly suggest increased melting of high-latitude glaciers in the last ten years and this is leading to substantially increased contribution to the total sea level rise. Further, it is now speculated that this glacier mass contribution would be accelerating in the next few decades leading to much higher value of SLR than what the IPCC has projected in its 2007 Climate Change Documents. In an Associated Press (AP) news item dated 25 February 2009, Colin Summerhayes, executive director of the UK-based Committee on Antarctic Research suggested that *glaciers in Antarctica are melting faster than previously thought and such accelerated melting could cause sea levels to climb by 3 to 5 feet (~75 cm to 130 cm) by the end of the twenty-first century.* In a scientific commentary by Meier et al (2007), the authors conclude that *contribution by small glaciers (other than two ice caps, Greenland and Antarctic) has accelerated over the past decade and this accelerated melting could cause 0.1to 0.25m of additional sea level rise by 2100.* In another recent paper, the authors (Bahr et al 2009) conclude that *if the climate continues to warm along current trends, a minimum of 373 +/- 21 mm of sea level rise over next 100 years is expected from glaciers and ice caps.*

The idea that '*world-wide glaciers and ice caps are melting faster than before*' seems to be gaining traction among the scientific community. What is however, not clear at this point in time is: how fast is sea level rising, at global and/or regional level? Studies reported in recent literature using the satellite-based altimeter data (known as Topex/Poseidon) suggest the sea level rise between 2.8 to 3.2 mm/year, but other studies using tide gauge data or a combination satellite and tide gauge data show a significantly smaller value. In a recent paper, Holgate (2007) analyzed nine long and nearly continuous sea level records over one hundred years (1903-2003) and obtained a mean value of SLR as 1.74 mm/yr with higher rates of sea level rise in the early part of twentieth century compared to the latter part. **Figure 3** shows cumulative sea level rise through the twentieth century as obtained by Holgate. In a most comprehensive recent study, Wunsch et al (2007) generate over 100 million data points using a 23-layer general circulation model with a 1^0 horizontal resolution. The general circulation model uses different types of observed data, like ocean salinity, sea surface temperature, satellite altimetry and Argo float profiles over a period 1993-2004. Based on careful analysis of such a large database, the authors (Wunsch et al) obtain a global mean value of SLR as 1.6 mm/year for the period 1993-2004. The authors also identify several uncertainties and regional variations in the altimetric data and conclude that *"it remains possible that the database is insufficient to compute sea level trends with the accuracy necessary to discuss the impact of global warming – as disappointing as the this conclusion may be"*

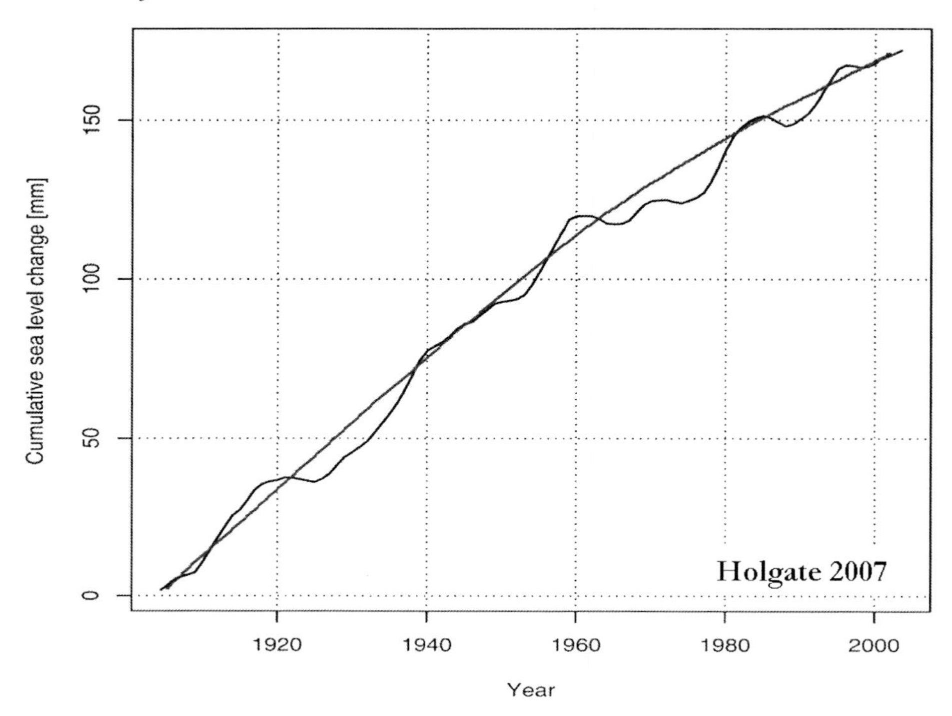

Fig. 3. Cumulative sea level rise 1900-2003. The total sea level rise in the first half of twentieth century was about 10 cm, while for the latter half the rise was about 7 cm. (Holgate 2007)

5. Sea level rise using GRACE data

As mentioned earlier, the GRACE (Gravity Recovery & Climate Experiment) satellite launched in March 2002 has provided valuable data so far on the rate of mass loss from polar ice sheets and also on the post-glacial rebound of the earth's surface. The GRACE data appear to provide confirmation of the ongoing glacial rebound of the North American Continent caused by deglaciation of the Laurentide, Innuition and Cordilleran ice sheets that began about 21000 yr BP. (Paulson et al 2007; Peltier & Drummond 2008). Such a confirmation allows an improved estimate of sea level rise by removing GIA contamination of sea level rise due to recent glacial melts. Two important recent papers, one by Cazenave et al (2008) and the other by Peltier (2009) have provided useful estimates of sea level changes due to various processes. These findings of Cazenave et al & Peltier can be summarized as follows:

From 2003 to 2008, the total ocean mass contribution to sea level rise is now estimated as about 2.27 mm/yr. Of this, the contribution from the two ice sheets (Greenland and Antarctica) is estimated at about 1.0 mm/yr, while the contribution from remaining (small) glaciers and ice caps is estimated at about 1.1 mm/yr; additional contribution of about 0.17 mm/yr is from land waters due to future desiccation (of land areas). The thermal contribution is estimated as the difference between satellite altimteric sea level and the GRACE-based ocean mass component and comes out to be about 0.37 mm/yr. The total sea level rise in the most recent period 2003-2008, is now estimated to be about between 2.5 to 2.65 mm/yr, which is less than the satellite altimetric value of 3.1 mm/yr. Cazenave et al further observe that the thermal contribution during the 1993-2003 period was estimated to be about 1.2 mm/yr which has now declined to just about 0.37 mm/yr in the most recent period 2003-2008. Once again, the glacier (and ice cap) melting is identified as the dominant contributor to the present sea level rise, while the thermal contribution appears to have declined, possibly in response to declining sea surface temperatures (SST) in the last five years. This aspect of sea level rise and its linkage to temperature change of the 20th century is discussed below.

6. Sea level rise in the context of 20th century temperature change:

The earth's temperature history for the twentieth century (Figure 1) shows two warming periods, one during the early part of the twentieth century (1910-1945) and the second since about 1977 till 1998, after which the mean temperature appears to have 'leveled off'. It is also instructive to note that there was a distinct period of cooling 1945-1977, when the earth's mean temperature declined by about 0.25C. The sea level variation during the twentieth century is most enigmatic and deserves a careful assessment. According to Holgate (2007), the total rise was about 17 cm (see **Figure 3**) with higher rise (~10 cm) during the first half and lower rise (~7cm) during the second half of the 20th century. The decadal rise in sea level as shown in **Figure 4** reveals interesting variations with lowest decadal rise (-1-49 mm/yr) centered on 1964, while the highest rise is centered on 1980 (5.31 mm/yr). The latter part of **Figure 4** suggests increasing rates of sea level rise, with larger fluctuations of low and high values. Can these decadal variations be linked to land-ocean temperature changes of the last thirty years?

Let us consider global SST variations as shown in **Figure 5.** Here the global SST anomalies (with respect to a mean value for the 20th century) are plotted as 12-month moving averages and the SSTs appear to be increasing since 1980, reaching a peak value by early 1998. This peak SST value on a global scale is most certainly linked to the intense El Nino event of 1997/98 which helped raise SST over other ocean basins as well (see Arun Kumar et al 2001). The global SSTs have declined rather sharply since the peak of 1998 and this may help explain a significant decline in the thermal contribution to sea level rise during the last five years in particular. Future sea level rise due to thermal expansion (steric component) is expected to be smaller in view of low climate sensitivity values obtained in several recent studies.

The warming of the earth's climate during the early part of twentieth century (1910-1945) and its possible impact on glacier melt appears to have been neglected in the present debate, by the media and by the scientific community as well. A close look at Figure 1 shows that the earlier warming (1910-1945) was quite steep from about 1920 to 1935, especially in the Arctic region. According to Chylek et al (2005), the Arctic warmed at a faster rate during

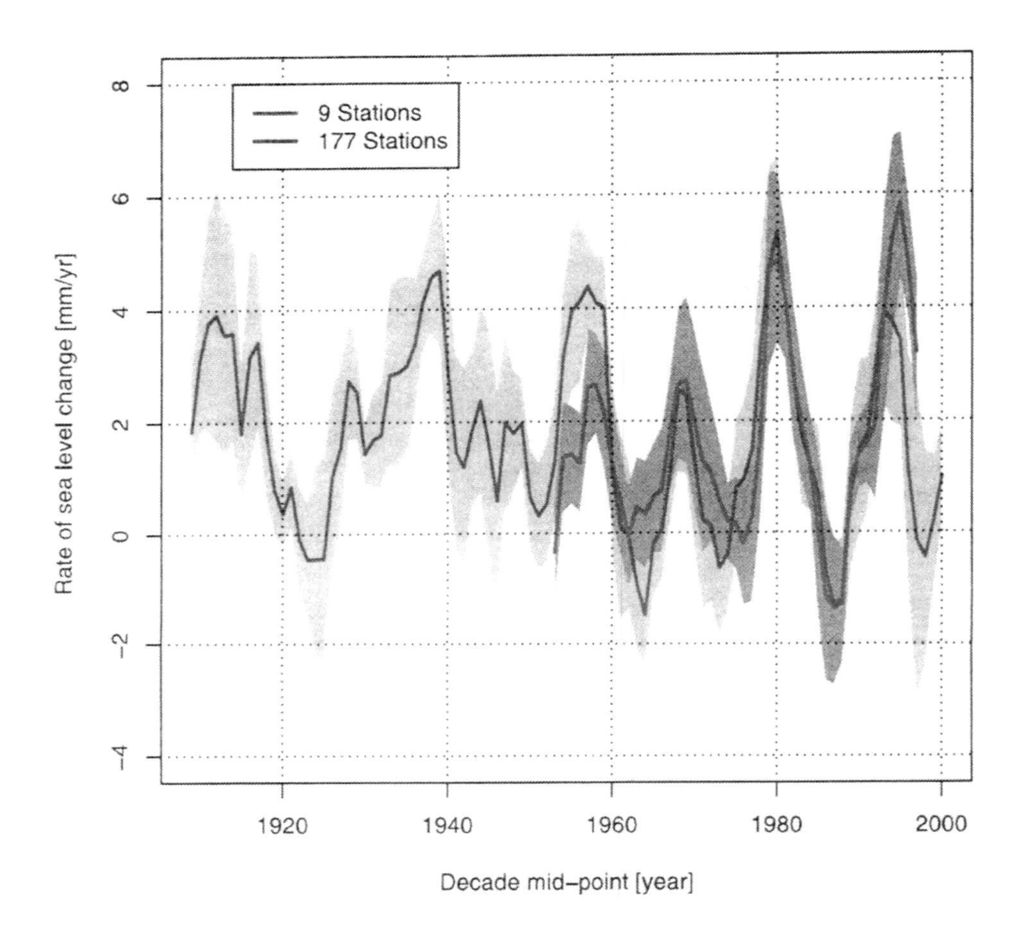

Fig. 4. Global mean decadal rates of sea level changes based on nine long tide gauge records (blue), compared with rates from 177 stations (red). All rates are corrected for glacial isostatic adjustment and inverse barometric effects. Shaded region indicates +/- 1 standard error (from Holgate 2007)

1920-1930 than it did during 1995-2005 and in the early part of the new millennium. According to Dr Igor Polyakov of the University of Alaska USA, who has collected an excellent set of data for the Arctic Basin from 1860 till present, the Arctic Basin was at its warmest in 1935 and 1936. In another recent study, Vinther et al (2006) have extended Greenland temperature records to 1874 using long-term temperature data available with the Danish Meteorological Institute. Vinther et al show that the decade 1930s and the 1940s were the warmest decades in Greenland and 1941 was the warmest year in Greenland in the 135-year temperature data of Greenland.

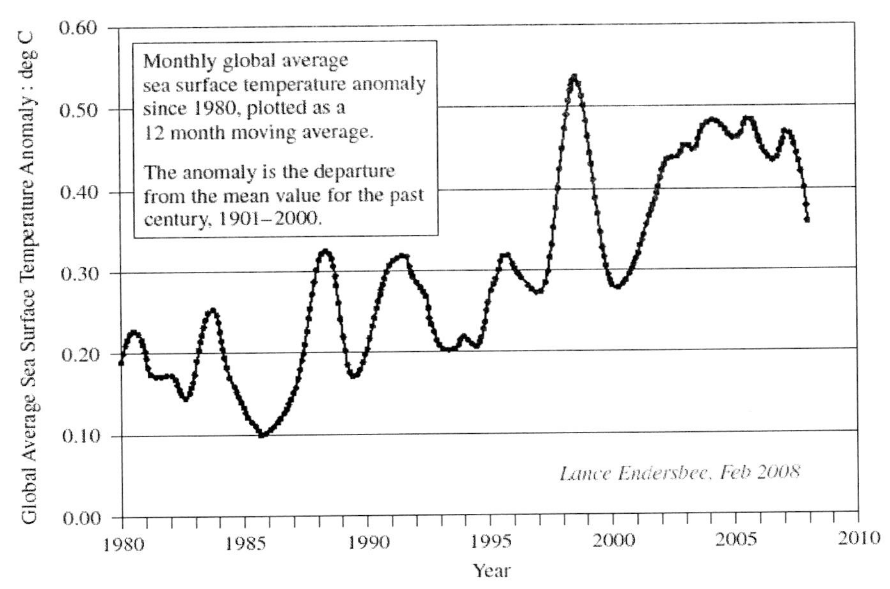

Fig. 5. Global average sea surface temperature anomalies plotted as a 12-month moving average. (Source: (late) Prof Lance Endersbee, Australia)

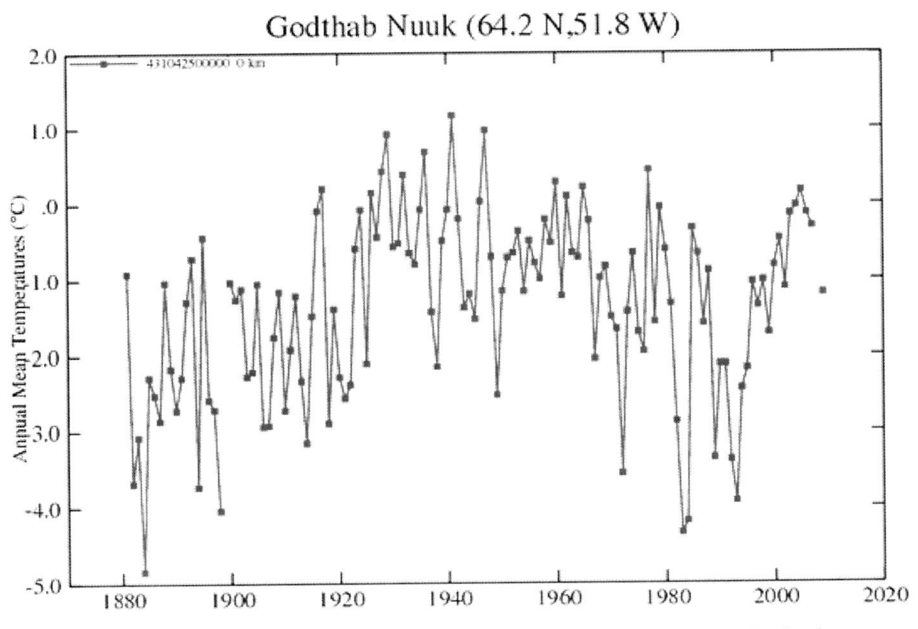

Fig. 6. Mean surface temperature history for Greenland 1880-2007. Note the highest temperature in 1941.

Figure 6 shows the temperature history of Greenland from 1860 and reveals clearly how Greenland was at its warmest in 1941. Since then Greenland temperature has declined till about 1980 and has climbed up to about the same level it was in 1941. A scientific commentary on the Arctic conditions in the 1920s reads: "*The Arctic Ocean is warming up, icebergs are growing scarcer and in some places the seals are finding the water too hot. Reports all point to a radical change in climate conditions and hitherto unheard of temperatures in the Arctic zone. Great masses of ice have been replaced by morains of earth and stones, while at many points well-known glaciers have entirely disappeared" (US Weather Bureau 1922).*

It is obvious from above that significant glacier and ice cap melting was going on in the Arctic during the 1920s and the 1930s. However, no satellite or other remote-sensed technology was available to obtain reliable estimates of glacier melts and rates of melting during the period 1920-1940, when most of the Artic was very warm and Greenland witnessed perhaps the 'hottest climate' in a 150-year instrumented temperature data. The impact of glacier melts of the 1920s and 1930s on subsequent sea level rise can be assessed by noting that total sea level rise since 1940 till 2008 is about 12 cm. Of this total rise, only about 8 cm (80 mm) can be attributed to glacier melt while the remaining is attributable to thermal expansion of water. Using this argument, we can estimate present glacier melt contribution to future sea level rise as about 12 cm or about 120 mm at most, in the next 100 years or so. This estimate is significantly less than the estimate obtained in some of the recent studies (e.g., Meier et al 2007; Bahr et al 2009). Adding the thermal contribution of about 110mm over next 100 years, we obtain a total sea level rise of just about 230 mm or 23 cm by 2100 (Khandekar 2009).

7. Observed sea level variations at selected locations

Here we present examples of observed sea level variations in a few selected locations. These locations have been well-publicized in the present climate change debate as *being threatened by escalating sea levels and flooding.* Among the locations most often discussed in the media and also in scientific community: **1.** The Maldives **2.** Tuvalu & Vanuatu islands in the south Pacific and **3.** Bangladesh and the Bay of Bengal region. Let us look at each of these three locations and assess the sea level change situation.

- **The Maldives**: These group of islands (about 1200 or more) in the equatorial Indian Ocean (~ $2\text{-}5^0$ N, $72\text{-}74^0$ E), about 1000 km southwest from the southern tip of India, have become an icon in the current debate on escalating sea level rise and the Islands *disappearing* in the next few decades. The question of sea level rise in the Maldives has been extensively studied by Prof N-A Morner of the University of Stockholm Sweden, who was the President of the INQUA Commission for Sea Level Changes & Coastal Evolution. Morner & his associates have prepared a number of reports on sea level changes and the morphology of these islands (Morner 2004a,b; 2007). Morner's findings on sea level changes and morphology of the Maldives can be summarized as: *The Maldives lie right in the centre of the earth's deepest geoid depression of about -100m. Sea level records on the islands show several rapid oscillations, due to local and regional factors, which are not linked to global sea level changes. The people of Maldives survived a higher sea level about 800 to 1000 yr BP, when the sea level was estimated to be about 50-60 cm higher than present. In the 1970s the sea level there fell by 20-30 cm, probably from an increased evaporation linked to Monsoon circulation changes. In the equatorial Indian Ocean, the dynamic sea level is significantly lowered with respect to the geoid by strong regional evaporation. Many other sharp spikes in sea level are linked to dynamic changes in sea level there with ocean circulation playing* *a significant role in combination with changes in air pressure, evaporation, precipitation etc.. The sea level at the Maldives is NOT affected by global sea level changes.*

- **Tuvalu and Vanuatu islands**: These two islands are located in the south Pacific about 1000 km north & west of the Fiji Islands. These islands have become iconic in the global warming debate with the people of Tuvalu threatening a legal action against the USA for "*ongoing flooding on the island*", while the small community living on the island of Vanuatu are often being identified as "*The first climate change refugees*" because of possibility of the island being submerged in a few years. (Vanuatu New Port Villa Press Online, December 21 2005). The sea level changes at Tuvalu and Vanuatu are shown in **Figure 7.** The Tuvalu record extends from 1978 to 2007, while the Vanuatu sea level tide-gauge covers the period 1993-2006. Both these records show no significant changes in recent years. The Tuvalu record shows irregular oscillations, with two of the lowest values (in early 1983 and early 1998) coinciding with the strong El Nino events of 1982/83 and 1997/98. It is widely recognized now that major El Nino events can produce significant changes in sea levels over the entire equatorial Pacific Basin from the Indonesian Archipelago in the east to Ecuador on the west coast of South America. The record at Vanuatu does not show any El Nino signature as the Vanuatu island is slightly south of 15^0 latitude, where El Nino-induced sea level changes are expected to be minimal. Besides El Nino-induced changes, the two records do not show any increasing or decreasing trend which can be linked to global warming or climate change.

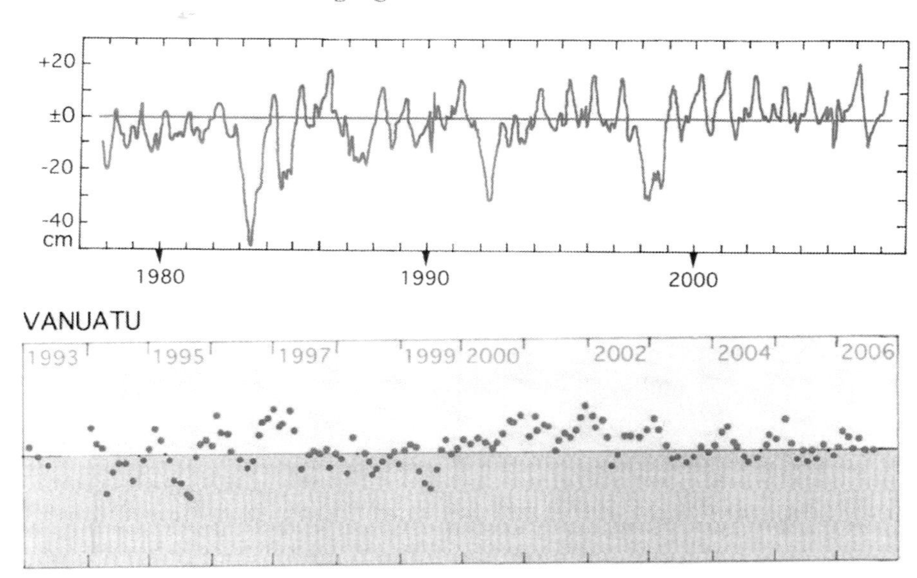

Fig. 7. Sea level records for the two south Pacific islands, Tuvalu and Vanuatu

- **Bangladesh & the adjacent area of Bay of Bengal:** Bangladesh is one of the most densely populated countries with a total present population of about 140 million, in a land area of just about 144000 km^2 A significant percentage of its population lives in the

Ganges River (*Holy River Ganga worshipped in India)* Delta, known as the *Sundarbans, the world's largest marsh area.* The Delta region is only a few m above the sea level and is subject to frequent flooding during the summer monsoon season and also when tropical cyclones from the Bay of Bengal hit the Delta region. The storm surges generated by tropical cyclones have killed hundreds of thousands of people over last four centuries (Murty and Neralla 1996). In recent times, an intense tropical cyclone struck the Delta region on November 13 1970, killing an estimated 200,000 thousand people, largest fatalities in a single weather-related disaster! There are no long-term tide gauge data available in the Delta region. The most recent paper by Morner (2010), analyzes sea level change over the Delta using two sea level records from nearby India, one at Visakhapatnam on east coast of India and the other at Mumbai on the west coast. The sea level changes at these two locations are shown in **Figure 8.** The Figure shows locations of Mumbai and Visakhapatnam in conjunction with Kotka, a location in the Delta region where Morner's latest study analyzes sea level changes. Based on a careful assessment of sea level changes in the Bay of Bengal region, Morner concludes that *there is NO global sea level rise signature in the Delta region of Bangladesh at this point in time. Flooding during the Monsoon season and associated river flooding remains a major problem. However, sea level rise is not a major problem for the Sundarbans region of Bangladesh today.*

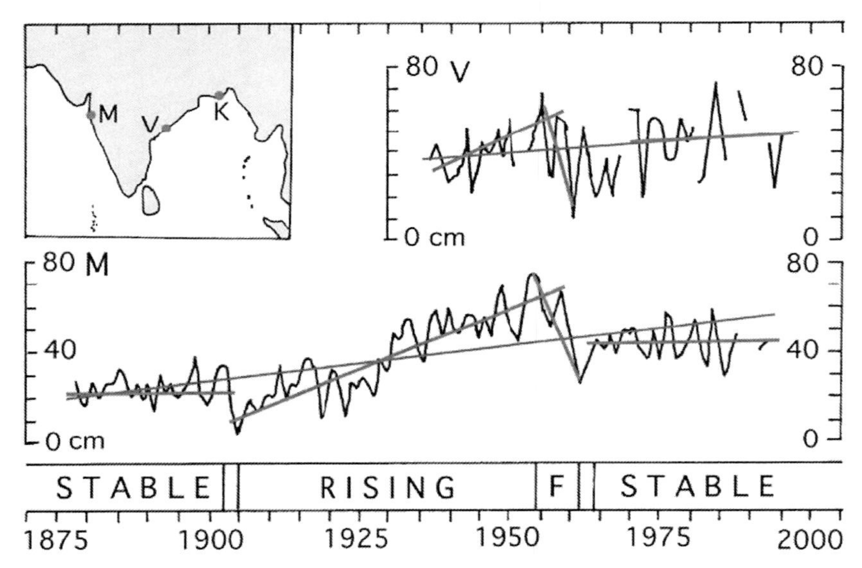

Fig. 8. Sea level trends at Mumbai (M) and Visakhapatnam (V). K denotes a location in the Sundarbans region of Bangladesh where sea level changes are analyzed (Morner 2010).

8. Concluding remarks

The present debate on global warming, climate change and glacier melt has highlighted the possibility of escalated sea level rise in the next few decades. A careful analysis of latest studies suggests that since about 2007, global sea level rise is estimated as between 2.3 to 2.8 mm/yr. A climatological analysis of past sea level changes in the twentieth century allows an estimate for future sea level rise over the next 100 years of about 23 cm +/- 5 cm with a 95% confidence. In view of declining global SSTs and possible continued cooling of the earth's mean temperature over the next decade (Keenlyside et al 2008), the best guess value for sea level rise from now until 2025 is estimated to be just about 30 mm +/- 10 mm. This value appears modest and does not pose a major threat to some of the most vulnerable areas of the world like the Maldives, Bangladesh and the south Pacific islands of Tuvalu and Vanuatu.

9. Acknowledgements

It is a pleasure to acknowledge useful communications with Prof Morner of the University of Stockholm, Sweden. Thanks are also due to my wife Shalan for her technical help in reproducing the Figures used in this Chapter.

10. References

Arun Kumar et al 2001: The sustained North American warming of 1997 & 1998. *Journal of Climate,14, p.345-353*

Chylek P, M K Dubey & G Lesins 2006: Greenland warming of 1920-1930 and 1995-2005. *Geophysical Research Letters 33 L11707 doi:10.1029/2006GL026510*

Chylek P, M K Dubey & U Lohmann 2008: Aerosol radiative forcing and climate sensitivity deduced from the Last Glacial Maximum to Holocene transition. *Geophysical Research Letters 35 L04804 doi;10.1029/2007GL02759*

Bahr D B, M Dyurgerov & M Meier 2009: Sea-level rise from glaciers and ice caps: A lower bound. *Geophysical Research Letters 36 L030501 doi:10.1029/2008GL036309*

Cazenave Anny 2006: How fast are the ice sheets melting? *Science 314 p.1250-1252*

Cazenave Anny et al 2008: Sea level budget over 2003-2008: A reevaluation from GRACE space gravimetry, satellite altimetry and Argo. *Global & Planetary Change, doi:10.1016/j.gloplacha.2008.10.004*

Douglas B C and W R Peltier 2002: The puzzle of global sea level rise. *Physics Today March 2002 p. 35-40 American Institute of Physics USA*

Flemming K et al 1998: Refining the eustatic sea level curve since the Last Glacial Maximum using far- and intermediate-field sites. *Earth and Planetary Science Letters 163 p.327-342*

Holgate S J 2007: On the decadal rates of sea level changes during the twentieth century. *Geophysical Research Letters 34 L01602 doi:10.1029/2006GL028492*

Holgate S J and P L Woodworth 2004: Evidence for enhanced sea level rise during the 1990s. *Geophysical Research Letters 31 L07305 doi:10.1029/2004GL019626*

Jevrejeva S et al 2006: Nonlinear trends and multiyear cycles in sea level records. *Journal of Geophysical Research 111 C09012 doi:10.1029/2005JC003229*

Keenlyside N S et al 2008: Advancing decadal-scale climate prediction in the North Atlantic sector. *Nature 453 p.84-88 doi:10.1038/nature06921*

Khandekar M L, T S Murty & P Chittibabu 2005: The global warming debate: A review of the state of science. *Pure & Applied Geophysics 162 p. 1557-1586.*

Khandekar M L, 2009: Global warming and sea level rise. *Energy & Environment 20 p. 1067-1074*

Knutti R et al 2008: A review of uncertainties in global temperature projections over the twenty-first century. *Journal of Climate 21 p.2651-2663*

Lambeck K 2002: Sea-level changes from mid-Holocene to recent times: An Australian example with global implications. *Ice Sheets, Sea Level and the Dynamic earth, Mitrovica & Vermeersen (Eds), Geodynamic Series, Vol 29 p. 33-50*

Lindzen R S 2007: Taking global warming seriously. *Energy & Environment 18 p. 937-950*

Meehl G A (plus 80 other auhors) 2007: Global Climate Projections. *Climate Change 2007: The Physical Science Basis-Contribution of Working Group I to the Fourth Assessment Report of the Intergovernmental Panel on Climate Change (IPCC), S Solomon et al (eds) p. 748-845 Cambridge University Press*

Meier M et al 2007: Glaciers dominate eustatic sea-level rise in the 21st century. *Science 317 p.1064-1067.*

Mitrovica J X, N Gomez & P Clark 2009: The sea-level fingerprint of west Antarctic collapse. Science 323 p.753

Morner N-C 2004a: New perspectives for the future of the Maldives. *Global & Planetary Change 40 p.177-182*

Morner N-C 2004b:Sea level changes: Are low-lying areas and coastal areas under threat? *The impact of climate change: an appraisal for future p.29-35 International Policy Press UK*

Morner N-C *2007:* Sea level changes and tsunamis. Environmental stress and migration over the seas. *Internationales Asienforum 38 p.353-374*

Morner N-C 2010: Sea level changes in Bangladesh: new observational facts. *Energy & Environment 21 p. 235-249.*

Munk W 2002: Twentieth century sea level: An enigma. *Proc. Of the National Academy of Sciences (USA) V 99 p. 6550-6565*

Munk W 2003: Ocean freshening, sea-level rising. *Science 300*, p.2041-2043.

Murty T S and V R Neralla 1996: El-Nino and storm surges in the Bay of Bengal. *Land-based and marine hazards M I El-Sabh et al (eds) Kluwer Pub. p. 97-111*

Paulson A, S Zhong & J Wahr 2007; Inference on mantle viscosity from GRACE and relative sea level data. *Geophysical Journal International 171 p.497-508*

Peltier W R 2009: Closure of the budget of global sea level rise over GRACE era: the importance and magnitude of the required corrections for global isostatic adjustment. *Quaternary Science Reviews 28 p. 1658-1674*

Peltier W R 1996: Global sea level rise and glacial isostatic adjustment: an analysis of data from the east coast of America. *Geophysical Research Letters 23 p.717-720*

Peltier W R 1998: Postglacial variations in the level of sea: implications for climate dynamics and earth geophysics. *Reviews of Geophysics 36 p.603-689*

Peltier W R 2001: Global glacial isostatic adjustment and modern instrumental records of relative sea level history. *Sea Level Rise: History & Consequences. B C Douglas M Kearney & S Leatherman (eds) Academic Press. International Geophysical Series Volume 75 p. 65-95*

Peltier W R and R Drummond 2008: Rheological stratification of the lithosphere:a direct reference upon the geodetically observed pattern of the glacial isostatic adjustment of the North American continent. *Geophysical Research Letters, 35,L16314 doi:10.1029/2008GL034586*

Revelle R and H E Suess 1958: Carbon dioxide exchange between atmosphere and ocean and the question of an increase of atmospheric CO_2 during the past decades. *Tellus 9 p. 18-27*

Schwartz S E 2007: Heat capacity, time constant and sensitivity of earth's climate system. *Journal of Geophysical Research 112 D24S05 doi;10.1029/2007JD008746*

Wunsch C, R M Ponte and P Heimbach 2007: Decadal trends in sea level patterns:1993-2004. *Journal of Climate, 20, p.5889-5911*

5

Potential Changes in Hydrologic Hazards under Global Climate Change

Koji Dairaku
National Research Institute for Earth Science and Disaster Prevention, Japan

1. Introduction

The water cycle is vital to life and ecosystems. The water cycle regulating climate stability and variability has complex interactions within the climate system and plays a basic role in it, therefore changes in the water cycle have arguably the most substantial impact on climate change.

Water-related natural disasters have increased in the last decade due to population growth, land use and land cover change such as urbanization or cultivation, and climate change. Historically, stable climatic conditions have been assumed for water resource management, planning, and civil engineering design. Although some predictions remain uncertain, the Intergovernmental Panel on Climate Change (IPCC) concluded that precipitation in high latitudes is increasing as decreases occur in most subtropical land regions. Extremes such as hot extremes, heat waves, and heavy precipitation will continue to become more frequent and tropical cyclones -- typhoons and hurricanes -- will become more intense [1]. Changes in the water cycle and extreme threats to life and ecosystems, such as droughts, floods, and soil erosion, constitute a basic global-warming concern. Water resource planners must increasingly quantify the risk of extremes [2].

Global climate changes may lead to changes in rainfall events by enhancing atmospheric moisture content [3]. Saturation vapor pressure governed by the Clausius-Clapeyron equation is observed and projected to increase by about 7% K^{-1} [4, 5]. A large portion of water vapor exists in the lower troposphere. Therefore column-integrated water vapor is primarily weighted by the lower troposphere. On the other hand, water vapor in the upper troposphere is relatively low and not directly constrained by the thermodynamic arguments. Soden et al. [6] reported a distinct radiative signature of upper tropospheric moistening by satellite measurements from 1982 to 2004. It also supports the robustness of the projected increase in column-integrated water vapor with global warming. Moisture is supplied to moderate or heavy precipitation mainly by transport, not from local evaporation. Three to five times the radius of the precipitating region expands to collect moisture over the area [4], therefore convective rainfall drawing in increased moisture from the surrounding atmosphere is expected to increase at the rate of water vapor, thus increasing the risk of extreme events such as flooding and heavy snowfall.

Emanuel [7] reported that the longevity and intensity of tropical cyclones have increased over the last 30 years. Webster et al. [8] concluded that the number of recent tropical cyclones of categories 4 and 5 in the last 35 years has increased. Due to their rare occurrence and the large natural variability of tropical storms, however, it is difficult to attribute change to global warning signals of climate change. A relatively high resolution (T106) atmospheric general circulation model (AGCM) projected the increase in precipitation of tropical cyclones in response to increased atmospheric moisture content despite decreased their frequency and intensity [9]. A higher resolution AGCM (20km-mesh) suggests that decrease of frequency in tropical cyclone (but increase in the North Atlantic), increase of the number of intense tropical cyclones, and increase in maximum surface wind speed of the most intense tropical cyclone [10]. Because they fail to provide compelling evidence, these reports and research are inconclusive. The events recently experienced are high-impact, amply evident of the urgency we face in learning to adapt to climate changes and cope with climate variability.

Prospects are increasing for changes in extremes such as droughts and floods profoundly impacting on society and the economy. These prospects are only now being adequately faced or addressed after release of the fourth assessment report of the IPCC [1] in studies on the impacts of climate change. Changes in extremes affect agriculture, hydrology, water resources, and the economies of many Asian countries where billions of people live. To adapt to climate changes in the face of drought, flood, and soil erosion, which frequently seriously threaten life and ecosystems, predictions and risk assessments in this area are required more frequently by policymakers.

Hydrologic predictions accounting for global climate changes mainly use GCMs [3, 11, 12, 13, 14]. Despite numerous studies, the regional responses of hydrologic changes -- atmosphere-ocean-land interactions, precipitation, and extreme events such as droughts and floods -- resulting from climate change remain unclear and involve many uncertain factors, e.g., the large amplitude of natural variability and anthropogenic influences such as black carbon, deforestation, and irrigation. Coarse spatial resolutions with grid spacing of approximately 300 km and uncertain physical processes limit the representation of terrestrial water/energy interaction and the variability and extremes in such systems as the Asian monsoon. Only very restricted regional-scale estimates are available to planners. Relatively high-resolution GCMs are now being developed and ensemble projections producing useful probability distributions can be provided to risk assessment models by improved computer capabilities.

We address two recent attempts of hydrologic projection under global climate change – (i) projected extreme events in the Asian summer monsoon with relatively high-resolution ensemble simulations and (ii) a case study of flood risk assessment in a watershed in Japan using multi-model ensemble projections based on atmosphere-ocean coupling general circulation models (AOGCMs) contributed to IPCC AR4.

2. Extreme events in the Asian summer monsoon with high-resolution GCM

The response of the Asian summer monsoon to global warming is of critical concern to Asia's large population. Despite numerous studies, however, the response remains unclear. Past studies have investigated factors that control the Asian summer monsoon, showing greater South Asian summer monsoon precipitation and interannual variability caused by enhanced land-sea thermal contrast [15] or by warmer ocean sea surface temperatures (SSTs) that provide more moisture [16] during global warming. Past studies have also shown a northward shift in lower tropospheric monsoon circulation and increased summer monsoon rainfall due to increased water vapor in the warmer atmosphere [17, 18, 19, 20, 21].

May [22] suggested weakened monsoon circulation and greater precipitation. Mitchell and Johns [23] theorized that additional forcing by aerosols could weaken monsoon circulation. Low-resolution climate models with grid spacing of approximately 300 km are particularly limited in forecasting possible changes in variability and extremes of daily rainfall during the Asian summer monsoon, and few modeling studies have been conducted [17, 24]. Many researchers have argued that increases in mean South Asian summer monsoon rainfall arise primarily from enhanced atmospheric moisture. Increases in water vapor can explain only part of the change, however, especially changes in daily precipitation [25, 26]. Emori and Brown [25] showed dynamic enhancement in lower latitudes mainly over the oceans and suppression over subtropics. Thermodynamic effects, however, were enhanced almost globally except for subtropical ocean, in mean change, corresponding to the increase in atmospheric moisture content. Taking results from high-resolution ensemble simulation with the AGCM [27, 28] on the earth simulator, we studied daily precipitation during the Asian summer monsoon.

2.1 Model and experiments

We ran time-slice experiments -- a five-member ensemble for the control and a seven-member ensemble for 2 × CO_2 -- using the Center for Climate System Research/National Institute for Environmental Studies/Frontier Research Center for Global Change (CCSR/NIES/FRCGC) AGCM. This model featured T106 spectral truncation in the horizontal (approximately 1.1° on an equivalent grid) and 56 vertical layers.
Each ensemble member covered 20 years. Ensemble simulation enhanced the statistical reliability of changes diagnosed in precipitation, particularly extreme events. Lower boundary forcing for control runs included SST and sea-ice concentration data for 1979–1998 provided by the Hadley Centre Global Sea Ice and Sea Surface Temperature (HadISST) dataset [29]. Each ensemble member was run with different initial conditions taken from a multiyear spinup run. Additional forcing for the 2 × CO_2 runs included seasonally varying SST anomalies derived from transient climate change experiments from different coupled GCMs -- CCSR, NCAR-CSM, NCAR-PCM, MPI, GFDL, CSIRO, and CCC. Output from these models was obtained from the IPCC Data Distribution Center (DDC), and additional forcing was added to control run forcing as "warming patterns." CO_2 concentrations were 345 ppmv for controls and 690 ppmv for 2 x CO_2 experiments.
The control ensemble showed uncertainty due to internal atmospheric variability, and results from 2 x CO_2 ensembles showed uncertainties from both internal variability and projected externally forced SST change.

2.2 Change in mean South Asian summer monsoon features

Precipitation and typical large-scale circulation patterns, e.g., strong easterlies in the upper troposphere and strong westerlies in the lower troposphere (Fig. 1a) agreed with observations during the summer monsoon season from June to September (JJAS, not shown). A general increase in the 2-meter air temperature by 2–6 K was projected in the region during JJAS. Greater warming occurred over land areas, such as the Tibetan Plateau, than over the ocean (Fig. 1b). Figure 1c shows a northward shift in lower tropospheric wind systems and enhanced precipitation over land in South Asia -- changes that agree with those in previous studies [17, 18, 19, 20]. Greater evaporation in this area enhanced the local water vapor supply (not shown). Larger moisture content in warmer air enhanced the water vapor flux, which increased Asian summer monsoon rainfall over land (Figs. 1c and 1d). Changes in precipitation indexes -- area-averaged rainfall near India (8.5°–30°N, 65°–100°E) and over land only [30] -- were 3.36% and 9.93%. Two dynamic indexes related to vertical zonal wind shear (U-shear) associated with Walker circulation [31] and vertical meridional wind shear (V-shear) associated with local Hadley circulation [32] could indicate projected change in the large-scale monsoon flow. Changes in U-shear were -7.46% and in V-shear -9.31%. These changes in weakening are attributable to shifts in the mean location of convergence over India and the tropical Indian Ocean. Current geostationary indexes may not, however, be appropriate [18] if positions of circulation systems change as a result of global warming.

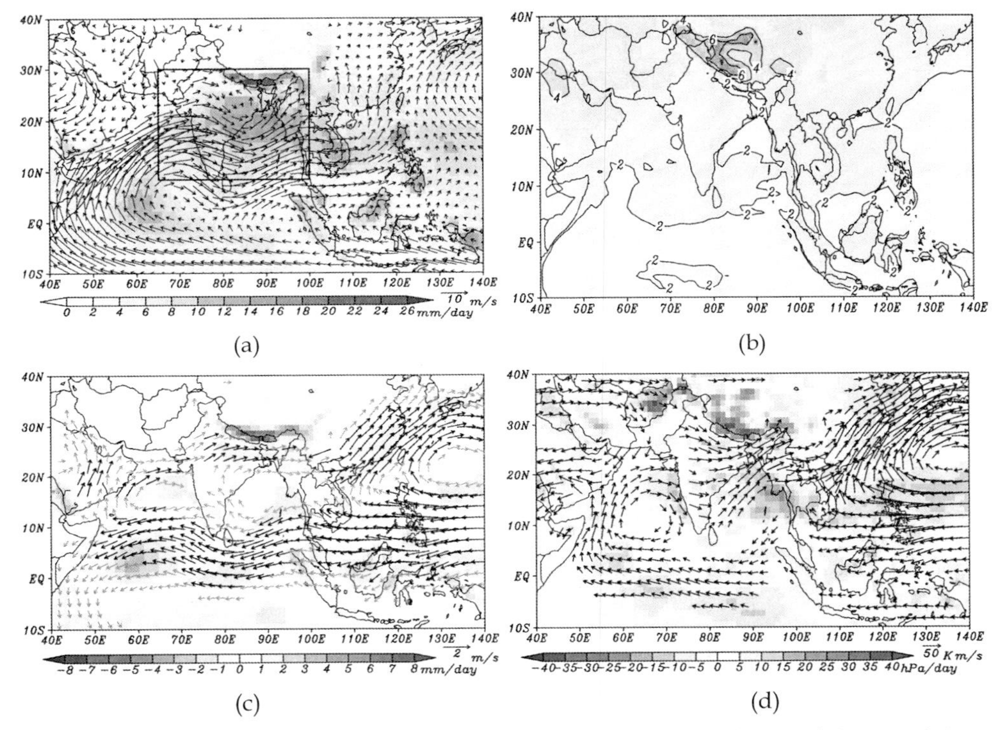

Fig. 1. (a) Simulated present mean precipitation and horizontal wind at 850 hPa over Asia during JJAS. Projected mean changes in (b) 2 m air temperature, (c) precipitation and horizontal wind at 850 hPa, and (d) vertically integrated water vapor flux and upward vertical velocity at 500 hPa.

The rectangle in (a) outlines the region (8.5°–30°N, 65°–100°E) used in Figs. 2-5. The temperature change in (b) exceeds the 99% confidence level. Shading in (c) shows precipitation changes statistically significant beyond the 95% level. Gray and black vectors in (c) show changes in wind speed exceeding 0.5 and 1.0 m/s; the reference vector is 2 m/s. Vectors in (d) show changes in water vapor flux exceeding 2.0×10^4 m/s; the reference vector in (d) is 5.0×10^4 m/s.

2.3 Change in extreme precipitation

Figure 2 shows the frequency distribution of daily precipitation over land in South Asia (8.5°-30°N, 65°-100°E) and China (20°-40°N, 100°-130°E) during JJAS. Simulated daily rainfall frequency in control experiments agreed relatively well with Global Precipitation Climatology Project (GPCP 1DD) data, although strong (approximately 20-30 mm/day) and extreme precipitation (>60 mm/day) events were overestimated. In China, simulated precipitation agrees well with observations. The limited validity of simulated extreme events in South Asia is partly due to the lack of observation because station measurements of daily precipitation are available only over limited land regions with dense population. Model validity should be refined by further improving parameterization. The intensification of extreme precipitation (>50 mm/day) occurred in 2 x CO_2 experiments, and the change exceeded the uncertainty range of one ensemble standard deviation in both South Asia and China (thin dashed lines in figure).

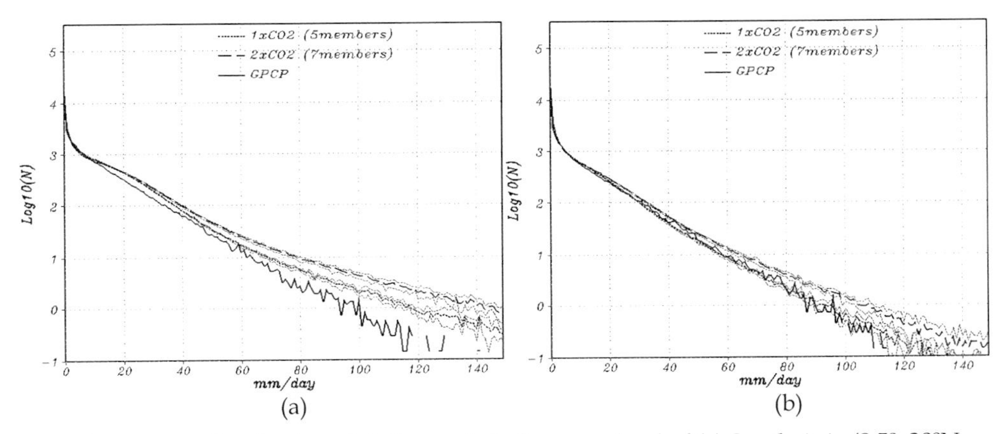

Fig. 2. Frequency distribution of daily precipitation over land of (a) South Asia (8.5°-30°N, 65°-100°E) and (b) China (20°-40°N, 100°-130°E) during JJAS. Solid lines show observations from the GPCP 1DD (1997-2003). The short dashed line and long dashed line denote the ensemble average from the control and 2 × CO_2 experiments. Thin dashed lines show the standard deviation of each ensemble.

Projected increase in mean South Asian summer monsoon rainfall is attributable primarily to enhanced atmospheric moisture as argued in previous studies. The question remains, however, whether the increase in water vapor alone is sufficient to explain the change in monsoon precipitation. Only part of changes in daily precipitation could be explained by the increase in water vapor. An investigation of the influence of dynamic and thermodynamic changes on possible daily precipitation is thus addressed.

Figure 3 shows expected daily precipitation to daily 500 hPa vertical velocity (dynamic regime). In contrast to the usual definition of pressure velocity, however, the positive values of ω showed ascent, and negative ω showed subsidence. The relative frequency of extreme upward motion was very small, but heavy precipitation would accompany a strong upward motion regime. A substantial enhancement of expected precipitation intensity for a given dynamic regime was noticeable and exceeded the uncertainty range of one ensemble standard deviation (thin dashed lines). Changes in extreme precipitation (17.4%; average of

500–1000 hPa/day) were largely consistent with precipitable water increases (18.9%) in the region of South Asia. It is almost same in China. The changes were quite likely due to increases in atmospheric moisture content, i.e., thermodynamic changes.

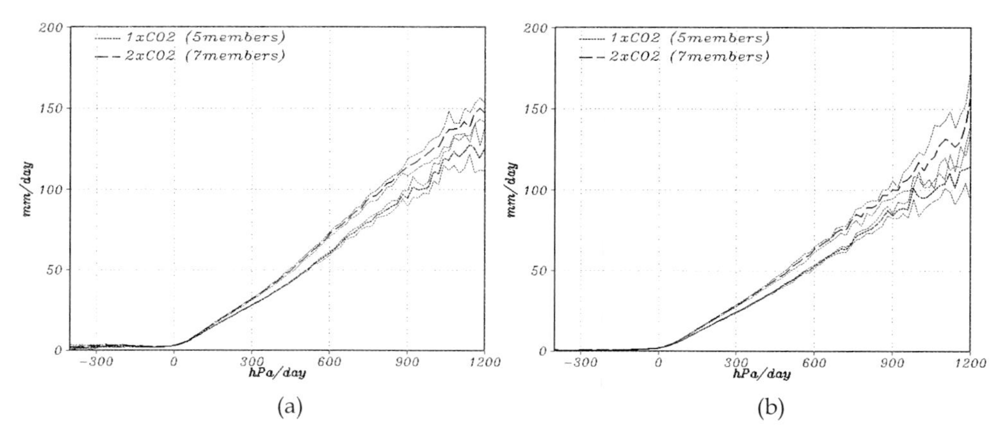

Fig. 3. As in Fig. 2, except for the expected daily precipitation intensity to daily 500 hPa vertical velocity.

Figure 4 shows spatial distribution of dynamic and thermodynamic contribution for projected precipitation changes in South Asia [25, 26]. Precipitation changes resulting from changes in dynamic circulation are *dynamic change*. *Thermodynamic change (nondynamic)* includes changes in precipitation features for given dynamic conditions and corresponds to intrinsic changes in precipitation mainly as a result of changes in atmospheric moisture content. *Covariation* is the change in these two effects. Covariation is not shown because the effect was negligible. Increased moisture convergence from enhanced atmospheric moisture is a thermodynamic effect. Increased moisture convergence from intensified convergence is a dynamic effect.

Figure 4 (a) and (b) shows dynamic and thermodynamic changes to mean precipitation change. Dynamic change reduces precipitation in a large portion of South Asia. Thermodynamic change generally enhances precipitation. Dynamic change, thermodynamic change, and covariance of mean precipitation change (9.93%) over land [8.5°–30°N, 65°–100°E] were 0.30%, 9.64%, and -0.01%. In regions including land and ocean, contributions of these components to mean precipitation change (3.36%) were -1.87%, 5.55%, and -0.32%. Mean precipitation changes in global warming scenarios over land during the South Asian summer monsoon primarily arose from thermodynamic effects. Dynamic changes over land had only a small influence. Dynamic changes over regions including land and ocean that reduced the intensification of precipitation played a relatively large role. The dynamic component decreased (increased) daily precipitation due to decreased (increased) relatively strong (weak) upward motion due to the shifted mean (to ascent) and reduced variability in vertical motions (not shown). The mean spatial distribution of dynamic changes was related to reduced variability in vertical motions and was partly a result of an anticyclonic anomaly associated with a northward shift in circulation systems over the Arabian Sea and the Bay of Bengal (Fig. 1[c]). Thermodynamic changes, in contrast, increased daily precipitation in most dynamic regimes and were the primary factor enhancing precipitation.

Figures 4 (c) and (d) indicate these contributions to extreme events -- the vertical velocity exceeds 500 hPa/day; it approximately corresponds to the rainfall event of about 50 mm/day in South Asia [8.5°–30°N, 65°–100°E] as in Figs. 2 and 3. The dynamic component in extreme events (>500 hPa/day) contributed more to precipitation changes than the thermodynamic component in a large portion of the regions. Compared to thermodynamic change to total precipitation change (Fig. 4[b]), thermodynamic change to extreme precipitation was the secondary factor in enhancing precipitation. On the southern edge of the Tibetan plateau, for instance, dynamic change rather than thermodynamic change contributes to precipitation.

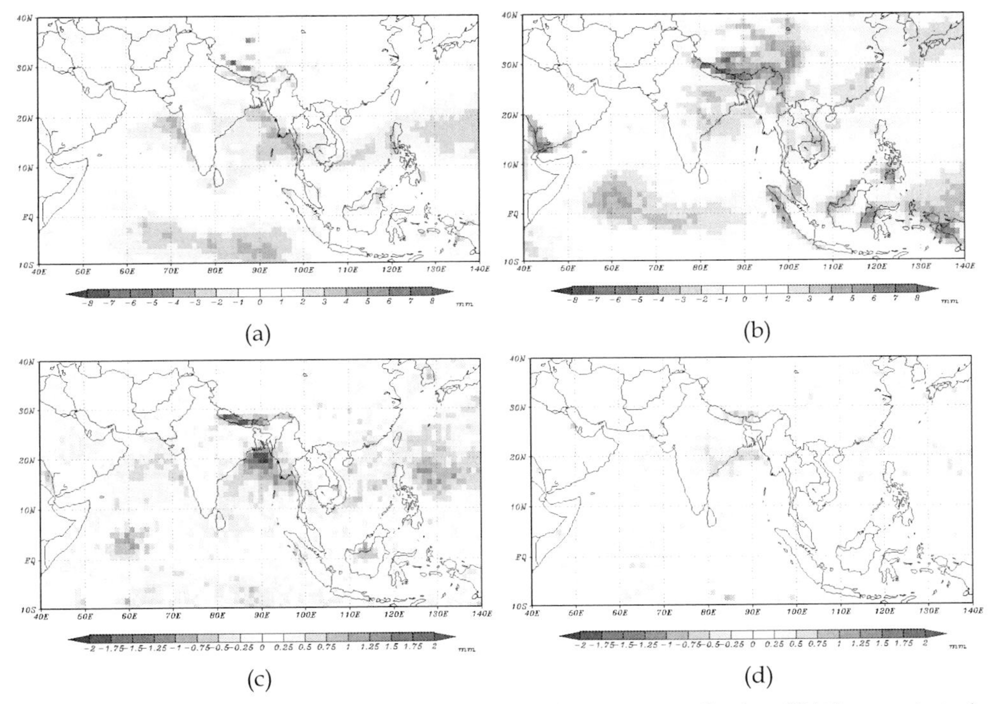

Fig. 4. Spatial distribution of dynamic and thermodynamic contribution (JJAS) to projected precipitation change. (a) Total dynamic component. (b) Total thermodynamic component. (c) Dynamic component to extreme events -- vertical velocity exceeds 500 hPa/day, (d) as in (c), but for the thermodynamic component.

2.4 Discussion

Results of projected extreme events in the Asian summer monsoon with relatively high-resolution ensemble simulation showed changes in the South Asian summer monsoon resulting from climate change. Model results under global warming conditions suggested more warming over land than over the ocean, a northward shift of lower tropospheric monsoon circulation, and an increase in mean precipitation during the Asian summer monsoon. The northward shift of circulation systems weakens large-scale monsoon indexes.

This shift may arise from enhanced land-ocean thermal contrasts in South Asia and a modulation of Hadley-Walker circulation due to projected El Niño-like warming patterns in the tropical Pacific, although such a pattern is not present in all model projections.

The number of extreme daily precipitation events has increased significantly. Increases in mean and relatively strong precipitation were attributed to greater atmospheric moisture content -- a thermodynamic change. Changes in relatively strong precipitation resulting from climate change were consistent with increases in precipitable water. Although dynamic changes limited the intensification of mean precipitation, enhanced extreme precipitation over land in South Asia arose from dynamic rather than thermodynamic change.

Where moist convection predominated in the tropics, a net radiative cooling anomaly in the upper troposphere and a net radiative heating anomaly at the surface in response to increased CO_2 increased upward latent heat transport. Resulting upper tropospheric warming increased the dry static stability (not shown), and enhanced stability suppressed the variability in vertical motions over most of the region investigated. The enhancement of extreme upward motion that occurred near mountainous areas, such as the southern edge of the Tibetan Plateau, regardless of changes in dry static stability, could be associated with a northward shift in monsoon circulation systems and could lead to more frequent extreme rainfall events on a regional basis -- a dynamic effect.

3. Flood risk assessment in a watershed in Japan using multi-model ensemble projections based on 12 AOGCMs

A key parameter in basin water management is local precipitation, and climate-change effects on heavy precipitation must be accurately estimated to prevent flood disasters. Relatively high-resolution GCMs are being developed and ensemble simulation by models are becoming available, but more frequently required basin-scale estimates to adapt to climate changes in the face of natural disaster are not sufficiently available to policy makers. Downscaling approaches can be used to produce finer scale information from GCMs with coarse spatial resolution via (i) statistical downscaling and (ii) dynamical downscaling [33, 34, 35, 36, 37].

The sections that follow provide a case study of flood risk assessment in a watershed in Japan using multi-model ensemble projections based on 12 state-of-the-art AOGCMs provided by the IPCC-DDC [1] using statistical downscaling approach [38].

3.1 Study area and analysis data

The Tama River basin, an urbanized watershed near Tokyo, Japan, (139.76E, 35.69N) (Fig. 5), is 1240 km^2 in area for a main stream 138 km long. The basic and estimated high-water discharges are 8700 and 6500 m^3/s, determined using 2-day precipitation with a 200-year return period.

We studied changes in the 200-year quantile in Tokyo caused by global warming using results from the 12 GCMs based on simulated precipitation data in 1981-2000 (2000), 2046-2065 (2050), 2081-2100 (2100), 2181-2200 (2200), and 2281-2300 (2300) under SRES A1B and B1 scenario conditions.

Table 1 lists GCMs and their simulated precipitation in 2000. Annual average precipitation, 2-day maximum precipitation, 40th largest 2-day precipitation, and extreme precipitation were calculated and compared to observed data. Regrettably, model bias is relatively large and no model can agree well with observation.

Heavy precipitation was defined as that in a partial duration series (PDS) [39] composed of 40 largest 2-day precipitations for 20 years.

In other words, the PDS was the time series exceeding the threshold which was set to the 40th largest 2-day precipitation. The frequency distribution for precipitation in the PDS is

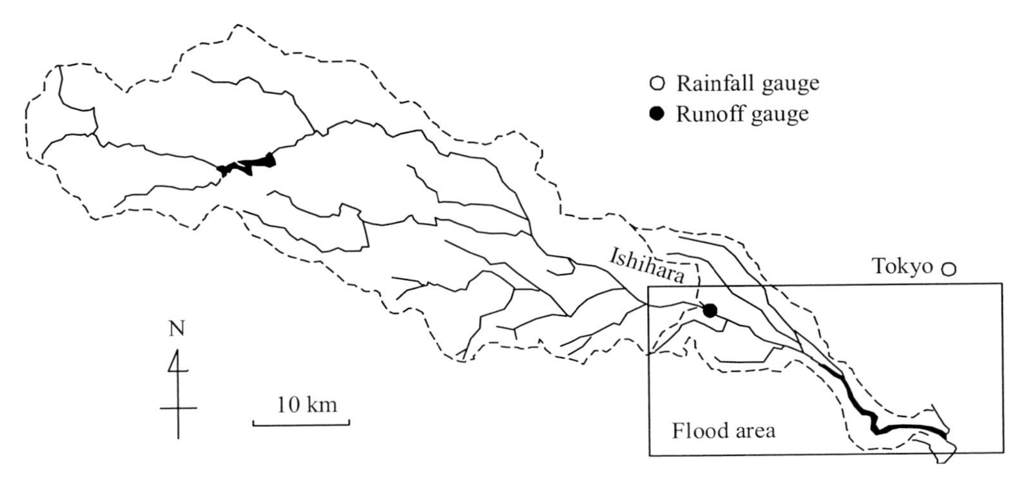

Fig. 5. Tama River basin overview.

Model	IPCC ID	Resolution long. × lat. (degree)		Precipitation (1981-2000) Annual Average (mm/year)	2-day Maximum (mm/2-day)	2-day 40th (mm/2-day)	Quantile (1981-2000) 100-year (mm/2-day)	200-year (mm/2-day)
Observed (Tokyo)				1479	294	86	376	415
Model emsemble				1733	151	66	170	184
cccma_cgcm3_1	CGCM3.1(T47)	3.8	3.7	1834	170	71	216	236
cnrm_cm3	CNRM-CM3	2.8	2.8	2227	214	86	266	290
csiro_mk3_0	CSIRO-Mk3.0	1.9	1.9	1600	124	61	139	149
gfdl_cm2_0	GFDL-CM2.0	2.5	2.0	1671	175	66	201	219
giss_aom	GISS-AOM	4.0	3.0	1795	68	49	89	94
giss_model_e_r	GISS-ER	5.0	4.0	1065	66	41	72	76
iap_fgoals1_0_g	FGOALS-g1.0	2.8	2.8	2077	144	77	162	173
ipsl_cm4	IPSL-CM4	3.8	2.5	2106	363	104	344	376
miroc3_2_hires	MIROC3.2(hires)	1.1	1.1	1823	143	70	147	157
miroc3_2_medres	MIROC3.2(medres)	2.8	2.8	1863	108	63	128	136
miub_echo_g	ECHO-G	3.8	3.7	1138	110	46	130	141
ncar_pcm1	PCM	2.8	2.8	1601	123	57	145	157

Table 1. GCMs with resolutions and simulated precipitation in present climate.

set as dimensionless using maximum (x_{max}) and threshold precipitation (x_0) in each model (Figure 6). The ensemble average of dimensionless precipitation frequency in 2000 agrees with that observed, and its probability density function is approximated by an exponential distribution. We also clarified that the frequency distribution does not change in 2050, 2100, 2200, or 2300.

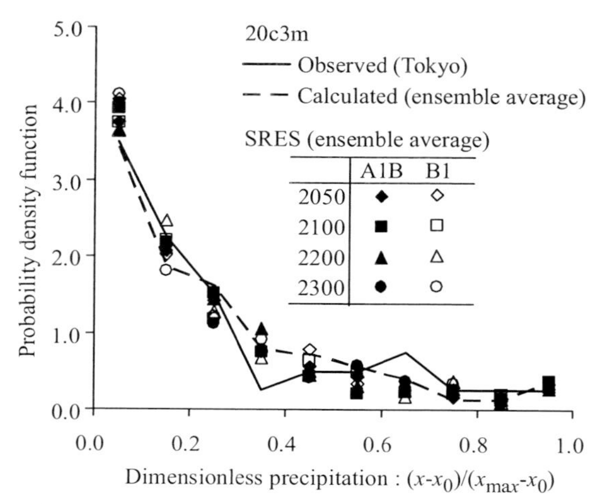

Fig. 6. Frequency distribution of precipitation in the PDS.

3.2 Changes in 200-year quantile caused by global warming

Figures 7 show changes in average precipitation in PDS caused by global warming. The values in 2000 are set to 1 in each model, and the ratio is used to calculate the ensemble average. The ensemble average ratio of change to the present one is 1.09-1.20 in the A1B scenario and 1.03-1.07 in the B1 scenario. Almost all model output in the A1B scenario indicates that future precipitation will exceed that in the present (Fig. 7(a)). Some model output indicates a trend toward a slight decrease in the B1 scenario (Fig. 7(b)).
In changes in the projected 200-year quantile caused by global warming (Figures 8), the ratio of the ensemble average of this quantile to the present one is 1.07-1.20 in the A1B scenario, indicating that heavy precipitation will slightly increase but not a statistically significant trend. The ratio remains stable at 1.0 in the B1 scenario, however, possibly because of less enhanced atmospheric moisture content associated with greenhouse gas concentration lower than that in the A1B scenario.

3.3 Global warming impact on flood risk

To assess changes in the estimated high-water discharge in the Tama River basin in the A1B scenario, we conducted rainfall runoff analyses under present geophysical conditions using the kinematic runoff model and unit hydrograph method to calculate direct discharge and base flow at Ishihara (Fig. 5) [38].
The kinematic runoff model [40] considers topography, land cover, channel networks, and storage facilities. The basin was divided into subbasins, each of which was modeled using two slopes and a channel. Slope and channel flows are approximated by a kinematic wave. Effective rainfall was calculated using a cumulated-retained curve. Flood risk was evaluated

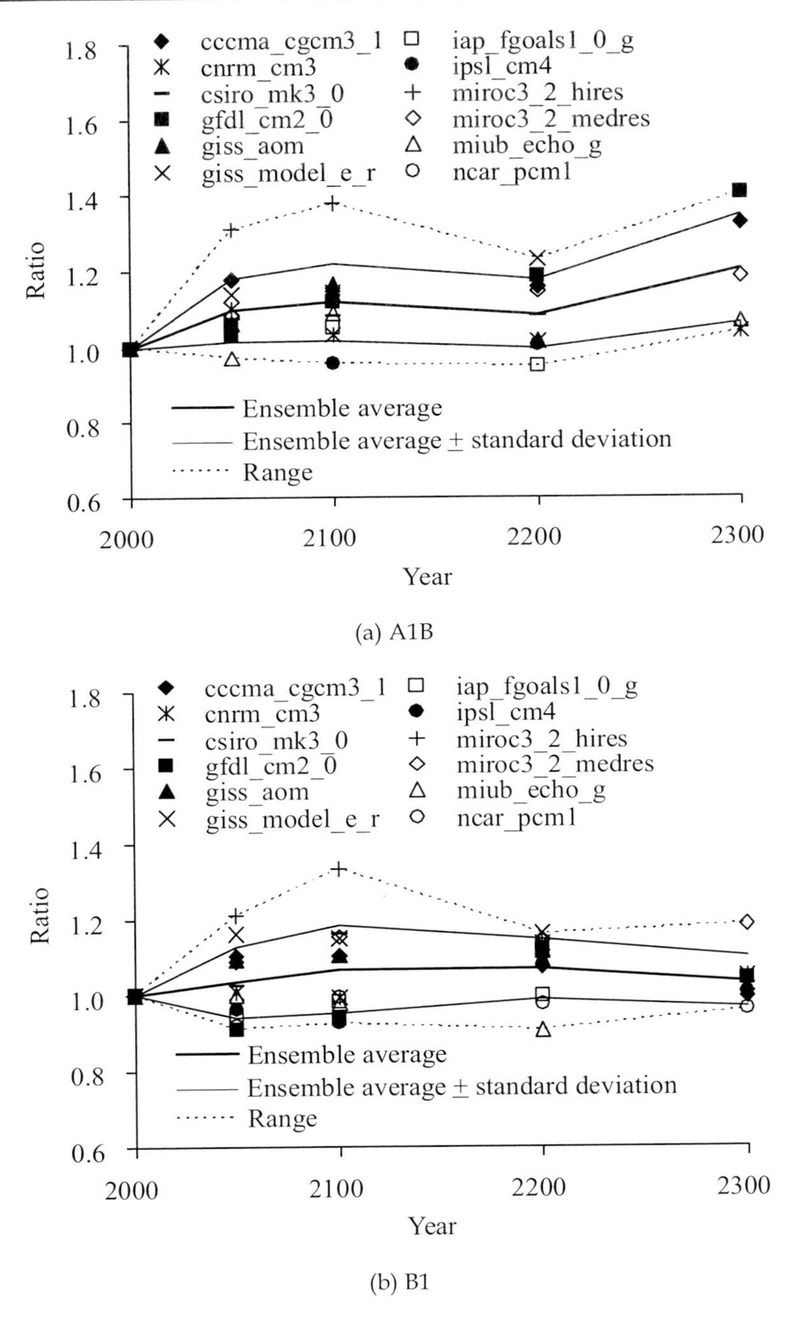

Fig. 7. Changes in average precipitation in the PDS caused by global warming.

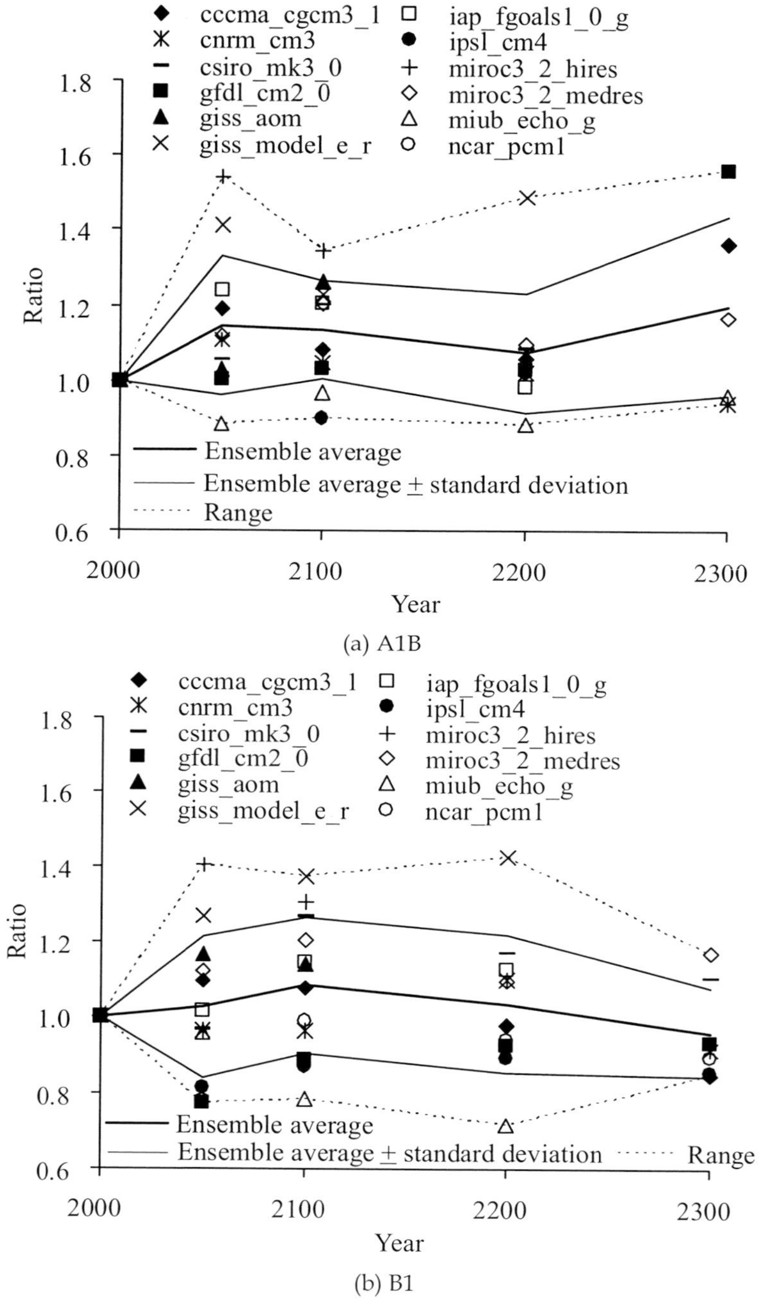

(a) A1B

(b) B1

Fig. 8. Changes in the 200-year quantile caused by global warming.

using numerical simulation for precipitation with a 200-year return period. The downstream area at Ishihara was defined as the inundation flow analysis area (Fig. 5). Tama River flow was analyzed one-dimensionally applying St. Venant equations, and flood plain inundation was analyzed two-dimensionally. Flows in the river and flood plain were combined using a weir discharge formula [41].

The 200-year quantiles in 2000 (present), 2050, 2100, 2200, and 2300 were set at 457, 523, 519, 491, and 548 mm/2-day based on the ensemble average in Fig. 8(a). The 200-year quantile in 2000 (present) corresponds to the 63, 72, 106, and 58-year quantiles in 2050, 2100, 2200, and 2300. Although extreme precipitation varies quite greatly due to large multi-decadal natural variability and the nonlinear response of hydrological cycles to global warming, we concluded that the 200-year quantile extreme event in the present climate is projected to occur in much shorter return periods in the A1B scenario.

Hyetograph (Figure 9) was defined as observed hourly precipitation from 10:00 on August 30 to 10:00 on September 1, 1949 -- one of the largest 2-day precipitations -- and multiplied by a constant so that 2-day precipitation equals the 200-year quantile in each period. Simulated changes in high water discharge and flood volume in the A1B scenario show ratios of the estimated high-water discharge to the present one to be 1.10-1.26 and those of the flood volume to be 1.46-2.31 (Figure 9). Flood volume increases dramatically compared to the increase in precipitation (Figure 10).

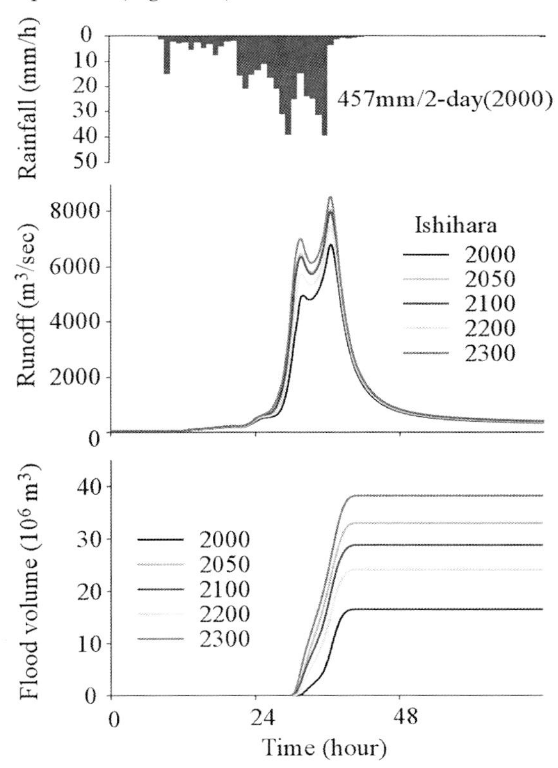

Fig. 9. Changes in hydrograph and flood volume in the A1B scenario.

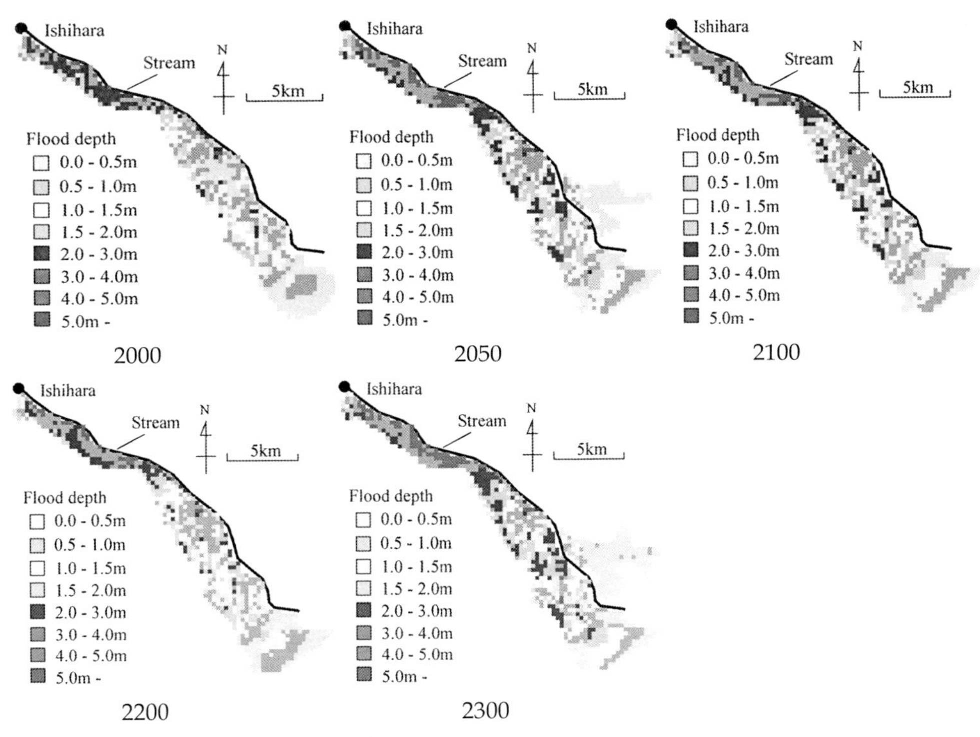

Fig. 10. Distribution of flood depth.

We used the multi-model ensemble average as a scenario of heavy precipitation for assessing the impact of climate change on risk of flood inundation. Even though heavy precipitation is slightly increased, the simulated results indicate the risk of flood in the basin is much higher than the present one in the A1B global warming scenario.

4. Summary

Two recent attempts at hydrologic projection in Asia were addressed. Time-slice ensemble experiments using a high-resolution (T106) AGCM on the earth simulator indicated changes in the South Asian summer monsoon resulting from climate change. Model results under global warming conditions suggested more warming over land than over the ocean, a northward shift of lower tropospheric monsoon circulation, and an increase in mean precipitation during the Asian summer monsoon. The number of extreme daily precipitation events increased significantly. Increases in mean and extreme precipitation were attributed to greater atmospheric moisture content -- a thermodynamic change. In contrast, dynamic changes limited the intensification of mean precipitation. Enhanced extreme precipitation over land in South Asia arose from dynamic rather than thermodynamic changes.

Results above obtained from high-resolution time-slice ensemble simulation are fairly robust. Ocean-atmosphere coupling is a basic feature of the Asian summer monsoon, and significant discrepancies exist between forced and coupled experiments [42, 43, 44]. Because dynamical downscaling by a regional climate model depends strongly on the results of parent GCMs, the robustness of results in the present study must be assessed using ensemble experiments based on high-resolution AOGCMs or AGCMs that are coupled to a slab ocean model.

Section 3 describes the impact of global warming on heavy precipitation features and flood risk, using 2-day precipitation of 12 AOGCMs. PDS-based frequency analysis indicated that multi-model ensemble average 200-year quantiles in Tokyo from 2050 to 2300 under IPCC SRES-A1B scenario climate conditions were 1.07-1.20 times as large as that under present climate conditions. The 200-year quantile extreme events in the present are projected to occur in much shorter return periods in the A1B scenario. Studying these influences on runoff discharge and flood risk in the Tama River basin using numerical simulation, we found that high-water discharge is projected to rise by 10%-26% and flood volume increase by 46%-131% in precipitation with a 200-year return period. Even though the increase of extreme precipitation as a result of global warming is not substantial, the risk of flooding in the basin is thus projected to be much higher than the present.

Climate-related disasters are serious problems in Asia. Advances in the understanding of meteorology and in the development of monitoring and forecasting systems have enhanced early warning systems, contributing immensely to reducing fatalities resulting from typhoons, cyclones, and floods. The frequency of extreme events causing water-related disasters has, however, been increasing in the last decade and may be increased in the future due to anthropogenic activity. The most advanced and trustworthy regional risk assessment for climate change is an urgent issue, and relatively high-resolution global climate models are not yet capabile of determining regional-scale feedback, especially between atmosphere and complex heterogeneous land surfaces such as topography and terrestrial ecosystems. Spatial resolution of less than 30 km grid spacing must thus be added and multi-model ensembles by RCMs and GCMs be conducted that include biophysical and biogeochemical processes to accurately assess critical interactions within systems.

5. Acknowledgments

The first part of the work was supported in part by the Global Environment Research Fund of Japan's Ministry of the Environment. Model simulations were made by the Earth Simulator at the Japan Agency for Marine-Earth Science and Technology for the Category 1 Research Revolution 2002 (RR2002) project of MEXT. We thank K-1 Japan project members for their support and feedback. The second part of this work was conducted as one of the research activities of the research project "Study on future changes in the global hydrologic cycle related disasters" of National Research Institute for Earth Science and Disaster Prevention. This research was partially supported by the resarch project on the disaster risk information platform by national research institute for earth science and disaster prevention, Japan. We also acknowledge the international modeling groups for providing their data for analysis, the PCMDI for collecting and archiving the model data.

6. References

[1] IPCC, Climate Change 2007: The physical science basis. Summary for policymakers, contribution of working group I to the fourth assessment report of the Intergovernmental Panel on Climate Change, 2007.

[2] Schnur, R., 2002, The investment forecast. *Nature*, 415, 483-484.
[3] Palmer, T.N., and Rälsänen, J., 2002, Quantifying the risk of extreme seasonal precipitation events in a changing climate. *Nature*, 415, 512-514.
[4] Trenberth, K.E., Dai A., Rasmussen, R.M., and Parsons, D.B., 2003, The changing character of precipitation. *Bulletin of the American Meteorological Society*, 84, 1205-1217.
[5] Held, I.M., and Soden B.J., 2006, Robust responses of the hydrological cycle to global warming. *Journal of climate*, 19, 5686-5699.
[6] Soden, B.J., Jackson, D.L., Ramaswamy, V., Schwarzkopf, M.D., Huang, X., 2005, The radiative signature of upper tropospheric moistening. *Science*, 310, 841-844.
[7] Emanuel, K., 2005, Increasing destructiveness of tropical cyclones over the past 30 years. *Nature*, 436, 686-688.
[8] Webster, P.J., Holland, G.J., Curry, J.A., and Chang, H.-R., 2005, Changes in topical cyclone number, duration, and intensity in a warming environment. *Science*, 309, 1844-1846.
[9] Hasegawa, A., and Emori, S., 2005, Tropical cyclones and associated precipitation over the Western North Pacific: T106 atmospheric GCM simulation for present-day and doubled CO2 climates, *Scientific Online Letters on the Atmosphere*, 1, 145-148.
[10] Oouchi, K., Yoshimura, J., Yoshimura, H., Mizuta, R., Kusunoki, S., and Noda, A., 2006, Tropical cyclone climatology in a global-warming climate as simulated in a 20 km-mesh global atmospheric model: Frequency and wind intensity analyses. *Journal of the Meteorological Society of Japan*, 84(2), 259-276.
[11] Coe, M.T., 2000, Modeling terrestrial hydrological systems at the continental scale: testing the accuracy of an atmospheric GCM. *Journal of Climate*, 13, 686-704.
[12] Koster, R.D., Suarez, M.J., and Heiser, M., 2000, Variance and predictability of precipitation at seasonal-to-interannual timescales. *Journal of Hydrometeorology*, 1, 26-46.
[13] Vörösmarty, C.J., Green, P., Salisbury, J., and Lammers, R.B., 2000, Global water resources: vulnerability from climate change and population growth. *Science*, 289, 284-288.
[14] Milly, P.C.D., Wetherald, R.T., Dunne, K.A., and Delworth, T.L., 2002, Increasing risk of great foods in a changing climate. *Nature*, 415, 514-517.
[15] Meehl, G.A., and Washington, W.M., 1993, South Asian summer monsoon variability in a model with doubled atmospheric carbon dioxide concentration. *Science*, 260, 1101-1104.
[16] Meehl, G.A., and Arblaster, J.M., 2003, Mechanisms for projected future changes in South Asian monsoon precipitation. *Climate Dynamics*, 21, 659-675.
[17] Bhaskaran, B., Mitchell, J.F.B., Lavery, J.R., and Lal, M., 1995, Climatic response of the Indian subcontinent to doubled CO2 concentrations. *International Journal of Climatology*, 15, 873-892.
[18] Kitoh, A., Yukimoto, S., Noda, A., and Motoi, T., 1997, Simulated changes in the Asian summer monsoon at times of increased atmospheric CO2. *Journal of the Meteorological Society of Japan*, 75, 1019-1031.
[19] Hu, Z.-Z., Latif, M., Roeckner, E., and Bengtsson, L., 2000, Intensified Asian summer monsoon and its variability in a coupled model forced by increasing greenhouse gas concentrations. *Geophysical Research Letters*, 27, 2681-2684.

[20] Ashrit, R.G., Douville, H., and Rupa Kumar, K., 2003, Response of the Indian monsoon and ENSO-monsoon teleconnection to enhanced greenhouse effect in the CNRM coupled model. *Journal of the Meteorological Society of Japan*, 81, 779-803.

[21] Douville, H., Royer, J.-F., Polcher, J., Cox, P.M., Gedeney, N., Stephenson, D.B., and Valdes, P.J., 2000, Impact of CO2 doubling on the Asian summer monsoon: Robust versus model-dependent responses. *Journal of the Meteorological Society of Japan*, 78, 421-439.

[22] May, W., 2004, Potential future changes in the Indian summer monsoon due to greenhouse warming: analysis of mechanisms in a global time-slice experiment. *Climate Dynamics*, 22, 389-414.

[23] Mitchell, J.F.B., and Johns, T.C., 1997, On modification of global warming by sulfate aerosols. *Journal of Climate*, 10, 245-267.

[24] May, W., 2004, Simulation of the variability and extremes of daily rainfall during the Indian summer monsoon for present and future times in a global time-slice experiment. *Climate Dynamics*, 22, 183-204.

[25] Emori, S., and Brown, S.J., 2005, Dynamic and thermodynamic changes in mean and extreme precipitation under changed climate. *Geophysical Research Letters*, 32, L17706.

[26] Dairaku, K., and Emori, S., 2006, Dynamic and thermodynamic influences on intensified daily rainfall during the Asian summer monsoon under doubled atmospheric CO2 conditions. *Geophysical Research Letters*, 33, L01704.

[27] Numaguti A., Takahashi M., Nakajima T., and Sumi A., 1997, Description of CCSR/NIES atmospheric general circulation model. *CGER's Supercomputer Monograph Report*. 3, pp 1-48. Center for Global Environmental Research, National Institute for Environmental Studies.

[28] Emori, S., Hasegawa, A., Suzuki, T., and Dairaku, K., 2005, Validation, parameterization dependence and future projection of daily precipitation simulated with a high-resolution atmospheric GCM. *Geophysical Research Letters*, 32, L06708.

[29] Rayner, N.A., Parker, D.E., Horton, E.B., Folland, C.K., Alexander, L.V., Rowell, D.P., Kent, E.C., and Kaplan, A., 2003, Global analyses of sea surface temperature, sea ice, and night marine air temperature since the late nineteenth century. *Journal of Geophysical Research*, 108, 4407.

[30] Sontakke, N.A., Plant, G.B., and Singh, N., 1993, Construction of all India rainfall series for the period 1844-1991. *Journal of Climate*, 6, 1807-1811.

[31] Webster, P.J., and YANG, S., 1992, Monsoon and ENSO: Selectively interactive systems. *The Quarterly Journal of the Royal Meteorological Society*, 118, 877-926.

[32] Goswami, B.N., Krishnamurthy, V., and Annamalai, H., 1999, A broad scale circulation index for interannual variability of the Indian summer monsoon. *The Quarterly Journal of the Royal Meteorological Society*, 125, 611-633.

[33] Dairaku, K., Emori, S., and Nozawa, T., 2005, Hydrological projection under the global warming in Asia with a regional climate model nested in a general circulation model. *Annual Journal of Hydraulic Engineering, JSCE*, 49(1), 397-402. (in Japanese with an English Summary)

[34] Dairaku, K., Emori, S., 2007, Potential hydrological change resulting from greenhouse warming: Climate change and water-related disasters of severe tropical storms in

East Asia, *Research Signpost "Geophysics"*, Tomonori Matsuura, Ryuichi Kawamura Eds., pp.105-123.

[35] Koji Dairaku, Seita Emori, Toru Nozawa(2008): Impacts of Global Warming on Hydrological Cycles in the Asian Monsoon Region, Advances in Atmospheric Sciences, 25, No. 6, pp.960-973

[36] Koji Dairaku, Seita Emori, Hironori Higashi(2008): Potential changes in extreme events under global climate change, Journal of Disaster Research, 3, No. 1, pp.39-50

[37] Castro, C.L., Pielke Sr, R.A., and Leoncini, G., 2005, Dynamical downscaling: Assessment of value retained and added using the Regional Atmospheric Modeling System (RAMS). *Journal of Geophysical Research*, 110, D05108.

[38] Higashi, H., 2007, Influences of climate change on the frequencies of storm rainfalls and flood disasters, *Research Signpost "Geophysics"*, Tomonori Matsuura, Ryuichi Kawamura Eds., pp.125-143.

[39] Stedinger, J.R., Vogel, R.M., and Foufoula-Georgiou, E., 1993, *Frequency analysis of extreme events, Handbook of Hydrology,* Maindment, D.J., ed. McGraw-Hill, ch. 18, 1-66.

[40] Iwagaki, Y., 1955, Fundamental studies on the runoff analysis by characteristics. *Bulletin of the Disaster Prevention Research Institute, Kyoto University*, 5(10), 1-25.

[41] Inoue, K., Toda, K., and Maeda, O., 2000, Inundation model in the region of river network system and its application to Mekong delta. *Annual Journal of Hydraulic Engineering,* JSCE, 44, 485-490.

[42] Douville, H., 2005, Limitations of time-slice experiments for predicting regional climate change over South Asia. *Climate Dynamics*, 24, 373-391.

[43] Inatsu, M., and Kimoto, M., 2005, Difference of boreal summer climate between coupled and atmosphere-only GCMs. *Scientific Online Letters on the Atmosphere*, 1, 105-108.

[44] Hasegawa A., Emori, S., 2007, Effect of air-sea coupling in the assessment of CO_2-induced intensification of tropical cyclone activity, *Geophysical Research Letters*, 34, L05701.

Section 3

6

On the Effect of Global Warming and the UAE Built Environment

Hassan Radhi
Faculty of Engineering,
UAE

1. Introduction

Climate changes have already been noted all over the world. The reasons for these changes are complex and there are disagreements in the scientific community about the causes. Some scientists believe that changes are part of natural variability while others point to human activity as the cause of increasing atmospheric concentrations of green house gases (GHGs) and the key driver of climate changes. Many scientific studies come to the conclusion that the expenditure of non-renewable energy has a direct impact on the climate, with potentially devastating results. This expenditure is said to be one of the main factors affecting the climate. It causes three major problems, namely air pollution, acid rain and greenhouse effects. The use of non-renewable energy has increased the carbon concentration in the atmosphere and has also increased the earth's temperature, which is known as "Global Warming". The Intergovernmental Panel of Climate Change [1] stated that there would be a steady increase in the ambient temperature during the end of the 21st century due to the large growth in carbon emissions. Much of this growth has come from energy generation, transport, industry and, above all, from building operation.
The energy generation represents the largest economic sector in the Gulf region. During the past few decades, the Gulf Council Corporation (GCC) countries, major oil producers, have witnessed an unprecedented economic and social transformation. Oil proceeds have been used to modernise infrastructure, create employment and improve social indicators. Due to the expenditure of oil, the GCC countries have fallen in the top countries of CO_2 emissions. On a global scale, all GCC countries fall in the top 25 countries of carbon dioxide emissions per capita, with UAE leading [2]. In addition, current reports on environmental policy in the GCC are very critical and have given them the image of being the worst environmental polluters worldwide, with UAE and Qatar at top. According to the Global Footprint Network [3], the UAE possesses the highest Ecological Footprint in the world. This issue in addition to the increase in energy demand have come to the agenda of the UAE government.

2. UAE agenda

Two important issues have become a hot topic in the UAE. Firstly, the current energy situation that shows a trend of growing demand. In one decade (1997-2007), the primary energy of this region increased by 55.8% with 15.3% change between 2007 and 2008. [4].

Secondly, the increase of CO_2 emissions. The statistics of the UAE show that the increase in CO_2 emissions is within the range of 33% and 35% between 1997 and 2006 [5]. The Environment Agency of Abu Dhabi stated that the UAE activities in pursuing developments, such as fossil fuel combustion, industrial processing, land-use change and waste management have caused the release of greenhouse gas (GHG) emissions into the atmosphere. Consequently, temperatures in the UAE regions could significantly increase. This increase will influence the economy, built environment and above all the micro-climate of the UAE.

2.1 Current and future climate

The United Arab Emirates (UAE) is a federation of seven Emirates located in the Gulf region (see Figure 1). It spans approximately 83,600 km^2 and can be divided into 3 major ecological areas: coastal areas, mountainous areas and desert areas. Over four-fifths of the UAE is classified as desert, especially in the western parts of the country. The general characteristics of the UAE's climate resemble those of arid and semi-arid zones. Figure 2 shows a brief analysis of climatic elements of the UAE provided by the Directorate of Meteorology of Abu Dhabi. The analysis shows two main seasons characterise the UAE's climate. Winter lasts from November through March, a period when temperatures seldom drop below 6 °C. Summers are very dry with temperatures rising to about 48 °C in coastal cities - with accompanying humidity levels reaching as high as 90%. In the southern arid regions such as Al-ain city, temperatures can reach to 50 °C. The UAE is blessed with a high solar radiation level. The highest monthly averages of total and direct radiation are 613 W/m^2 and 546 W/m^2 in May and October respectively, while the highest monthly average of diffuse radiation is 273 W/m^2 in July. Wind from a north-west direction throughout the year is the characteristic of the UAE. The wind speed average shows slight variation, being generally low from November to January with a monthly average of 3.5 m/s, while from February to October it is well above 4.2 m/s, reaching a monthly average of 4.6 m/s in May.
Hot arid regions, such as the UAE, are sensitive to climate changes and the effects they produce. The Environment Agency of Abu Dhabi and the Ministry of Energy [6] studied different scenarios of climate changes and stated that temperatures in the UAE regions could increase while precipitation levels could significantly decline by the end of the 21st century. This scenario was simulated and the output was generated at the regional level and then scaled to eight cities within the UAE including Abu Dhabi, Dubai, Sharjah, Al-Ain, Ras al-Khaymah, Khawr Fakkan, Umm al-Qaywayn, and Ajman. The result shows that the annual average temperatures in 2050 are projected to be between about 1.6 °C and 2.9 °C warmer than they were over the period 1961–1990 and between 2.3 °C and 5.9 °C warmer by 2100. It is clear that the climate of the UAE is tending to get warmer. This tendency is expected to impact the built environment, energy use in buildings and its associated CO_2 emissions.

2.2 Energy consumption and CO_2 emissions

The discovery of oil in 1958 in Abu Dhabi and 1966 in Dubai transformed the economy dramatically, enabling the country to move away from a subsistence economy toward a modern, industrial base. In some respects, however, it seems, the energy plans of the UAE is following the example of developed nations whose economic growth occurred through the use of technologies and expenditure of fossil fuels and electricity. The rapid and increasing economic expenditure with huge architectural projects and population growth rates and a fairly low energy cost are increasing the UAE's energy consumption, making it one of the highest energy consumers per capita in the world [7]. Generally, energy in the UAE is consumed in five broad sectors defined by four end-uses, including residential, commercial, industrial and agriculture sectors. If electricity generation is included, the five sectors account for all energy consumption in the economy of the UAE.

On an international level, the consumption of energy for the building sector is a significant factor in the economy of many countries. Recent studies show such trends in different parts of the world. In the United States, for example, 41% of the total national energy production and nearly 70% of electricity production is used in buildings, as well as 28% in transportation, which is at least partly influenced by urban design [8]. In the United Kingdom, the building sector consumes about 50% of all the country's energy [9]. In Brazil, 48% of the national energy is consumed in buildings [10], while in China, building sector currently accounts for 23% of the country total energy use [11]. The same situation can be seen in the UAE. Figure 3 shows the energy consumption per sector in Al-Ain and Dubai [12-13]. Clearly, buildings, particularly those in the residential sector, have the largest impact on this growth, as 30% and 46% of the total energy in Al-ain and Dubai is consumed in this sector. Unlike many developed nations, however, the UAE always reacted to its growth in energy consumption by adding new generation capacity. Whereas, the developed countries are focusing on demand-side-policies to reduce the energy consumption as can be seen in Japan, which is considered as the most energy efficiency economy in the world due to innovative policy instruments such as the top runner approach [14].

As the fraction of the total energy increases, the production of CO_2 emitted becomes greater. Figure 4 shows the increase in CO_2 emissions relative to the use of energy. It is important to note that the production and consumption of energy are the dominant source of GHG emissions in the UAE. The UAE statistic data show that about 4% of the CO_2 production is caused by the direct emissions of buildings, 43% by electricity generation and 45% by manufacturing and construction [15]. The remaining is caused by other resources.

3. Global warming and the UAE buildings

The increasing emission of CO_2 and its contribution to global warming has become a growing concern for building industry and regulation bodies in the UAE. There are two reasons: firstly, CO_2 is the main by-product of the generation from fossil fuels of energy. As buildings are one of the largest consumers of energy then they are also the largest contributor to the increase in the atmospheric CO_2 and hence global warming and climate change. Secondly, building operation is likely to be especially affected by global warming. Clearly, by using none renewable fossil fuels, buildings contribute to the CO_2 emissions leading to warming the globe. In turns, global warming influences the energy consumption of buildings leading to increase the production of CO_2 emissions.

To evaluate the interaction between buildings and global warming, the following methodology was used. Statistically-based weather data files were generated in order to reflect the increases in air-temperatures. Each file represented a weather input of a sophisticated simulation program [16]. A typical residential building was used as a simulation model in order to represent the mainstream residential buildings in the UAE. The model was then validated using measurement data from field study and audit reports. Based on the output of simulation, a regression model was developed in order to estimate the CO_2 emission. This evaluation first estimated the variation in heating and cooling degree-days, as they were the

most straightforward indicators on building energy demands. It then predicted the variation on heating and cooling energy demands of the typical residential building to help illustrating the consequences at the national level. To estimate the CO_2 emissions, the electricity consumption was multiplied by the conversion factor of fuels in the UAE.
The first part of this section explores the contribution of UAE building sector to global warming. The second part studies the impact of global warming on UAE building design and operation in the UAE. The third part forecasts the future transformations in energy and CO_2 emissions of the UAE building sector.

3.1 UAE building sector and its contribution to global warming

The energy consumption of buildings and its associated CO_2 emissions are influenced by the interaction between three major factors including building design and materials, occupant behaviour and above all climate. To reach the energy efficiency target, sequential processes should be followed. These processes start with an optimum climatic design and end with an efficient operation of building system by the occupants. The optimum design positively impacts the building systems, particularly the HVAC and lighting systems. It may reduce the building loads and equipment size and consequently the cost and energy use. However, to obtain the maximum benefits of this design the occupants should operate the building systems in an efficient way because they can directly alter the system performance through controllers. For example, the energy consumption for heating and cooling depends on internal temperature and ventilation and these parameters are controlled by the occupants.
In the ground, however, there is no question that the majority of buildings in the UAE are designed, built and operated without attention being paid to the environmental and energy system. Today, under the umbrella of a worldwide international style of buildings, and in an attempt to embark on a new trend of modern architecture, huge glass façades facing the sun have appeared in cities such as Dubai, Abu Dhabi and Al-Ain. In energy terms, this strategy is generally applied to gain the most solar radiation possible in order to heat up buildings and utilise daylight and therefore, it is often used for cold climates. For hot climates, such as that of the UAE, using this strategy may lead to a different scenario with respect to cooling load. To apply this strategy in hot climates, the energy design should utilise the availability of useful daylight by striking a balance between light and heat gain. Nevertheless, this is not the case in the UAE. Huge projects have been constructed with enormous glazed façades facing the southeast and southwest without protection against overheating and sun glare in the summer [17]. Furthermore, some construction materials have low impact on CO_2 emissions that result from raw material acquisition, manufacture, transportation, installation, maintenance and recycling, but provide a moderate reduction in terms of operational energy, and vice versa. Others positively impact the embodied energy and environmental performance and can optimise the cooling and heating energy performance. Replacing or at least reducing the use of some construction material such as concrete, reinforcing steels, formwork, and gypsum board have a direct impact on CO_2 emissions. Some materials and construction systems can decrease the amount of CO_2 emissions by around 6.9% [18]. In most projects in the UAE, however, materials are evaluated and selected based on aesthetics and cost and not on their energy and environmental performance [19]. It is, therefore, not surprising that 70% of the yearly electric energy use is consumed by building systems. Figure 5 shows the energy end-uses of a typical residential building in the UAE, where the electricity consumed by the HVAC

system is the most significant, particularly for cooling energy. The growth in electricity consumption for cooling buildings in the UAE region has increased ten times (from 5 to 50 Billion kWh) over the past two decades [20].

A key function of building design is to modify the indoor environment to be more suitable for habitation than the outdoor. If the building fails to meet this objective due to one or more reasons, such as insufficient design and materials selection or variations in climate parameters that probably make it impossible for any certain level of comfortable indoor environment to be achieved through passive means. Then, it is necessary to rely upon mechanical means to achieve the comfort level. As a result, additional electricity will be used by the HVAC system to provide a comfortable internal temperature for human being. Most people feel comfortable at indoors temperature ranging from 22 °C to 24 °C along with a range of 40–60% relative humidity. For a residential building it would normally be designed with comfort temperature selected from range 20 °C to 24 °C. With a heating system one figure would be chosen, but with air-conditioning system two figures would be selected, the higher one for summer (cooling) conditions. These figures are taken to apply generally for cold climates such as North America and Europe, and for warm countries higher figures would often be used, and in the harsh climates of the UAE, where the average maximum air-temperature reaches above 50 °C, an internal temperature of 26 °C and 27 °C would be considered comfortable. A significant amount of electricity and between 26.8% and 33.6% savings in cost can be achieved by raising the set point temperature from 24 °C to 26 °C in similar climate [21]. Nevertheless, this is not the case in most cities in the UAE, as the point temperatures are often set below 24 °C. This attitude can be related to two main reasons, first, low electricity prices and second, the support of the government where the citizen pays 0.05 AED (1 AED = 0.27 USD) for each kilowatt-hours and in some cases the government pays for the consumption [22]. These two reasons have reduced the people immediate interest in electricity conservation.

Clearly, harsh climatic conditions, building design and occupant behaviour in the UAE are contributing negatively to the increase in energy consumption and its associated CO_2 emissions. As mentioned earlier that about 4% of the CO_2 production in the UAE is caused by the direct emissions of buildings, 43% by electricity generation and 45% by manufacturing and construction. Electricity use in building sector is within the range of 50% to 73% with an average of 60%. The net energy consumption of the UAE reached 52.6 Billion kWh and the total annual CO_2 emission got the level of 137.8 million metric tonnes [5]. These figures coupled with percentages in Figure 3 give a rough estimate of CO_2 emissions per sector in the UAE. Around 5.50 million metric tonnes is caused by the direct emissions of buildings, 35.5 million metric tonnes from electricity use by building sector and 62.0 million metric tonnes by material manufacturing and building construction.

To reduce the above figures, it is necessary to eliminate the reasons behind the CO_2 emissions. First, reduce the inefficient used of energy by educating people and providing a good energy management. Secondly, modify the impact of climate with minimum electricity use through climatic design. In this way it is possible to reduce the negative contribution of buildings to the global warming.

3.2 Impact of global warming on building design and operation

Changes in the external air-temperature will have significant consequences upon building thermal performance, particularly cooling and heating energy. The severity of the outside air-temperature related to cooling and heating energy consumption can be measured using the so-called degree-days. Figure 6 shows the impact of air-temperature on the cooling and heating degree-days in the UAE. It is clear that there is a significant change, which positively influences the heating degree-days, but negatively influences the cooling degree-days. Cooling degree days can increase between 16% and 27% by 2050. This increase can reach between 22% and 42% by 2100. The growth in cooling degree-days implies that to reach a comfortable internal environment in the hot summer of the UAE, a dramatic change will occur in the amount of electricity used by air-conditioning systems. Table 1 illustrates the simulated impact of global warming on the cooling and heating demands. As can be seen, there is a brief drop in heating energy demand with different rates ranging from 9.5% to 37.1% due to the increase in air-temperature by 1.6 °C and 5.9 °C respectively. When this applies to the cooling and ventilation energy, a different scenario occurs. There is a sharp increase in the cooling energy which reaches a peak of 23.5% due to 5.9 °C increase. This increase represents a clear indication that global warming will lead to a negative impact on the total electricity demand, where changing from the current climate has reduced the heating energy demand at the expense of a rise in annual cooling energy demand, and therefore, additional total energy has been consumed. From the total energy increase; there has been in effect a further CO_2 increase, with electric cooling energy consumption.

3.3 Forecasting future transformations in energy consumption and CO_2 emissions of the UAE building sector

To forecast future transformations in the energy consumption and CO_2 emissions of the UAE residential sector, a simple regression model was constructed in the light of current building design and operation as well as the future weather conditions. The primary analysis of the constructed model is based on a weighted ordinary least squares regression. This type of regression is used to know the relationship between several independent or predictor variables and a dependent or criterion variable. In the current case, the cooling energy is the dependent and variables in the right side of the equation are the independents.

$$CE = C_0 + C_1\, Tao + C_2\, WWR + C_3\, U - v_{(w)} + C_4\, U - v_{(g)} + C_5\, SC_{(g)} \qquad (2)$$

The result of regressing the simulated cooling energy (*CE*) as obtained from Table 1 onto the outside temperature (*Tao*) and building design parameters including U-value of the wall, $U\text{-}v^{(w)}$, window-to-wall ratio, WWR, U-value of the glazing, $U\text{-}v^{(g)}$, shading coefficient of the glazing, $SC^{(g)}$, as found in the representative residential building is shown in Table 2. The coefficient of determination, or R^2 of the CE, is 0.97 which would indicate a strong relationship between the CE variables and the outside temperature, U-value of the wall, WWR, U-value of the glazing and the SC of the glazing. The amount of CO_2 emissions (*E*) is subjected to the cooling energy, operational schedule (*Op_sch*) and the conversion factor of fuel (*Cf*). Therefore, a simple linear equation was developed and used to calculate the CO_2 emission reduction due to the examined weather and none weather dependants. The following equation was used.

$$C_{emission} = (CEI \times Op\_sch) \times Cf$$

With the current building design and operation in the UAE, the residential sector accounts for 2646 GWh, or almost 46% of the total regional consumption. The global warming is likely to increase the energy used for cooling buildings by 23.5% if the UAE warms by 5.9 °C leading to a growth in electricity consumption to almost (current consumption + 12.5%) 2977

GWh, and consequently the total CO_2 emissions will grow to almost 7.6 million metric tonnes. The net Emirati CO_2 emissions could increase at around 138.4 million metric tonnes over the next few decades.
When energy efficiency techniques are applied to the current building design and operation, different scenario is occurred. Table 3 illustrates cooling energy savings due to each efficiency technique under different scenarios. The energy breakdown of the representative building show that electricity used for space cooling is approximately 65% or 97.5 MWh. As illustrated, adding thermal insulation to the case building due to 1.6 °C increase reduces the cooling demand by 19.3%. Considering the large amount of cooling energy demand this figure is significant. The minimum reduction is 15.5% due to 5.9 °C increase. At the same time, replacing the glazing type from single glazing to double low-energy glazing produces a significant savings in cooling energy demand as can be seen in the fall of energy consumption which reaches 10.5% due to 5.9 °C increase. As a great amount of cooling energy can be saved by glazing type, an appropriate design of window area offers a considerable opportunity to control electricity used by the AC system. As seen, reducing the WWR reduces the cooling energy by 3.7% and 9.0% under current climate and 5.9 °C increase. These figures indicate that thermal insulation performs best, followed by glazing type and then window area in descending order.
As these techniques stop the heat flow from the outside and reduce the cooling load and energy consumption leading to decreasing the CO_2 emissions, authorities and energy code bodies in the UAE should develop such techniques and make the relevant part of the building design and regulations more stringent, and emphasise that the goal of saving energy is to reduce CO_2 emissions into the atmosphere. This can be done by using CO_2 emissions as one of the principal criteria by which the design of a building is judged. It can be implemented jointly with measures on specific envelope elements, system components and energy use patterns in order to ensure the dissemination of the most efficient building

4. The UAE strategy towards sustainability in the built environment

Indeed, the less a country depends on finite resources such as natural gas and oil, the stronger and more stable the economy will remain in the face of energy cost increases or reduced supplies. From an environmental point of view, the expenditure of non-renewable energy has a direct impact on the natural environment. Thereby, following the example of developed world without any consideration to the local environment may lead to critical economic and environmental consequences. To avoid such consequences, two major changes in patterns are proposed, first, effective measures to protect the depleted resources and second, valid policies to replace fossil fuels with non-fossil fuels.

4.1 Policies and legislations to reduce the energy demand

There has recently been a consensus to legislate for energy efficiency in the UAE. The government has realised the benefits of energy efficiency not from the point of view of the balance between energy supply and demand, but rather from a socio-economic and environmental standpoint. As the building sector is a major consumer of energy, the UAE government has concentrated on this sector and recognised the important role that efficiency codes play in reducing the amount of energy consumption, especially that of the HVAC systems. The thermal insulation code was first applied. The green building codes were then introduced. The new building energy codes conform to the most demanding global standards and have been developed in tandem with the International Code Council (ICC), responsible for advising US regulators on their exacting regime. Therefore, the UAE building codes is considered as the first step towards developing consistent sustainable policies in the UAE and the region. The UAE government, also, launched the Estidama Program and the Pearls green building rating system which would become integrated into the building code and therefore enforceable, as well as the launch of the Emirates Green Buildings Council [23]. The rating system introduced by Estidama Program can be considered as an important step towards low carbon emission buildings. Rating the performance of buildings against itself and other buildings plays a key role in protecting the environment, reducing energy consumption and checking on energy efficiency. Its most significant contribution is that it provides a target for improvement.

A survey [24] concerns with the environmental sustainability in the UAE showed the residential buildings before the codes as poor energy and carbon emission performers, while a benchmarking study [25] categorised most educational buildings in Abu Dhabi Emirate as poor energy and environmental performers when compared to international benchmarks. Those studies indicated the inefficient building design and the poor energy management as the main reasons behind the high energy consumption and CO2 emissions of those buildings. An evaluation of the new building codes and their impact on energy and CO_2 emissions [22, 26] showed that using such codes can reduce the CO_2 emissions of buildings by 50%.

4.2 Initiatives to utilised renewable energy

Although the UAE has no consistent policy frameworks for sustainable technologies and renewable energies, it has planned economic development programmers dedicated to establishing new economic sectors focused on alternative energy and sustainable technologies. For instant, two promising projects are planned to be completed in the next few years: first, a $350 millions solar power plant and second, a $2 Billions hydrogen-fuelled power plant [22]. Such projects, in general, can contribute to the sustainable development including economic, environmental and technological well being. They will not only contribute towards employment generation, but also reduce significant amount of GHG emissions which would have taken place in ordinary power plant scenario with natural gas and fuel oil based generation. In the latter project, the CO_2 will be kept underground which represent one of the world first carbon capture and storage projects. Moreover, solar energy based power generation system will be a robust and clean technology involving the latest state of art renewable energy options to be used for the purpose of electricity generation.

Utilisation of clean and renewable energy has become a trend in the UAE, not only through the establishment of sustainable power stations, but also through the construction of low energy and free carbon emission built environment. There are some remarkable projects going on in the UAE. The most notable project among these is Masdar City. Although the concept is not new, Masdar City is planned to be a carbon-neutral, zero-waste city with the aim of being one of the world's most sustainable urban development powered by renewable energy [27]. This huge project incorporates various sustainability techniques and renewable energy technologies. It is planned to host two important institutions. First, the headquarter of the International Renewable Energy Agency (IRENA) which will be the first global agency based in the Middle East. Second, the Masdar Institute of Science and Technology which will offer MSc and PhD programmes in alternative energy and sustainable

technologies as well as give opportunities to do various research activities in sustainable design [28].
To this end, protecting the depleted resources and switching towards more efficient use of energy coupled with replacing fossil fuels with non-fossil fuels would have a number of benefits for the UAE:

- The UAE would be given a better reputation in the regional and international policy arena.
- The reduction in the use of fossil fuels will lead to an increase in the exported oil and natural gas.
- The UAE would gain another important benefit from none-fossil fuels such as solar and wind energy. Consequently, it will be prepared for the post-oil era.
- Reducing the use of fossil fuel and the use of renewable energy will limit the effect of global warning on the UAE and on other countries in the Gulf region.

5. Summary

As energy scarcity and global warming are threatening human sustainability, governments and organisations must spend much effort in reducing the energy consumption and CO_2 emissions. Buildings are one of the largest consumers of energy then they are also the largest contributor to the increase in the atmospheric CO_2 and hence global warming and climate change. At the same time, building operation is likely to be especially affected by global warming. A rise in the ambient air-temperature can lead to a significant increase in electricity consumption and its associated CO_2 emissions. Global warming is likely to increase the energy used for cooling residential buildings by 23.5% if the UAE warms by 5.9 °C. At the regional level, the energy consumption can be increased at around 5.4%. Consequently, the CO_2 emissions can increase to almost 7.6 million metric tonnes. The net Emirati CO_2 emissions could increase at around 138.4 million metric tonnes over the next few decades.
To cope with global warming and the increase of CO_2 emissions, two major changes in patterns are suggested in the UAE: first, effective measures to protect the depleted resources and second, valid policies to replace fossil fuels with non-fossil. The former can be seen in the new building energy regulations. Implementing these regulations can reduce the CO_2 emissions by 50%. The latter can be seen in establishing a new economic sector. This sector focuses on alternative energy and sustainable technologies through the installation of new power plants that use renewable resources in power generation. In addition, the construction of low energy and free carbon emission built environment such as Masdar City. Such a project can served as the foundation for an extension of activities in the field of low carbon emission buildings and renewable resources with the goal of reducing the impact of global warming on our life, economy and above all our built environment.

6. References

[1] Climate change. Synthesis report, intergovernmental panel of climate change. See, http://www.ipcc.ch/ipccreports/ar4-syr.htm; 2007.
[2] United Nations Statistic Division, 2007. Environmental Indicators, Climate Change, New York.
[3] Global footprint network 2010. Available at:

http://www.footprintnetwork.org/en/index.php/GFN/page/footprint_for_nations/
[4] BP, 2009. BP Statistical Review of World Energy, June 2008, London. Can be found at: http://www.bp.com/liveassets/bp_internet/globalbp/globalbp_uk_english/reports_and_publications/statistical_energy_review_2008/STAGING/local_assets/2009_downloads/statistical_review_of_world_energy_full_report_2009.pdf
[5] Energy Information Administration. UAE energy profile 2008. Available from: http://www.eia.doe.gov/cabs/UAE/Electricity.html.
[6] Ministry of Energy. Initial National Communication to the United Nations Framework Convention on Climate Change. United Arab Emirates; 2006.
[7] Kazim AM. Assessments of primary energy consumption and its environmental consequences in the United Arab Emirates. Renewable and Sustainable Energy Reviews 2007;11:426–46.
[8] EIA. Annual Energy Review 2008. June 2009. Can be found at: www.eia.doe.org
[9] Steemers, K. (2003) Energy and the city - density, buildings and transport. Energy and Buildings 35 (1): 3-14.
[10] Westphal, F. S., & Lamberts, R. (2004) The use of simplified weather data to estimate thermal loads of non-residential buildings. Energy and Buildings 36 (8): 847-854
[11] Yao R, Li B, Steemers K. Energy policy and standards for built environment in China. Renewable Energy 2005; 30 (13): 1973-1988.
[12] Al-Ain Distribution Company. Special water and electricity report. Al-Ain, United Arab Emirates; 2008.
[13] Dubai Electricity and Water Authority. Thermal insulation. Available from: http://www.dewa.gov.ae/community/ThermalInsulation/thermalInsIntro.asp.
[14] Nordqvist J. Evaluation of Japan top runner programme- within the framework of the Aid-EE project. 2006. Can be found at: http://www.aid-ee.org/documents/018TopRunner-Japan.PDF
[15] World Resources Institute. Climate and atmosphere–UAE 2006. Available from: http://earthtrends.wri.org/pdf_library/country_profiles/cli_cou_784.pdf.
[16] Visual DOE. User manual. USA: Architectural Energy Corporation; 2004.
[17] Radhi H and Sharples S. (2008) Developing energy standards for low energy buildings in the Gulf States, Architectural Science Review 51(4): 369-381
[18] Hong WK, Kim JM, Park SC, Lee SG, Kim SI, Yoon KJ, et al. A new apartment construction technology with effective CO2 emission reduction capabilities. Energy 2009; doi:10.1016/j.energy.2009.05.036.
[19] Radhi H. (2010). On the optimal selection of wall cladding system to reduce direct and indirect CO2 emissions, Energy 35:1412-1424.
[20] Annual Statistical Report. Dubai Electricity and Water Authority (DEWA); 2003.
[21] Al-Sanea SA, Zedan MF. Optimized monthly-fixed thermostat-setting scheme for maximum energy-savings and thermal comfort in air-conditioned spaces. Applied Energy 2008;85(5):326–46.
[22] Radhi H. Evaluating the potential impact of global warming on the UAE residential buildings - A contribution to reduce the CO2 emissions. Building and Environment 2009; 44: 2451- 2462.
[23] Emerites Green Building Council 2010. Available at http://www.esoul.gohsphere.com/default.aspx

[24] AboulNaga MM, Elsheshtawy YH. Environmental sustainability assessment of buildings in hot climates: the case of the UAE. Renewable Energy 24 (2001) 553–563
[25] Radhi H and Al-Shaali R. Energy and CO2 Emissions Benchmarks: A Step towards Performance Standards for Educational Buildings in Al-Ain City. Special report. UAE university 2010
[26] Radhi H. (2010). On the optimal selection of wall cladding system to reduce direct and indirect CO2 emissions, Energy 35:1412-1424.
[27] Reiche D. Renewable Energy Policies in the Gulf countries: A case study of the carbon-neutral "Masdar City" in Abu Dhabi. Energy Policy 2010, 38: 378–382
[28] Masdar city. Abu Dhabi Future energy company (Masdar) 2010 can be found at: http://www.masdar.ae/en/home/index.aspx

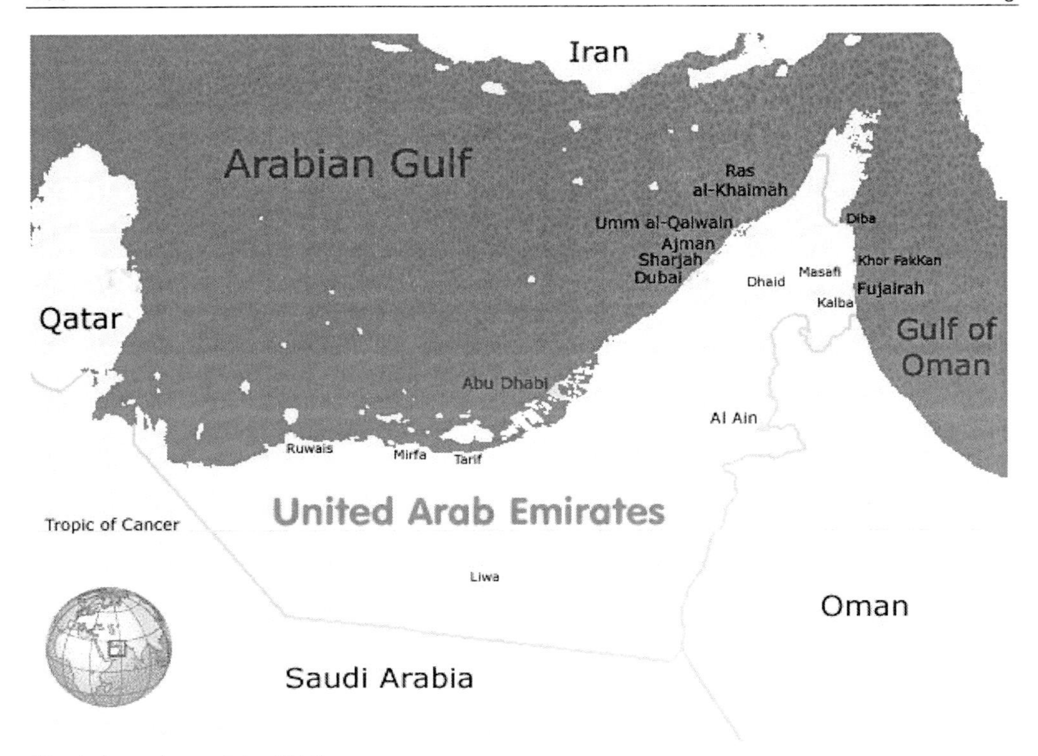

Fig. 1. Locations of the UAE

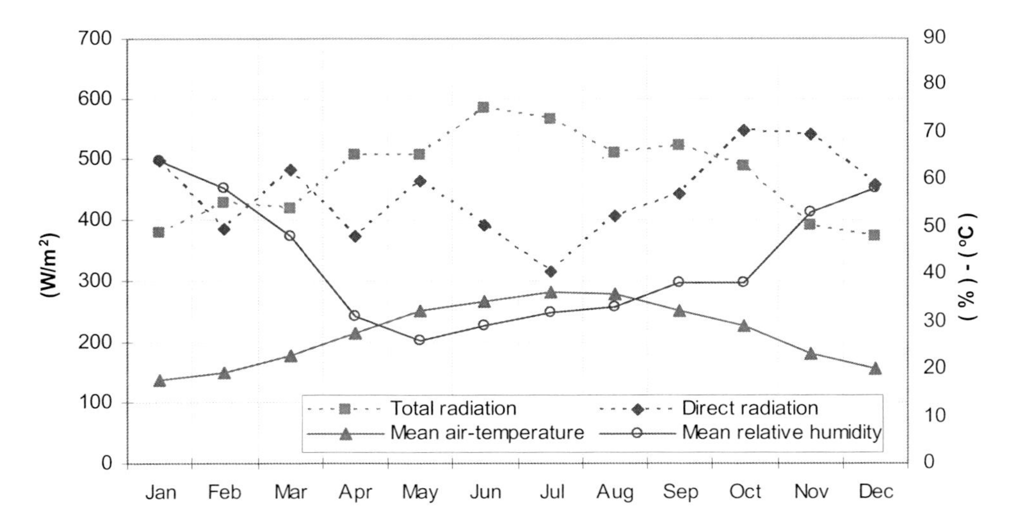

Fig. 2. Analysis of UAE climate

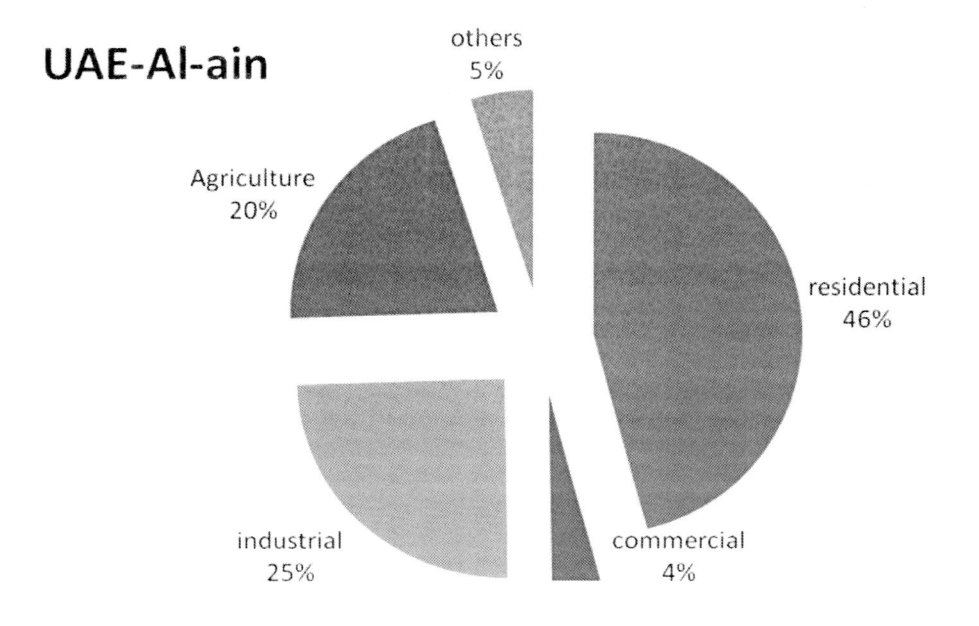

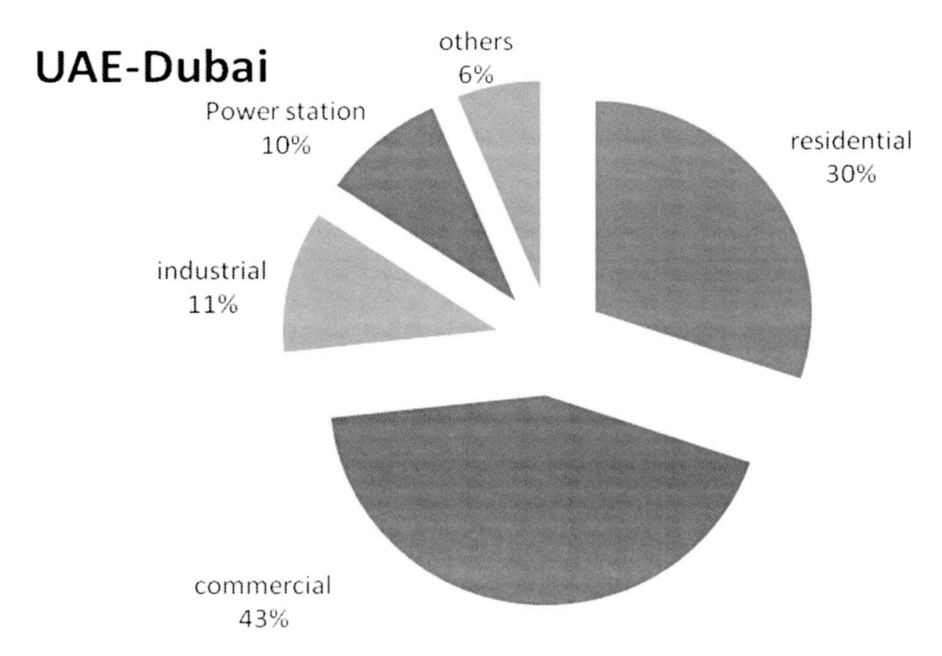

Fig. 3. Energy consumption per sector

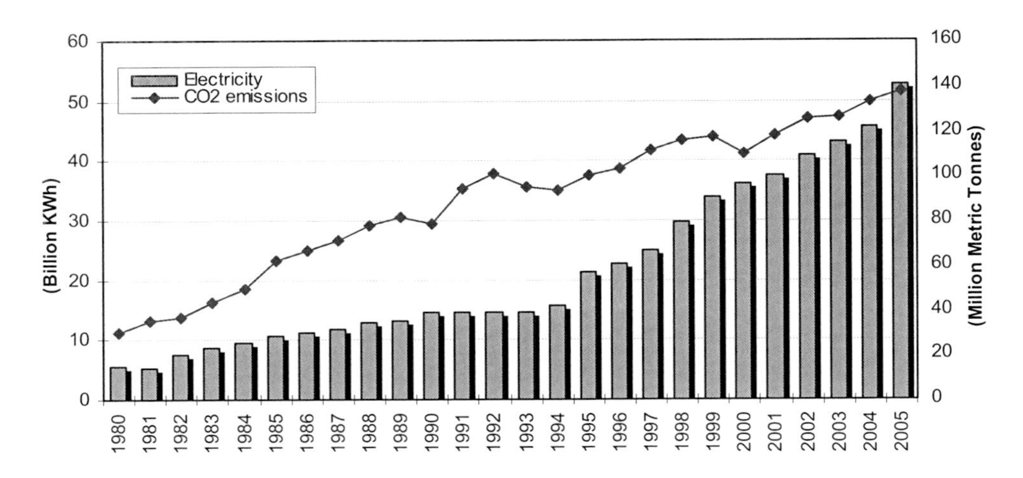

Fig. 4. Increase in CO_2 emissions relative to the use of energy

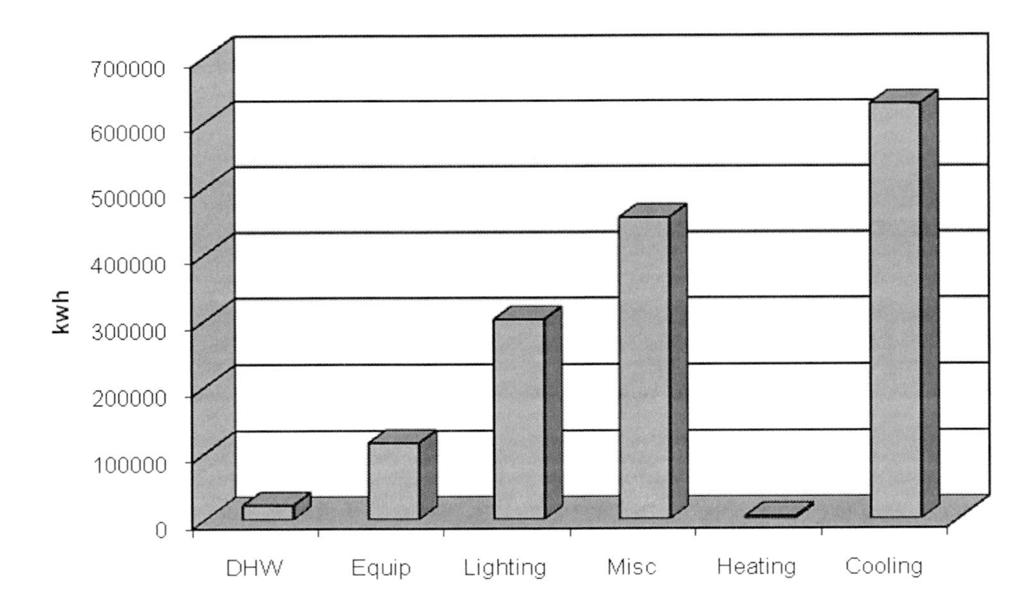

Fig. 5. Energy end-uses in the typical building

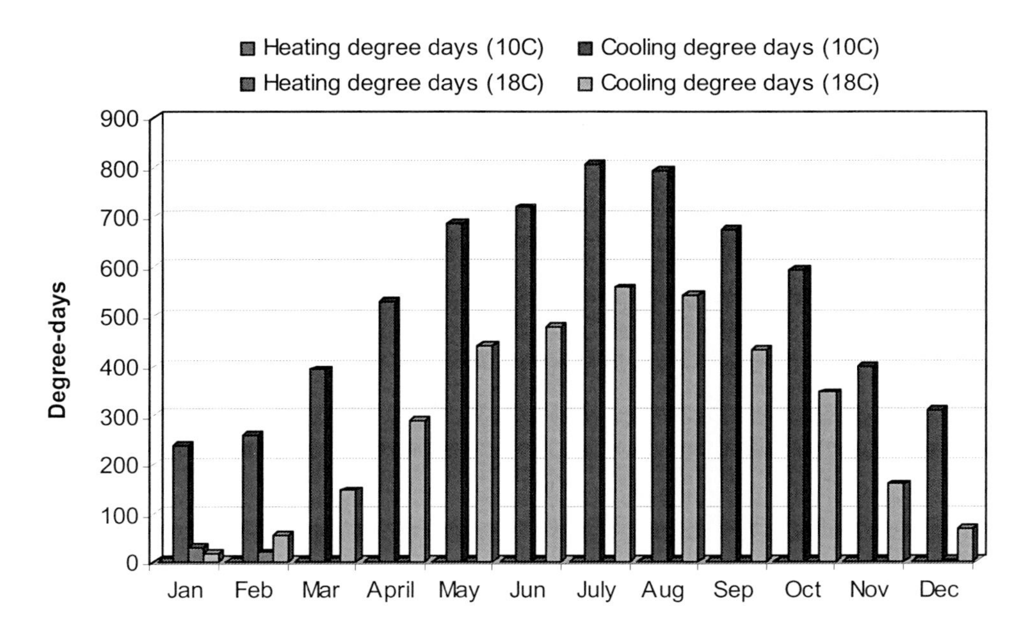

Fig. 6. Monthly heating and cooling degree days

	Heating (KWH)	Cooling (KWH)	Fans (KWH)	Electricity (KWH)	CO_2 emissions ($Kg/m^2/yr$)
Baseline (consumption)	6369	73049	11886	122920	176
I.6 °C (%)	-9.5	7.3	3.9	4.1	183
2.9 °C (%)	-14.2	11.7	5.8	6.7	188
2.3 °C (%)	-17.4	16.7	6.8	9.5	193
5.9 °C (%)	-37.1	23.5	12.3	12.9	197
(-) reduction					

Table 1. Increase in electricity and CO_2 emissions due to global warming

Cooling requirement (CR)						
	C_o	C_1	C_2	C_3	C_4	C_5
	-258	11.8	20.2	249	2.9	27.6
	7.4	0.2	0.7	8.6	0.5	4.9
R^2	0.97					
F	952					

Table 2. Regressing the energy cooling energy requirement

Climate	Baseline	1.6 °C	2.9 °C	2.3 °C	5.9 °C
Consumption (KWH)					
Cooling	75462	80434	83390	86811	96203
Electricity	126836	131393	134173	137397	145486
Reduction due to thermal insulation (%)					
Cooling	19.3	19.7	19.9	19.7	15.5
Electricity	15.5	15.9	16	15.9	13.1
Reduction due to glazing system (%)					
Cooling	5.4	5.4	5.5	5.5	10.5
Electricity	4.5	4.6	4.7	4.7	8.1
Reduction due to glazing area WWR (%)					
Cooling	3.7	3.8	3.9	3.9	9
Electricity	3.2	3.2	3.3	3.3	6.8
(-) increase in energy demand					

Table 3. Performance of design technologies under different scenarios

7

Transport Planning and Global Warming

Pedro Pérez, Emilio Ortega, Belén Martín,
Isabel Otero and Andrés Monzón
TRANSyT-UPM, Centre for Transport Research, Universidad Politécnica de Madrid
Spain

1. Introduction

Transport energy consumption in industrialised countries is based primarily on fossil fuels, and is associated with the main negative impacts of transport: climate change, air pollution, congestion and accidents (Sperling, 2004). The emissions of many pollutants are being moderated due to improvements in engines and fuels, but the consequences for health are a growing concern, and particularly the risks posed by nitrogen oxides and particles, which are closely associated to transport. CO_2 emissions (the gas considered mainly responsible for the greenhouse effect) are also increasing, and this phenomenon can be seen most intensely in the transport sector.

The European Commission's 2001 White Paper on transport (and the 2006 revised edition) declared that the sustainability of the transport energy model must include the control of transport demand and an improvement in the efficiency of transport modes. It is this area which offers the greatest potential for establishing an effective strategy of action. This requires a greater commitment to the processes of transport deregulation –in order to make consumers aware of price considerations–, the establishment of mechanisms to ensure that these prices reflect actual costs, and the promotion of energy savings. This approach was underlined in the 2005 Green Paper on energy end-use efficiency and energy services, which suggests that overall consumption in the European Union can be reduced by up to 20% without compromising economic profitability. This was subsequently ratified by the European Council's March 2007 Action Plan which established this as an objective for the year 2020. The European Parliament and Council has also approved Directive 2006/32/EC concerning end-use energy efficiency, as well as revising a proposal for a directive for the development of clean and energy-efficient road vehicles.

However, measures require some time after their implementation in order to take effect, and they must be supported by changes in lifestyle which will effectively influence transport use over the forthcoming decades (Rodenburg et al., 2002). A reduction in transport GHG emissions can be achieved by reducing the need for transport, improving the energy efficiency of the different modes of transport and fuels, and balancing modal distribution (Schipper et al., 1997; Steenhof et al., 2006).

The measures that can be applied in the transport sector to promote savings and improvements in energy efficiency are well known in general terms (Rodenburg et al., 2002; Cuddihy et al., 2005). These include everything from correctly setting energy prices, and reflecting these prices in the cost of services, including external costs; economic and tax

incentives which favour a reduction in energy intensity; the optimisation of travel in order to increase occupancy; joint planning of transport infrastructures and land uses so as to reduce average distances; development of new low-carbon fuels and low-consumption engines; and making more use of communications technologies as a resource.
The United Nations' Intergovernmental Panel on Climate Change (IPCC) and other institutions in this area consider that energy savings and efficiency will be a key element in guaranteeing sustainable development in forthcoming decades, until such a time as any current or future technological innovations can be implemented on a massive scale (Kahn Ribeiro et al., 2007). The United Nations Convention on Climate Change outlines the main technologies and commercial practices available to the sector to mitigate GHG emissions: these include energy-efficient vehicles, hybrid vehicles, clean diesel vehicles, biofuels, modal change from roads to railways and public transport, and non-motorised transport (UN-FCCC, 2007). It also details the technologies and practices which are expected to be available on the market by 2030: second-generation biofuels, more energy-efficient aircraft, more advanced hybrid and electric vehicles with more powerful and reliable batteries. All these measures can serve as the basis for a low-emission economy, and this will be possible only if low-emission fuels are used to supply the different forms of motorised energy necessary for transportation, and the complete chain of energy transformations which make that energy available to the end users (Van Wee et al., 2005). Thus the consumption of one unit of energy for railway traction involves the consumption of 2.5 units of primary energy. Energy savings and efficiency are therefore key in securing an energy supply which is low in CO_2.
Another aspect is the reduction in concentrations of air pollutants, for which the European Union is establishing guidelines for all European countries. Many of the directives in this area include measures which coincide with those for improving energy efficiency, and particularly regarding fuels and vehicles.

2. The energy and environmental behaviour of transport

2.1 Transport energy consumption

In the Kyoto protocol, the European Union undertook to reduce GHG emissions in its area by 8% over 1990 levels between 2008 and 2012. The significant increase in GHG emissions for the transport sector cannot be explained simply by demographic growth, nor even by economic growth, both of which have grown at a lower rates. This indicates that productive processes are increasing their consumption of transport, contrary to Community targets which aim to generate economic growth with lower increases in transport flows of passengers and freight (European Environmental Agency, 2008). In Spain, for example, the energy intensity of road transport has gone from 0.46 tonnes of oil equivalent (toe) per inhabitant in 1990 to 0.71 in 2008 (an increase of 54%). Similarly, the energy intensity of road transport (at constant 1995 prices) has gone from 0.045 ton per million euros in 1990 to 0.052 in 2008 (15% growth).
Energy consumption and CO_2 emissions can be estimated based on transport data by using the methodology and the factors developed by the Intergovernmental Panel on Climate Change (1995). These emissions are directly proportional to the carbon content of the fuel used in transport (expressed in kilotonnes of equivalent CO_2 per pegajoule, $ktCO_2$ eq./PJ). Most of the carbon is converted into CO_2 during combustion, although a part is released as CO, CH_4 or hydrocarbons without methane which oxidise into CO_2 over time. The fuel oil used in maritime transport has the highest carbon content, followed by diesel, kerosene (air transport)

and petrol. Also to be taken into account is the carbon consumed in the electrical power used by modes of rail transport, which depends on the mix of fuels used in electricity production (Hernández-Martínez, 2006). Coal has the highest carbon content, followed by petroleum and natural gas. For nuclear power and renewable sources such as biomass, hydraulic, solar, wind and geothermal energy, net carbon emissions equal to 0 are assumed (Schipper et al., 1997).

The energy consumption of different modes of transport is influenced by direct and indirect factors. Direct factors involve the use of the vehicle, and indirect factors concern the construction and maintenance of infrastructures, and the production and maintenance of vehicles (Van Wee et al., 2005). The direct factors can be divided into logistical, technical and operational factors. The most important logistical factors are rate of occupancy and load, and network density. Technical factors include characteristics such as the weight, capacity, engine, fuel and the aerodynamic coefficient of the vehicle. Operational factors refer to the way in which the vehicle is used, and include the speed and the driving dynamic.

The energy consumption required to move the vehicle and for the use of auxiliary features (i.e. lighting and heating systems, air conditioning) and energy loss (in the engine and in the transmission) comprise the direct consumption of primary energy. The production and distribution of fuel and electricity also consume energy ("well-to-tank", WTT). Thus the consumption of 1 PJ of electricity in Spanish trains involves an average consumption of 2.5 PJ of primary energy in the form of fossil fuels in an electrical power plant (40% efficiency), as well as associated emissions of over 185 ktCO_2 eq.

On average, CO_2 emissions occasioned during the production, distribution and consumption of electrical energy are around 80% of the emissions produced during the extraction, distribution and consumption of fossil fuels. This percentage varies according to the energy mix required to produce the energy, and particularly with the technology used during the production and distribution of the electricity associated to the fuels used (renewable energies, nuclear, coal, petroleum, gas, etc.). Thus the technology used by fossil fuels has attained a higher level of development than the technology used in electricity, which is currently still under improvement.

Measures to encourage energy savings and efficiency can therefore significantly reduce the volume of energy required, making it possible to achieve an energy supply which is low in CO_2. Refineries need an average of 1.14 PJ of petroleum to produce 1.0 PJ of a particular fuel, and its distribution requires an additional 0.02 PJ per 1.0 PJ (Pilo et al., 2006). Diesel and petrol are produced from conventional crude oil with an efficiency of 85-90%, depending on the situation of the oil well and the production process (Kaul & Edinger, 2004). The review in this work includes both categories of direct primary energy consumption: well-to-tank and tank-to-wheel (TTW).

For diesel trains, the WTT and TTW losses are between 10-15% and 68-70% respectively (Pilo et al., 2006). For electric trains, WTT and TTW losses are between 57-63% and 13-19%. Electric railways operate using a regenerative dynamic brake which saves 50% of TTW energy during the braking process. In road vehicles, TTW losses vary between 60% and 79% depending on whether the traction is diesel or petrol. In general, electric traction vehicles have lower TTW losses and higher WTT losses than petroleum traction vehicles. The average efficiency of energy production in Spain is 47%, due to the composition of the energy mix, which is based on coal and nuclear technologies.

As well as the direct consumption of energy, the production and maintenance of vehicles and transport infrastructures are important factors in total energy consumption. This is known as indirect energy use. The construction costs represent non-recurrent consumption and are not usually included in statistics on transport energy consumption. On the other hand, direct energy consumption occurs in a recurrent manner throughout the operative life of the motorways, and is now included in the consumption statistics.

The energy consumption model must include direct primary energy and indirect energy. The contribution of indirect energy to consumption varies according to the different transport modes, types of vehicle and categories of infrastructure (Saari et al., 2007; van Wee et al., 2005). Van Wee et al., (2000) estimates that about four times more energy is used to operate a road vehicle (during the useful life of the vehicle) than in manufacturing it. In Spanish cities, the direct energy consumption of modes of rail transport is 40-50%, and in road transport modes it is 30-45% of the direct consumption of primary energy (Zamorano et al., 2004).

Similarly, Lenzen (1999) estimated indirect energy consumption according to the mode of transport, and this figure varied between 18% (heavy articulated lorries) and 44% (railway) for freight transport. It is therefore important to take indirect energy into account when calculating energy consumption and efficiency. Thus although the energy efficiency of railways is about three times greater than that of road transport with regards fuel consumption, the production energy efficiency is the same as that of heavy articulated lorries, and the total efficiency is only two times greater.

In summary, both energy consumption and CO_2 emissions can be said to depend on the following factors:

1. Global demand for travellers and freight (tkm and vkm).
2. Modal distribution.
3. Energy intensity of each mode, or energy consumption for each unit transported (MJ/tkm-MJ/vkm energy). Another factor to be considered in parallel is the intensity of GHG emissions ($gCO_2eq.$/tkm vs. $gCO_2eq.$/vkm).
4. Energy mix of each mode.

2.2 Energy and environmental efficiency of transport modes

Once a global analysis has been made of the dimension of the problem from the environmental and energy viewpoint, the next step is to analyse the efficiency, as the absolute figures depend to a considerable extent on the increase in the units transported (passengers and freight) in each mode of transport. This is done by merging the data for transport activity with the data for energy consumption and emissions. The result is what is understood by efficiency (energy and/or environmental), and also expresses intensity (understood as consumption of resources).

Energy efficiency is determined by two factors: the energy required to move the vehicle and the use of the vehicle's capacity. The energy required to move the vehicle is determined by the fuel consumption, transport conditions (traffic and geography) and the vehicle's characteristics (model and size). The use of the vehicle's capacity depends on the levels of occupancy and load of each individual vehicle, the relative use of each type of vehicle, and the distribution of the different types of vehicles within the fleet of vehicles as a whole (Leonardi & Baumgartner., 2004).

In addition, the concept of environmental efficiency must be defined for each of the air pollutants, as well as for sound contamination. Environmental efficiency is measured in emissions of each pollutant for the same units of transport.

As an example, the following figure shows the average growth in the period from 1990-2007 of three indicators relating to passenger transport in the following modes: road, railway, air, boat, and underground railway. The following variables are analysed: growth in demand, GHG emissions, and energy intensity. As can be seen, energy consumption has grown less than demand, which means that in all cases -except in underground railways- the energy intensity (consumption per unit transported) has improved (negative value).

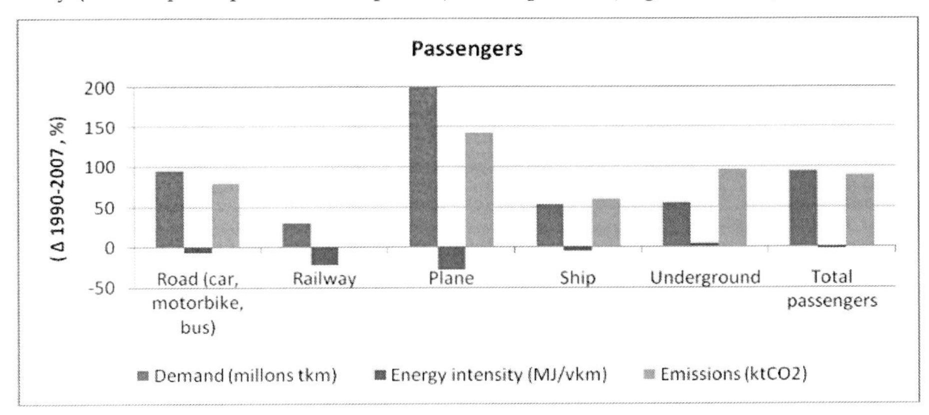

Fig. 1. Change in demand, energy intensity and CO_2 emissions for passengers, 1990-2007. Source: National Inventory of Emissions 2008, Annual Report on Transport and Postal Services 2008 and authors' own compilation.

The figure below shows the average growth in the same period of three indicators relating to freight transport in the following modes: road, railway, air, boat, and pipeline. The variables analysed, as in the case of passengers, are growth in demand, GHG emissions, and energy intensity.

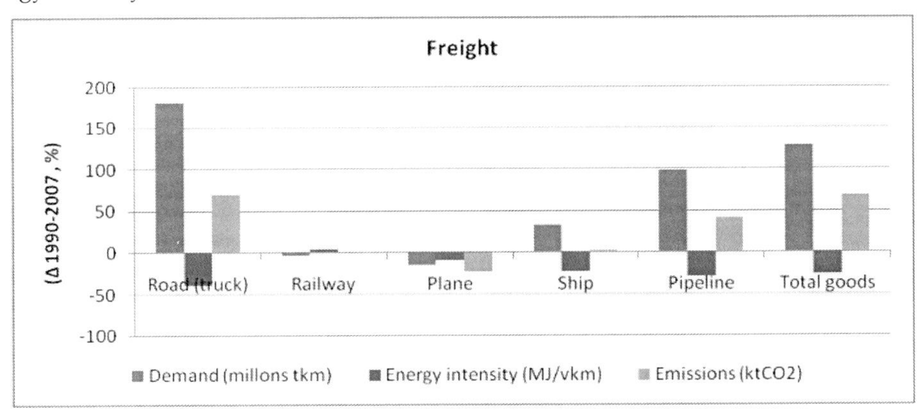

Fig. 2. Change in demand, energy intensity and CO_2 emissions for freight, 1990-2007. Source: National Inventory of Emissions 2008, Annual Report on Transport and Postal Services 2008 and authors' own compilation.

As can be seen from the figures above, the greatest growth in the percentage of demand occurred in road transport (freight) and in air transport (passengers), with a lower increase in GHG and energy consumption. This is most likely due to the considerable technological improvements made in the road transport of freight and air transport of passengers. On the

other hand the cost of fuel is a major item, and a reduction in these costs therefore increases the profitability of road and air services. Passenger trains, which are electrified on the most heavily used routes, have reduced their GHG emissions and upgraded their fleet, leading to significant improvements in energy efficiency. Passenger road transport has also seen a net change for the better, as GHG emissions have grown less than demand, although the reduction in consumption is lower, perhaps due to the fact that the energy efficiency was already fairly high. The increase in traffic congestion may also play a role, as well as the proportion of city travel and the use of ever larger vehicles.
On average, the intensity of CO_2 emissions on the road is 5 times greater than for railways in the case of passenger transport, and 4 times greater in the case of freight. For this reason the decreasing trend in railway transport partly explains the accelerated growth in emissions in our country. In Spain, the energy intensity of transport modes as a whole has decreased, as the energy intensity of the modes has decreased individually.

3. Proposed strategies for action in transport planning to reduce emissions

Transport infrastructures are a vital social and economic resource, as they structure space and determine mobility (Short & Kopp, 2005), providing access to current economic and social opportunities (Richardson, 2005). Investment in the construction and maintenance of transport infrastructures is of considerable importance and has repercussions which affect society as a whole (Hildén et al., 2004; Short & Kopp, 2005). It is therefore essential that the planning of transport systems should be correctly carried out (Hildén et al., 2004).
The increase in the use of transport systems has gone hand in hand with a heightened awareness of the impacts this increase has provoked (RCEP, 1994; Hine, 1998). A new requirement has therefore arisen which demands that this development should also be sustainable and integrated (EC, 1998; US Department of Transportation, 2000).
The decisions concerning major transport plans are taken in a previous strategic phase regarding the development of a particular region.
Strategic Environmental Assessment[1] has been used in recent years as a tool for introducing and integrating environmental, social and economic concerns into the decision-making process for policies, plans and programmes (Dalal-Clayton & Sadler, 1999).
The following are some proposals for possible strategies for mitigating emissions in the transport sector. These strategies should be taken into account in the initial phases of the decision-making process.
The transport sector can apply different measures and instruments to reduce GHG emissions (Accut & Dodgson, 1996). From among these measures, the United Nations Convention on Climate Change (UN-FCCC, 2007) highlights taxing the purchase and use of vehicles, vehicle registration and road use, are based on the vehicle's emissions. Other measures include special taxes on hydrocarbons, tolls for the use of roads and parking, and investment in public transport and other non-motorised forms of transport. The impact of taxation may be minor if incomes increase. It also proposes mixing biofuels and conventional fuels, the admission of CO_2 standards for road transport, and making fuel savings obligatory. These measures may have a limited impact as they are restricted to small

[1]Strategic Environmental Assessment (SEA) is "a systematic process for evaluating the environmental consequences of proposed policy, plan or programme initiatives in order to ensure they are fully included and appropriately addressed at the earliest stages of decision making" (Sadler & Verheem, 1996; Arce & Gullón, 2000; Fisher, 2003; Dalal-Clayton & Sadler, 1999).

sector of the total vehicle fleet. New developments include measures to influence mobility through the coordination of land use and the planning of infrastructures.
As the conditions and circumstances of a country change, it should be possible to measure the potential for mitigating GHG emissions for a specific period of time. This measurement should take past trends into account, as well as the current and future state of all the factors and indicators which determine the mitigation potential of a country. According to Pacala & Socolow (2004), we now have sufficient scientific, technical and industrial knowledge available to solve the problem of GHG emissions and climate change in the forthcoming decades, but this will require assuming the structural modifications and economic costs of such a change. This change in trend must be achieved by through a series of actions, aiming to enhance the synergy between them, and which are institutionally coordinated.
The possible mitigation strategies for CO_2 emissions can be grouped into the following categories:
A) Technological improvements in vehicles and fuels (TCC)
The last decade has seen a reduction in vehicle fuel consumption through voluntary agreements subscribed by vehicle manufacturers (ACEA agreements), and an improvement in fuels (AUTOIL programme). These reductions in energy are also thanks to improvements in engine performance, weight reduction, vehicles with a more aerodynamic design, the use of electric and hybrid engines, etc. (Orasch & Wirl. 1997; Advenier et al., 2002). It is also possible to act on the quality of the fuels by improving their strength, or by introducing biofuels and other energy sources, electric or mixed vehicles, hydrogen batteries, etc. (Hill et al., 2006). Governments can encourage these measures through the exemplary use of vehicle fleets which are run on alternative fuels, using reduced power, etc. (Schipper, 2007). Another measure is the application of a system of taxation based on vehicle emissions and not on engine power.
B) Change in the modal distribution of freight (DMM)
This involves reducing dependence on roads, although this may still be the most heavily used mode. However, railways must aim to reverse their current trend in order to attain a similar rate to that of other European countries. Maritime transport also has the potential for growth in the case of large volumes of freight. However the reinforcement of these different modes of transport must include improvements in the multimodal chain, as roads serve as a complementary mode to other forms of transport for the collection of freight and its final distribution. Thus the availability of multimodal logistical facilities will be a key factor in obtaining a significant change in modal distribution (Vasallo & Fagan, 2007; Janic, 2007).
C) Balance in the modal distribution of interurban passenger demand (MTI)
Similarly, quality railway services must be able to capture travel, both from private vehicles and -in the case of high-speed trains- from planes (López-Pita & Robusté, 2003). Bus services by road have also been shown to be a competitive and quality alternative when they are well run, and offer a network with adequate frequencies and destinations.
D) Balanced modal distribution of urban passenger demand, reduction in length and number of motorised journeys (MTU)
This involves the implementation of Urban Mobility Plans in cities in order to change people's mobility habits and reduce dependence on the car. Plans should also be designed for transportation to work centres (industrial estates, business parks etc), and to shopping and leisure centres so that the locations and facilities are accessible by public transport (IDAE, 2006a).

Travel on foot and by bicycle should be encouraged, with the recovery of urban spaces for non-motorised journeys.

Priority should be given to public transport by improving the quality of the following elements: bus lanes, priority in traffic lights, improved appeal through attractive system design and a high quality service. This will in each case involve providing the most suitable resources: buses, underground, trams, etc. Coordination and integration actions will be of key importance in this area: transportation hubs, combination tickets, user information, etc. (Ministry of the Environment. 2008; IDAE, 2006b).

Finally, there is a need for information and awareness campaigns so that citizens can make their decisions fully conscious of the effects of their choice of transport mode on the environment. Many trips can be avoided, others can be clustered or done in a cleaner way. In some cases it would be useful to indicate the economic and environmental costs of lifestyles which are dependent on mobility by cars (Schafer & Victor, 2000).

E) Efficient use of vehicles (UE)

This refers to the proper management of transport systems for both passengers and freight. One aspect of management involves controlling the use of the infrastructure, either by means of restrictions of time, tariffs, or any other aspect for particular vehicles, for example at times of congestion. GHG emissions increase with the speed of the vehicles; speed control is therefore a key factor in reducing emissions. Emissions also increase the longer the engine is running, so traffic jams must also be minimised (Hensher, 2006; Berger, 2007).

The second aspect refers to the fleet, which should be upgraded with suitable regularity and undergo a strict maintenance regime in order to reduce consumption (Van Weet al., 2000). The management of the fleet is also an important factor: variables such as size and power, frequencies, level of occupancy etc. must be optimised in order to improve energy behaviour in both passenger and freight transport. Efficient driving from the energy point of view has also been shown potentially to reduce emissions, as well as reducing the costs of fuel.

F) Impact of fuel prices on consumption

Another possible measure worth highlighting is the increase in the price of fuel. It would be logical to assume that an increase in the price of fuel would provoke a decrease in the use of transport modes, and particularly in road travel. However it can be observed in Spain that alterations in the price of fuel have had barely any effect on changes in transport activity and consumption of energy in the transport of freight and passengers. Figure 3 shows that despite the considerable increases in fuel prices in 1993 and 2000, and decreases in 2000 and 2004, there was no appreciable effect on the consumption curve.

The short-term impact of fuel prices is limited, partly due to the lack of alternatives, and partly to the mobility habits associated with particular modes such as the car. According to AEMA (2008), a 10% increase in fuel prices produces an average reduction of 2.5% in short-term (first year) fuel consumption for passenger road vehicles. The long-term impact is greater, as there are more alternatives available, such as for example changing one's place of work or residence, and using more fuel-efficient vehicles.

4. Scenarios in 2020 for CO_2 emissions from freight transport, and strategies for mitigation

In this chapter, a theoretical study is approached in order to know the CO_2 emissions after implementation of the actions envisaged in the Spanish Strategic Plan of infrastructures and Transport (Plan Estratégico de Infraestructuras y Transportes (PEIT)), which horizon year is 2020.

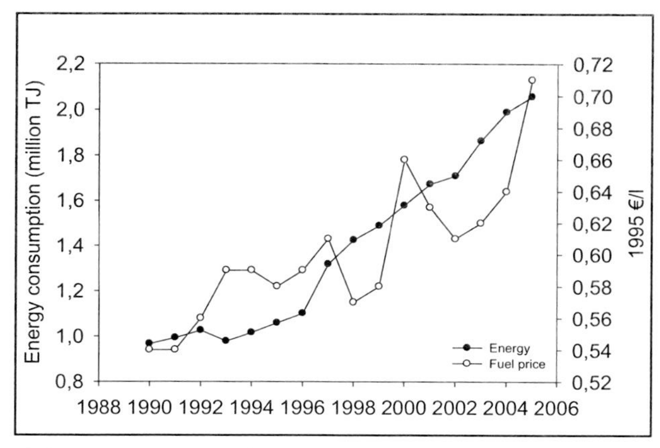

Fig. 3. Evolution of the consumption of energy for transport and fuel prices in Spain (1990-2005). Source: Source: Ministry of Public Works, 2008 Annual Report; Ministry of the Economy, Consumer Price Index, National Institute of Statistics 2008; Ministry of the Economy, Internal Revenue Service 2008.

4.1 Strategic plan for infrastructures and transport

The 2005-2020 Strategic Plan for Infrastructures and Transport (PEIT) aims to establish an efficient framework for transport in Spain by developing a transport system which integrates all the different modes. An additional objective is to use transport as an element which contributes to economic and territorial development, due to its function as a force for linking different areas of the country.

The actions included in the PEIT involve extending the road network to a total of 15,000 km of high-capacity highways, reducing the network's radial configuration and promoting a mesh-type road system, and reaching the target of 9,000 km of high-performance railway lines (Ministerio de Fomento, 2005). The plans for the road network essentially entail a reconversion of national roads into high-performance highways. However, the actions included in the PEIT regarding the railway network consist of creating infrastructures along new routes for the high-performance train system.

The objective established by the PEIT is that the railway system should progressively become the lynchpin in an articulated system of intermodal transport services for both passengers and freight.

4.2 Definition of scenarios

This section analyses the freight transport sector and offers a medium-term prediction under examples of various proposals for action scenarios. Three scenarios are designed to estimate the CO_2 emissions for the Spanish freight transport sector in 2020. The scenarios consist of descriptions of future patterns of behaviour in the sector, and include modal distribution, technologies and mix of different fuels (Pacala & Socolow, 2004). These scenarios are described below in detail, together with a summary of the basic parameters which define them and their estimated variation.

All the scenarios have the same estimates for the levels of future transport demand, based on the projections for Spanish freight transport up to 2007; this trend would represent an increase of 57% between 2007 and 2020, but with a different distribution of the transport modes and different types of vehicles. With the exception of electric trains, the scenarios include only the vehicles powered by internal combustion engines. The differences in CO_2 emissions between the scenarios stem from the different engine technologies, differences in aerodynamic and rolling resistance, and the variety of modal distributions (Orasch & Wirl, 1997; Advenier et al., 2002, Schipper, 2007). The scenarios represent different fuel conditions.

Scenario 1 - "Business As Usual" trend (BAU) assumes that the same trends in activity, energy intensity, fuel and modal distribution observed during the period from 1990-2007 will continue until 2020. There are minor mode transfers from the railways to the road, and the predominance of fossil fuels, -primarily diesel- remains unchanged. The energy intensity of road transport in Spain continues decreasing, due to ever stricter environmental regulations and technological improvements to engines. The use of capacity in the base scenario is 9.4 tons kilometer per vehicle kilometer with load (2007). This value has changed very little from the 9.0 recorded in 1997 (4%). The energy intensity of railways decreases by 2% between 2007 and 2020. Fossil fuels are used in all vehicles with the exception of electric trains.

Scenario 2 - Development of the Railway Sector in Spain (DSF) assumes an increase in the modal share of the railway, and an improvement in energy efficiency through the introduction of new technologies. Freight transport by rail will constitute almost 10% of total freight transport in 2020 (twice the proportion in 1990). This scenario is consistent with the EU's policy measures designed to promote cleaner modes of transport, as in almost all European countries, the railway sector is losing modal share (Vassallo & Fagan, 2007; Janic, 2007). The energy intensity of the railway decreases by up to 20% in this scenario, due to the introduction of new technologies for electric propulsion, which is included in the context of the new railway regulation (Izquierdo & Vassallo, 2004). The decrease in intensity is specific to the fuel used, and is a result of the improved energy use of electrical locomotives as compared to diesel engines. The consumption of fossil fuels decreases in favour of electricity.

Scenario 3 - Road Efficiency and Development of Biodiesel (ECB) assumes that the road maintains its predominance in the freight transport sector, at the same time as there are significant improvements in the efficiency of diesel propulsion engines, and advances in biodiesel engines. Diesel engines show an increase of 55% in efficiency with regards current levels, and significant advances in biodiesel engines makes them more competitive. This increase in efficiency follows the trend of the base scenario (the BAU scenario predicts a 45% increase in efficiency). An additional 10% improvement in efficiency could be obtained by operating high-productivity vehicles. Assuming that the fleet is upgraded every five years, at the start of 2020 all new lorries will need to be 55% more efficient. After exploring the sensitivity of this scenario to changes in the introduction of biodiesel on the market, biodiesel in 2020 will contribute almost 10% of the energy consumed on the roads. The choice of 10% biodiesel is justified by the EU's biofuels directive, which aims to increase the share of biofuels to 5.8% of the energy content of the total consumption of fuel in 2010 (Biofuels barometer, 2008). Biodiesel is studied by analysing the whole of its life cycle (the CO_2 emitted in producing the biofuel), and the real carbon factor for biodiesel (56.1 $ktCO_2eq./PJ$) is 24% less than for conventional diesel.

4.3 Results

The first point is that actions of this kind are required to achieve the objectives proposed in the E4 for the sector, despite the fact that these measures will not be sufficient to fulfil Kyoto targets, even in 2020.

The second observation is that from the point of view of GHG emissions, the DSF scenario based on enhancing railway travel, would be just as efficient as the ECB scenario with its emphasis on roads, using efficient vehicles and fuels. In the BAU scenario, CO_2 emissions exceed 40 $MtCO_2eq$. (an increase of 29% since 2007). This is the scenario with the most significant increase in emissions. The scenario with the greatest reductions is ECB, where emissions fall by 2.4% (1.22 million tons per year less than the BAU scenario, and 15 million tons less in the overall period from 2008-2020). In the case of the DSF scenario, the reduction in emissions is somewhat lower (1.20 million tons per year less than BAU, which represents a decrease of 1.9% since 2007).

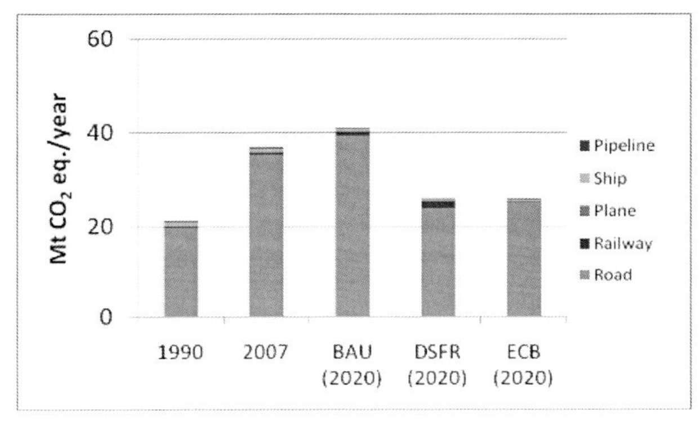

Fig. 4. Scenarios for GHG emissions from freight transport, 1990, 2007 and 2020 (BAU, DSF, ECB). Source: compiled by author.

In the DSF and ECB scenarios, CO_2 emissions are reduced to levels which are lower than the objectives in the E4 strategy; however, none of the scenarios fulfils the target of the Kyoto protocol. In the Kyoto protocol, Spain has undertaken to achieve an average annual increase in GHG -and thus in consumption- of 15% as compared to 1990 levels, between 2008 and 2012 (27.9 Mt CO_2 eq. per year). In the E4 strategy, Spain has committed itself to achieving 38.2 Mt CO_2 eq. a year in 2012, which means 4.5 Mt less than the 42.7 corresponding to the BAU scenario (Ministerio de la Presidencia, 2006).

In the ECB scenario, emissions of freight lorries are 1.5% lower than in 2007, a result of the modification of the different fuel parameters due to technological advances. Similarly, emissions from the railway and maritime sector are 37.3% and 24.3% lower respectively. Aircraft emissions are 38.4% lower, as a result of a decrease in air transport. In this scenario the increase in transport activity cancels out the reductions obtained through technological improvements. The emissions could be lower than the Kyoto target if there were no increase in road transport activity. With the DSF scenario, road emissions are 7% lower and railway emissions are 417.7% higher than in 2007, as a result of the increase in railway activity. With the technological level anticipated in the BAU scenario, emissions from all modes of transport (except air transport) are greater than in 2007.

5. Conclusion

The transport of freight and passengers has grown linked to economic growth, but at a higher rate, especially for passengers. These increases have been well above population

growth, indicating an increase in the number and length of trips per capita. These data indicate an increase in activity, but also a change in the territorial model, with a greater dispersion of activities, whether industrial, and commercial and residential.
In spite of the decrease in energy intensity and emissions of pollutants, the growth in the mobility of persons and freight suggests the possibility of an undesirable growth in emissions, which may also be unsustainable in the long term.
There are some possible mitigation strategies for CO_2 emissions. They can be grouped into six categories. Efficient use of vehicles is the option which offers the greatest potential, and is also vital in drastically reducing emissions of carbon and other pollutants. In second place are the technological improvements in vehicles and fuels. The recourse to multimodality is perhaps an option which requires a greater co-operation between operators, as well as technical and administrative coordination. Achieving a greater balance between modes and dedicating each mode to the segments of the market where they are most competitive would lead to a greater efficiency in emission factors and a reduction in costs, through the improved exploitation of the capacities of the existing networks.
From the point of view of GHG emissions, the enhancing railway travel would be just as efficient as the using efficient vehicles and fuels.
It is necessary that all these strategies be coordinate and integrated into the planning process.

6. References

Acutt, M. & Dodgson, J. (1996). Policy instruments and greenhouse emissions from transport in the UK, *Fiscal Studies*, 17-2, 65-82.

Advenier, P.; Boisson, P.; Delarue, C.; Douaud, A.; Girard, C. & Legendre, M. (2002). Energy Efficiency and CO_2 Emissions of Road Transportation. Comparative Analysis of Technologies and Fuels, *Energy & Environment*, 13, 631-646.

AEMA Agencia Europea Medio Ambiente (2008). Transport and environment. facing a dilemma. TERM 2007, In: *Environmental issues series No 3*, Janse, P., 52, Copenhagen.

Arce, R. & Gullón, N. (2000). The application of Strategic Environmental Assessment to sustainability assessment of infrastructure development. *Environmental Impact Assessment Review*, 20, 393–402.

Berger, W.J. (2007). Abschätzung der Auswirkungen einer Einführung von Tempolimit 80 km/h auf Landstraβen in Österreich, *Straβenverkehrstechnik*, 8, 409-416.

Biofuels barometer (2008). 7.7 MTOE consumed in EU in 2007, *Systèmes Solaires, le journal des energies renouvelables*, 185.

Cuddihy, J.; Kennedy, C. & Byer, P. (2005). Energy use in Canada. environmental impacts and opportunities in relationship to infrastructure systems, *Canadian Journal of Civil Engineering*, 32, 1-15.

Dalal-Clayton, D.B. & Sadler, B. (1999). Strategic Environmental Assessment. A Rapidly Evolving Approach. *Environmental Planning Issues*, 18.

EC (1998). *The Common Transport Policy. Sustainable Mobility. Perspectives for the Future*, Commission Communication to the Council, European Parliament, Economic and Social Committee and Committee of the Regions.

Fisher, B. (2003). Strategic Environmental Assessment in post-modern times. *Environmental Impact Assessment Review*, 23, 155–170.

Hensher, D.A. (2006). Integrating accident and travel delay externalities in an urban speed reduction context, *Transport Reviews*, 26, 521-534

Hernández-Martínez, F. (2006). *La producción de energía eléctrica en España. análisis económico de la actividad tras la liberación del sector eléctrico. Documento de Trabajo 290/2006,* Fundación de las Cajas de Ahorro, Madrid.

Hildén, M.; Furman, E. & Kaljonen, M. (2004). Views on planning and expectations of SEA. the case of transport planning. *Environmental Impact Assessment Review,* 24, 519–536.

Hill, J.; Nelson, E.; Tilman, D.; Polasky, S. & Tiffany, D. (2006). From the Cover. Environmental, economic, and energetic costs and benefits of biodiesel and ethanol biofuels, *Proceedings of the National Academy of Sciences* 103, pp. 11206-11210, National Academy of Sciences.

Hine, J.P. (1998). Roads, regulation and road user behaviour. *Journal of Transport Geography,* 6 (2), 143-158.

IDAE (2006b). *Guía práctica para la elaboración de planes de movilidad sostenible,* IDAE, Madrid.

IDAE (2006a). *Guía práctica para la elaboración de implantación de planes de transporte al centro de trabajo,* IDAE, Madrid.

Izquierdo, R. & Vasallo, J.M. (2004). *Nuevos sistemas de gestión y financiación de infraestructuras de transporte,* Colegio de Ingenieros de Caminos, Canales y Puertos, Madrid.

Janic, M. (2007). Modeling the full costs of an intermodal and road freight transport network, *Transportation Research Part D. Transport and Environment,* 12, 33-44.

Kahn Ribeiro, S.; Kobayashi, S.; Beuthe, M.; Gasca, J.; Greene, D.; Lee, D.S.; Muromachi, Y.; Newton, P.J.; Plotkin, S.; Sperling, D.; Wit, R. & Zhou, P.J. (2007). Transport and its infrastructure, In: *Climate Change 2007. Mitigation. Contribution of Working Group III to the Fourth Assessment Report of the Intergovernmental Panel on Climate Change,* Metz, B.; Davidson, O.R.; Bosch, P.R.; Dave, R. & Meyer, L.A., 323-385, Cambridge University Press, Cambridge.

Kaul, S. & Edinger, R. (2004). Efficiency versus cost of alternative fuels from renewable resources. outlining decision parameters, *Energy Policy,* 32 (7), 929-935.

Lenzen, M. (1999). Total requirements of energy and greenhouse gases for Australian transport, *Transportation Research Part D. Transport and Environment,* 4, 265-290.

Léonardi,J. & Baumgartner, M. (2004). CO_2 efficiency in road freight transportation. Status quo, measures and potential, *Transportation Research Part D. Transport and Environment,* 9, 451-464.

López-pita, A. & Robusté, F. (2003). The Madrid-Barcelona high-speed line, *Proceedings of the Institution of Civil Engineers-Transport,* 156, 3-8.

Ministerio de Fomento (2005). *PEIT. Plan Estratégico de Infraestructuras y Transporte 2005-2020.* Secretaría General Técnica. Ministerio de Fomento, Madrid.

Ministerio de la Presidencia (2006). Real Decreto 1370/2006, de 24 de noviembre, por el que se aprueba el Plan Nacional de Asignación de derechos de emisión de gases de efecto invernadero, 2008-2012, *Boletín Oficial del Estado,* Vol. 282, 41.320-41.440.

Ministerio de Medio Ambiente (2008). *Inventario de Gases de Efecto Invernadero de España-Edición 2008 (serie 1990-2007), sumario de resultados,* Subdirección General de Calidad Ambiental, Madrid.

Orasch, W. & Wirl, F. (1997). Technological efficiency and the demand for energy (road transport), *Energy Policy,* 25, 1129-1136.

Pacala, S. & Socolow, R. (2004). Stabilization wedges. solving the climate problem for the next 50 years with current technologies, *Science,* 305, 968-972.

Pilo, E.; Jiménez, J.A. & López, J.L. (2006). *Jornadas de eficiencia energética en el ferrocarril,* Fundación de los Ferrocarriles Españoles, Madrid.

RCEP (Royal Commission on Environmental Pollution) (1994). *Transport and the Environment, Eighteenth Report.* HMSO. London.

Richardson, B.C. (2005). Sustainable transport. analysis frameworks. *Journal of Transport Geography*, 13, 29–39.
Rodenburg, C.A.; Ubbels, B. & Nijkamp, P. (2002). Policy scenarios for achieving sustainable transportation in Europe, *Transport Reviews*, 22, 449-472.
Saari, A.; Lettenmeier, M.; Pusenius, K. & Hakkarainen, E. (2007). Influence of vehicle type and road category on natural resource consumption in road transport, *Transportation Research Part D*, 12 (1), 23-32.
Sadler, B. & Verheem, R. (1996). *Strategic Environmental Assessment. Status, Challenges and Future Directions.* Ministry of Housing, Spatial Planning and the Environment. International Study of Effectiveness of Environmental Assessment. No 53. The Netherlands.
Schafer, A. & Victor, D.G. (1999). Global passenger travel. implications for carbon dioxide emissions, *Energy*, 24, 657-679.
Schipper, L. (2007). «Automobile fuel; economy and CO_2 emissions in industrialized countries. troubling trends through 2005/2006», World resources institute, EMBARQ cities on the move, Washington, pp. 20.
Schipper, L.; Scholl, L. & Price, L. (1997). Energy use and carbon emissions from freight in 10 industrialized countries. an analysis of trends from 1973 to 1992, *Transportation Research Part D. Transport and Environment,* 2, 57-76.
Short, J. & Kopp, A. (2005). Transport infrastructure. Investment and planning. Policy and research aspects. *Transport Policy,* 12, 360–367.
Sperling, D. (2004). Environmental impacts due to urban transport, In: *Urban Transport and the Environment. An International Perspective,* Nakamura, H.; Hayashi, Y. & May, A.D., 99-189, Elsevier, Oxford.
Steenhof, P.; Woudsma, C. & Sparling, E. (2006). Greenhouse gas emissions and the surface transport of freight in Canada, *Transportation Research Part D. Transport and Environment*, 11, 369-376.
UN-FCCC (2007). *Synthesis of information relevant to the determination of the mitigation potential and to the identification of possible ranges of emission reduction objectives of Annex I Parties,* United Nations.
US Department of Transportation (2000). *Strategic Plan 2000-2003.* Department of Transportation. USA.
Van Wee, B., Moll, H.C. & Dirks, J. (2000). Environmental impact of scraping old cars, *Transportation Research Part D. Transport and Environment*, 5, 137-143.
Van Wee, B.; Janse, P. & Van Den Brink, R. (2005). Comparing energy use and environmental performance of land transport modes, *Transport Reviews,* 25, 3-24.
Vassallo, J.M. & Fagan, M. (2007). Nature or nuture. why do railroads carry greater freight share in the United States than in Europe?, *Transportation,* 34, 177-193.
Zamorano, C.; Biga, J.; & Sastre, J. (2004), *Manual para la planificación, financiación e implantación de sistemas de transporte urbano*, CRTM, Madrid.

Section 4

8

Global Warming and Hydropower in Turkey for a Clean and Sustainable Energy Future

I. Yuksel[1] and H. Arman[2]
[1]*Sakarya University, Faculty of Technology, Department of Construction, 54187 Sakarya*
[2]*Sakarya University, Engineering Faculty Department of Civil Engineering, Esentepe, Campus, 54187, Sakarya,*
[1,2]*Turkey*
[2]*Guest Professor: United Arab Emirates University, College of Science, Department of Geology, P.O. Box. 17551, Al-Ain, UAE.*

1. Introduction

More generally, global warming and climate change and sustainable development interact in a circular fashion. Climate change vulnerability, impacts and adaptation will influence prospects for sustainable development, and in turn, alternative development paths will not only determine greenhouse gas (GHG) emission levels that affect future climate change, but also influence future capacity to adapt to and mitigate climate change. Impacts of climate change are exacerbated by development status, adversely affecting especially the poor and vulnerable socio-economic groups. The capacity to adapt to climate change goes beyond wealth, to other key pre-requisites of good development planning, including institutions, governance, economic management and technology (Kaygusuz, 2001; Yuksel, 2008a).
Meanwhile, global warming and climate change poses an unprecedented threat to all human beings. While this problem is important in the long-run, most decision-makers recognise (especially in the developing countries), that there are many other critical sustainable development issues that affect human welfare more immediately. However, even in the short term, climate is an essential resource for development. For example, in many countries (especially the poorest ones), existing levels of climatic variability and extreme events pose significant risks for agriculture, economic infrastructure, and vulnerable households. Climatic hazards continue to take their human and economic toll even in wealthy countries. Such climate threats, which undermine development prospects today, need to be better addressed in the context of the long-run evolution of local and regional climates (PEWCLIMATE, 2002; Yuksel, 2008a).
Delivering sustainability demands that this access and security of supply be provided, while avoiding environmental impacts, which would compromise future social and economic development. Drawing on the wide-ranging discussions of the Congress, the World Energy Council draws some conclusions a few of these as follows (WEC, 2004; Yuksel and Sandalci, 2009):

- Climate change is a serious global concern, calling for changes in consumer behavior, but offering potential win-win opportunities. These include increased transfer of efficient technologies from industrialized to developing countries and incentives to investment through emerging voluntary and regulated emissions trading.
- Technological innovation and development is vital to reconciling expanded energy services for more equitable economic development with protection of the environment.
- Research and development (R&D) must be more strongly and consistently supported than has been the case. It is the pre-condition of the innovation which is needed. A starting point is the reduction of R&D redundancies through international cooperation.

However, developing the remaining hydropower potential offers many challenges and pressures from some environmental action groups over its impact has tended to increase over time. Hydropower throughout the world provides 17% of our electricity from an installed capacity of some 730 GW is currently under construction, making hydropower by far the most important renewable energy for electrical power production. The contribution of hydropower, especially small hydropower (SHP) to the worldwide electrical capacity is more of a similar scale to the other renewable energy sources (1-2% of total capacity), amounting to about 47 GW (53%) of this capacity is in developing countries (Yuksel, 2007; Yuksel and Sandalci, 2009).

2. Global warming

One major disadvantage of the two-actor matrix presented above is that it gives the false impression that Greens and Developmentalists are evenly matched in their struggle to shape energy politics in Turkey. The actual struggle, however, is far from being between two equals. Developmentalist ideology rules supreme in Turkey and energy politics is no exception to this rule. While energy-related environmental activism, as exemplified by the movements against the Gökova thermic power plant, the Akkuyu Nuclear Power Plant, and the Fırtına valley hydropower dam, is at the heart of environmental politics in Turkey, they either achieve short-lived victories (e.g. the reintroduction of the plans of nuclear power plants) or end-of-pipe solutions that do little to change the overall policy structures (e.g. installation of filters at Gökova). Yet, as several other contributors have argued in this collection, the state in Turkey remains highly sensitive to international forces and dynamics and has frequently improved its environmental policies and practices in response to outside pressures. Therefore, this concluding section discusses the potential impact of global warming and the Kyoto Protocol on the future of Turkish energy policies (Ogutcu, 2002; Kaygusuz, 2003a; Kaygusuz and Sari, 2003; Kaygusuz, 2004a,b; IEA, 2005; MENR, 2005; TEIAS, 2005; Yuksel 2010).

When the United Nations Framework Convention on Climate Change (UNFCCC) was adopted in 1992, all OECD members were included in the list of developed countries in Annex II. Turkey asked for an exception on the grounds that its relative underdevelopment from other OECD members justified special treatment. Such an exception was granted at the Seventh Conference of Parties in Marrakech in 2001, where Turkey was removed from the Annex II. Consequently, the parliament is expected to ratify the Convention. This exception is notable because the flexible implementation mechanisms of the Kyoto Protocol (assuming eventual ratification by Turkey) will open up new avenues for foreign investments for energy efficiency and clean technology projects (IEA, 2005; Yuksel, 2010).

Following the ratification of the Framework Convention and the Kyoto Protocol, Turkey has become eligible for trade in carbon credits under the provisions of the Clean Development

Mechanism. While the necessary institutional capacities and information systems remain to be developed, the government declared its willingness to comply with the general provisions of the UNFCCC. Unlike domestic energy procurement strategies, the global warming dimension of energy politics receives scant attention from civil society and environmental NGOs. Nevertheless, international pressure, especially through the European Union, is likely to lead Turkey to take real steps toward helping prevent global warming (Ogutcu, 2002; MENR, 2005; Yuksel 2010).

Toward this end, the preparation of the 8th Five-Year Development Plan included for the first time an Expert Committee on Climate Change. The committee's recommendations lean heavily toward market-based solutions, support the recent trend toward increased natural gas consumption and make a number of commonsensical suggestions (WECTNC, 2004; Yuksel 2010). A number of promising steps have been taken toward the implementation of these policies. The Electricity Market Act and the Natural Gas Market Law, both of 2001, increased competition and further private involvement. However, given the projected increase in energy demand and consumption, any meaningful reduction of future greenhouse gases in Turkey will necessitate significant investment in renewable energies beyond the current interest in hydropower.

Energy development in Turkey has been dominated by public investment and management. The current government, however, is keen to complete the process of liberalization, restructuring, and privatization in the energy sector. Turkey has made early and extensive use of financing models such as build-own-operate (BOO) and build-own-transfer (BOT). As yet, however, no decisive breakthrough has been achieved. This does not mean a complete withdrawal of the state from energy development. In fact, state involvement in formulating and implementing favorable policies for renewable energy development remains vital. To ensure timely and effective investment in renewable sources, however, the state needs to mobilize the extensive funds available to the private sector. A number of renewable energy projects, such as certain hydropower and solar thermal applications, are already commercially attractive to private interests.

Since possible results of the global warmth gradually started to form the most basic problem on environmental basis, "Framework Convention on Climate Changes" (FCCC) is constituted which was due on March 21, 1994 followed by its approval by 50 countries after being first approved in Rio Environment and Development Conference held in 1992. Aim of the Convention is to keep the concentration of greenhouse gas in the atmosphere at a constant level necessary to prevent its hazardous man caused impact on climate system. On the other hand, international society will come to a common decision in Conference of Parties (COP) held annually where all participating countries are closely involved in decision making process. The countries in Convention's Appendix-1 list decided by Kyoto Protocol to be due between 2008 and 2012 will be forced to reduce total emission level of gases (CO_2, CH_4, N_2O, HFCs) that have direct greenhouse effect 5% below the level in 1990 (Say, 2006; Yuksel, 2008a).

However, more often than not, they are placed in a dilemma when left to balance between economic growth and environment. Conflicts often rise between social, environmental and economic objectives (World Bank, 1992 and 2000). The headlong pursuit of economic growth is the cornerstone of developing countries. A top Turkish environmental official accepted that economic growth must take precedence over environmental protection for years to come because the former is not only of great importance to maintaining political stability but also to funding the environmental clean-up (Yuksel, 2008a).

3. Climate change

Sustainable development has been recognized as a key cross-cutting theme in the preparation of the Intergovernmental Panel on Climate Change (IPCC) fourth assessment report. Researchers could make pivotal contributions to the IPCC's work on sustainable development, with contributions to this volume highlighting some of the key issues requiring investigation and analysis. On the other hand, technologies and practices to reduce GHG emissions are continuously being developed. Many of these technologies focus on improving the efficiency of fossil fuel energy or electricity use and the development of low carbon energy sources, since the majority of GHG emissions are related to the use of energy. Energy intensity (energy consumed divided by gross domestic product, GDP) and carbon intensity (CO_2 emitted from burning fossil fuels divided by the amount of energy produced) have been declining for more than 100 years in developed countries without explicit government policies for decarbonization, and have the potential to decline further (IPCC, 2001; Yuksel, 2008a).

Perhaps the most contentious issue is the conceptual framework for addressing climate change within a sustainable development mandate. Various stakeholders are bound to have different views and analytical frameworks to support their positions. Given the extent to which the respective debates on climate change and sustainable development have evolved separately in the past, it will be a significant challenge to re-integrate climate change with development policy (Briden and Downing, 2002; Yuksel, 2008a).

A debate on policy requires a framework for evaluating risks and solutions. The choices revolve around the extent to which a framework seeks to explore and visualize alternatives or recommend desirable solutions, the representation of values, and the role of actors. The contributors to this volume do not evaluate their frameworks-often presuming that the structure they use (be it approaches based on cost-benefit analysis, integrated assessment or social analysis), are adequate to the challenge (Munasinghe and Swant, 2004; Yuksel, 2008a).

The more technical issues in analytical methodologies involve persistent challenges to researchers. For example:

- A narrowly focused cost-benefit analysis assumes that researchers can comprehensively estimate the monetary implications of mitigation policy and climate impacts in the economic, social and environmental domains. Few researchers believe that calculations of potential impacts of climate change are well known, and many are sceptical of the hubris involved in bridging the local/global nature and present/century time-scales of climate change. For example, currently available estimates of the social cost of carbon are inadequate in assessing secondary effects, climatic disasters and potential large-consequence risks.
- Integrated Assessment Models (IAMs) do not capture the role of decision-makers (i.e. their worldviews, goals and strategies), rather relying on rational economic criteria or statistical trends in a pressure-state-impacts-responses framework. For example, technological developments are often handled as continuous functions, such as a co-efficient for autonomous improvement in energy efficiency (Downing et al., 2003; Yuksel, 2008a).
- Scenarios are not very reliable frameworks for optimizing present decisions, although they are often used in this manner. Existing scenarios are seldom probabilistic and socio-economic projections tend to be static world-views with little correspondence to the punctuated, dynamic, event-response nature of reality. For example, few vulnerability/adaptation researchers consider scenarios of GHG emissions projections as adequate for understanding potential failures of climate policy.

Given the importance of the conceptual frameworks, there is surprisingly little research into what comprises a 'good' framework. Some research communities have attempted to systematically compare their own frameworks (e.g. the vigorous discussion among IAM teams). Even agreed criteria are missing for comparing such broad scoping frameworks and methodologies (Downing et al., 2003; Yuksel, 2008a).

While it is relatively easy to raise equity and values as key research-policy issues, there is a tendency by researchers to say in effect, "we provide the facts and let policy-makers negotiate on the values and make choices". This is an unduly conservative approach to research. Equally, it removes from climate policy research the rich traditions of some social sciences and the humanities.

Turkey's most recent Five-Year Development Plan, adopted in 2000, affects all policies in all economic sectors and has an indirect impact on greenhouse (GHG) emissions. The first Special Expert Committee on Climate Change was established as one of 98 consultative committees during preparation of this plan. The committee's recommendations were published by the Turkish prime minister as official policy for the current planning period (see Table 1).

• Privatizing energy resource production.
• Increasing the share of natural gas in consumption.
• Transferring electricity production and distribution to the private sector to make utility services more efficient.
• Encourage power savings by matching costs to prices and preventing theft.
• Developing new and renewable energy sources and ensuring their greater role in the market.
• Converting railway management to commercial orientation to ensure efficient, market oriented services.
• Investing in natural gas pipelines and storage facilities.
• A comprehensive strategy is needed for developing renewable energy sources offshore and this should cover assessment of environmental impacts.
• Combining heat and power plants should be regarded primarily as a source of heat.
• Increasing energy efficiency and ensuring energy savings.
• Improving the petroleum product quality for cut sulphur emissions.
• Using proper energy management model for the future of Turkey.

Source: WECTNC (2003)

Table 1. Emission mitigation potential in Turkey

These recommendations serve to guide government actions, but their actual implementation depends on the actions of various agencies and regulators. Under the Electricity Market Act adopted in 2001, the power sector will soon undergo profound reform, leading to the introduction of competition and increasing private involvement. The new Natural Gas Market Law, also adopted in 2001, establishes a competitive gas market and harmonizes Turkish legislation with European law. The Turkish Council of Ministers has adopted several measures to stabilize fuel prices. An automatic pricing formula was abolished and gasoline taxes were made consistent with European countries. For example, taxes comprised over 60% of the price of gasoline by late 2000. To increase energy efficiency in industrial sectors, energy conservation regulations were issued in 1995. These required industrial establishments with annual consumption above 84 terajoules to establish an internal energy management system, conduct energy audits, and appoint an energy manager in their plants. Some 1,250 plants accounting for 70% of Turkish industrial energy use are covered by this regulation (Kaygusuz, 2004b; Yuksel, 2008a).

Turkey's total carbon dioxide (CO_2) emissions amounted to 239 million tones (Mt) in 2006. Emissions grew by 5% compared to 2001 levels and by just over 50% compared to 1990 levels. Oil has historically been the most important source of emissions, followed by coal and gas. Oil represented 45% of total emissions in 2004, while coal represented 40% and gas 15%. The contribution of each fuel has however changed significantly owing to the increasingly important role of gas in the country's fuel mix starting from the mid-1980s (MENR, 2005; MENR, 2007; Yuksel and Sandalci, 2009).

According to recent projections, total primary energy supply (TPES) will almost double between 2006 and 2020, with coal accounting for an increasingly important share, rising from 24% in 2006 to 36% in 2020, principally replacing oil, which is expected to drop from 40% to 27%. Such trends will lead to a significant rise in CO_2 emissions, which are projected to reach nearly 600 Mt in 2020, over three times 2004 levels (MEF, 2007; MENR, 2007; IEA, 2008; Yuksel and Sandalci, 2009).

In 2006, public electricity and heat production were the largest contributors of CO_2 emissions, accounting for 30% of the country's total. The industry sector was the second largest, representing 28% of total emissions, followed by transport, which represented 20% and direct fossil fuel use in the residential sector with 8%. Other sectors, including other energy industries, account for 14% of total emissions. Since 1990, emissions from public electricity and heat production have grown more rapidly than in other sectors, increasing by 6%. Simultaneously, the shares of emissions from the residential and transport sectors both dropped by 7% and 3% respectively while the share of emissions from the manufacturing industries and construction sector remained stable (MENR, 2005; DIE, 2006; DPT, 2006; Yuksel and Sandalci, 2009).

4. Global warming and climate change policy in Turkey

Turkey was a member of the OECD when the UNFCCC was adopted in 1992, and was therefore included among the so-called Annex I and Annex II countries. Under the convention, Annex I countries have to take steps to reduce emissions and Annex II countries have to take steps to provide financial and technical assistance to developing countries. However, in comparison to other countries included in these annexes, Turkey was at a relatively early stage of industrialization and had a lower level of economic development as well as a lower means to assist developing countries. Turkey was not given a quantified emissions reduction or limitation objective in the Kyoto Protocol. Following a number of negotiations, in 2001 Turkey was finally removed from the list of Annex II countries but remained on the list of Annex I countries with an accompanying footnote specifying that Turkey should enjoy favorable conditions considering differentiated responsibilities. This led to an official acceptance of the UNFCCC by the Turkish Grand National Assembly in October 2003, followed by its enactment in May 2004. Turkey has not yet signed the Kyoto Protocol (Kaygusuz, 2003b; MENR, 2005; IEA, 2008; Kaygusuz, 2009; Yuksel and Sandalci, 2009).

Throughout this process, the government carried out a number of studies on the implications of climate change and its mitigation. The first efforts were undertaken by the

National Climate Coordination Group in preparation for the 1992 Rio Earth Summit. Following this, a National Climate Program was developed in the scope of the UNFCCC. In 1999, a specialized Commission on Climate Change was established by DPT in preparation of the Eighth Five-Year Development Plan (2001-2005). The Five-Year Development Plan was the first planning document to contain proposals for national policies and measures to reduce greenhouse gas (GHG) emissions, and funding for climate-friendly technologies (ESMAP, 2003; Yuksel and Sandalci, 2009).

Running counter to the technological and economic potential for GHG emissions reduction are rapid economic development and accelerating change in some socio-economic and behavioral trends that are increasing total energy use, especially in developed countries and high-income groups in developing countries. Dwelling units and vehicles in many countries are growing in size, and the intensity of electrical appliance use is increasing. Use of electrical office equipment in commercial buildings is increasing. In developed countries, and especially the USA, sales of larger, heavier, and less efficient vehicles are also increasing. In addition and usually related to technological innovation options, there are important possibilities in the area of social innovation. In all regions, many options are available for lifestyle choices that may improve quality of life, while at the same time decreasing resource consumption and associated GHG emissions (IPCC, 2001; Yuksel, 2008a).

Of course, the target readers for this volume are not only relevant researchers, but also those concerned with and responsible for climate and sustainable development policy at all levels. At an operational level, there is a need to increase awareness of climatic risks in sustainable development, to look for synergies in policy, and to demonstrate effective solutions. International negotiators bear the front-line responsibility for helping to effectively integrate global climate policies with national sustainable development strategies. The pace of achievements seems slow, particularly in the face of certain climate change. The challenges of resolving conflicts among the world-wide community of stakeholders and implementing a broad reaching sustainability paradigm are indeed formidable. Boundary organisations, linking vulnerable groups with civil society, government and private actors, should explore emerging opportunities.

What should researchers seek to achieve? The ultimate aim of the United Nations Framework Convention on Climate Change (UNFCCC) is to stabilize GHG concentrations at a level that prevents dangerous climate change. While there is no agreement on what such a "dangerous" level might be, stabilization of GHG concentrations at any level would require cuts in global GHG emissions substantially below their current levels. The only international agreement in place so far with legally binding emission targets is the Kyoto Protocol, whose targets, if fully implemented, would amount to only a 5% cut in GHG emissions in industrialized countries from 1990 levels by 2008–2012. Although the Protocol does signal an important change in policy and could, particularly through subsequent negotiating rounds, have long-term impacts on technology and economic development, by itself it is far from achieving the aim of the UNFCCC (1999) (Yuksel, 2008a). Annex I countries, need to show far more effective leadership in controlling their own emissions. The gap between where we are today and the policy target of a stable and acceptable climate system is the defining challenge of research into climate policy (Downing et al., 2003; Yuksel, 2008a).

Developing countries, while varying in size and population, political system, economic structure, bear many similarities. They are facing less favourable economic circumstances, worsening environmental degradation and challenges in curbing climate changes. The present paper only focuses on the issues of contradictory objectives, unrealistic standards and limited public participation.

Policy makers in developing countries are well aware of the importance of environmental protection. However, more often than not, they are placed in a dilemma when left to balance between economic growth and environment. Conflicts often rise between social, environmental and economic objectives (World Bank, 1992; Yuksel, 2008a). The headlong pursuit of economic growth is the cornerstone of developing countries. A top Turkish environmental official accepted that economic growth must take precedence over environmental protection for years to come because the former is not only of great importance to maintaining political stability but also to funding the environmental clean-up. This very contradictory objective in developing countries is well materialized in the implementation of "Polluter Pays Principles" (the PPP), the value of which is dramatically belabored. A good example can be found in the way the governments deal with state-owned enterprises (SOEs) in emissions abatement.

Environmental degradation harms human health, spoils amenities and reduces economic productivity, e.g. agriculture production (Arıkan and Kumbaroğlu, 2001; Yuksel, 2008a). However, protecting the environment is a vital part of improving economic productivity as well as improving the well-being of people today and tomorrow. The evidence shows that the gains from protecting the environment outweigh the costs in the long run (World Bank, 1992, 2000; Yuksel, 2008a). While there is still uncertainty as to the extent and the physical effects of climate change, the costs of not taking actions may well be greater than the costs of preventive actions taken now, especially when the absence of action today may lead to irreversibly undesirable environmental consequences. On the other hand, for developing countries, great importance should be attached to the acceleration of environmentally responsible development rather than following the past, and arguably the present, path of the industrial world in pursuit of "unrestricted economic growth without considerations to its effects on the natural environment".

On May 24, 2004 Turkey became the 189th party by signing Framework Convention on Climate Changes. In the first six months after Turkey became a party of FCCC, the country is obligated to first national declaration to United Nations General Secretariat until November 24, 2004. After this stage is completed Turkey will both have to fulfill new liabilities such as to present national greenhouse gas inventories and national declaration reports to Convention Secretariat regularly, and will also actively participate in efforts carried on global wide so that convention will achieve its ultimate goal. When we compare sectoral distribution of greenhouse effect emissions occurring due to fuel consumption to obtain consumption and projectional values, it is observed that some sectors increased their shares, and some had a significant decrease in their shares. In 2003, it is estimated that 36% of CO_2 emissions occurred due to energy, 34% due to industry, 15% due to transportation and 14% due to other (housing, agriculture and forestry) sectors and in 2020 40% will occur due to energy, 35% due to industry, 14% due to transportation and 11% due to other sectors (Yuksel, 2008a).

Since possible results of the global warmth gradually started to form the most basic problem on environmental basis, "Framework Convention on Climate Changes" (FCCC) is constituted which was due on March 21, 1994 followed by its approval by 50 countries after being first approved in Rio Environment and Development Conference held in 1992. Aim of the Convention is to keep the concentration of greenhouse gas in the atmosphere at a constant level necessary to prevent its hazardous man caused impact on climate system. On the other hand, international society will come to a common decision in Conference of Parties (COP) held annually where all participating countries are closely involved in decision making process. The countries in Convention's Appendix-1 list decided by Kyoto Protocol to be due between 2008 and 2012 will be forced to reduce total emission level of gases (CO_2, CH_4, N_2O, HFCs) that have direct greenhouse effect 5% below the level in 1990 (Say, 2006; Yuksel, 2008b,c).

However, more often than not, they are placed in a dilemma when left to balance between economic growth and environment. Conflicts often rise between social, environmental and economic objectives (World Bank, 1992 and 2000; Yuksel, 2008c). The headlong pursuit of economic growth is the cornerstone of developing countries. A top Turkish environmental official accepted that economic growth must take precedence over environmental protection for years to come because the former is not only of great importance to maintaining political stability but also to funding the environmental clean-up (Yuksel, 2008b,c).

5. Renewable and sustainable energy in Turkey

The government considers alternative transport fuels to be an important option in the longer term to mitigate energy security concerns and reduce GHG emissions. However, it deems current technologies to be expensive and a risky investment, while not offering significant life cycle GHG reduction benefits, especially if the fuel is derived from fossil fuels. On the other hand, liquefied petroleum gas (LPG) demand in transport increased between 1998 and 2000 owing to a government subsidy and a zero-taxation policy. The trend was subsequently reversed in 2001 with an increase in taxation and the removal of the subsidy. Nevertheless, LPG remains an important transport fuel as its share was 8.9% of the total oil product demand in the transport sector in 2002. LPG is used, for example, in taxis in the major cities (MEF, 2007; Yuksel and Sandalci, 2009).

Hydropower generation climbed from 2 Mtoe (23.1 TWh) in 1990 to 3.0 Mtoe (35.3 TWh) in 2004, growing on average by 3.8% per year. The economic hydropower potential has been estimated at 128 TWh per year, of which 35% has been exploited. The government has a strategy for developing the hydropower potential and expects a few hundred plants to be constructed over the long term adding more than 19 GW of capacity. Construction costs would be approximately US$ 30 billion. The government expects hydropower capacity to reach about 31 000 MW in 2020. Some 500 projects (with a total installed capacity over 20 400 MW), which are in different phases of the project cycle, are awaiting realization. On the other hand, Turkey has a lot of potential for small hydropower (< 10 MW), particularly in the eastern part of the country. At present the total installed capacity of small hydropower is 176 MW in 70 locations, with annual generation of 260 GWh. Ten units are under construction with a total installed capacity of 53 MW and estimated annual production of 133 GWh. Furthermore, 210 projects are under planning with a total capacity of 844 MW and annual production of about 3.6 TWh (DSI, 2005; Yuksel and Sandalci, 2009).

6. The Role of hydropower for renewable and sustainable energy

The hydropower industry is closely linked to both water management and renewable energy production, and so has a unique role to play in contributing to sustainable development in a world where billions of people lack access to safe drinking water and adequate energy supplies. On the other hand, approximately 1.6 billion people have no access to electricity and about 1.1 billion are without adequate water supply. However, resources for hydropower development are widely spread around the world. Potential exists in about 150 countries and about 70% of the economically feasible potential remains to be developed-mostly in developing countries where the needs are most urgent (IEA, 2002; IHA, 2003; WEC, 2001; Yuksel, 2008b,c).

Hydropower is available in a broad range of project scales and types. Projects can be designed to suit particular needs and specific site conditions. As hydropower does not consume or pollute the water it uses to generate power, it leaves this vital resource available for other uses. At the same time, the revenues generated through electricity sales can finance other infrastructure essential for human welfare. This can include drinking water supply systems, irrigation schemes for food production, infrastructures enhancing navigation, recreational facilities and ecotourism. Hydropower has very few greenhouse gas emissions compared with other large-scale energy options (see Table 2).

Technology	Energy pay back time in months	SO_2 emission In kg/GWh	NO_2 emission In kg/GWh	CO_2 in Ton/GWh
Coal fired	1.0-1.1	630-1370	630-1560	830-920
Gas (CCGT)	0.4	45-140	650-810	370-420
Large-hydro	5-6	18-21	34-40	7-8
Micro hydro	9-11	38-46	71-86	16-20
Small hydro	8-9	24-29	46-56	10-12
Wind turbine				
4.5 m/s	6-20	18-32	26-43	19-34
5.5 m/s	4-13	13-20	18-27	13-22
6.5 m/s	2-8	10-16	14-22	10-17
Photovoltaic				
Mono-crystalline	72-93	230-295	270-340	200-260
Multi-crystalline	58-74	260-330	250-310	190-250
Amorphous	51-66	135-175	160-200	170-220

Source: UNDP, 2000

Table 2. The comparison of energy amortization time and emissions of various energy technologies.

In addition, by storing water during rainy seasons and releasing it during dry ones, dams and reservoirs can help control water during floods and droughts. These essential functions protect human lives and other assets. This will be increasingly important in the context of global warming, which implies an expected rising variability in precipitation frequency and intensity. On the other hand, hydropower projects do not export impacts such as acid rain or atmospheric pollution. Environmental impacts are limited to changes in the watershed in which the dam is located. When well managed, these changes can sometimes result in enhancements, and other impacts can be avoided, mitigated. Hydropower can contribute to mitigating the widespread potential human impacts of climate change (IHA, 2003; WEC, 2001; Yuksel, 2008b,c).

Hydropower energy is a renewable, sustainable and clean energy in the other alternative energy sources. Moreover, it does not deprive future generations in terms of raw materials, or burdening them with pollutants or waste. Hydroelectric power plants utilize the basic national and renewable resource of the country. Although the initial investment cost of hydropower seems relatively high, the projects have the lowest production costs and do not depend on foreign capital and support, when considering environmental pollution and long-term economic evaluation (Paish, 2002; Yuksel, 2008b; Yuksel, 2008c).

7. Hydropower and dams for renewable and sustainable energy in Turkey

In 2005, primary energy production and consumption has reached 28 and 94.3 million tons of oil equivalents (Mtoe) respectively (see Table 3). The most significant developments in production are observed in coal production, hydropower, geothermal, and solar energy. Turkey's use of hydropower, geothermal and solar thermal energy has increased since 1990. However, the total share of renewable energy sources in total primary energy supply (TPES) has declined, owing to the declining use of non-commercial biomass and the growing role of natural gas in the system. Turkey has recently announced that it will reopen its nuclear programme in order to respond to the growing electricity demand while avoiding increasing dependence on energy imports (DPT, 2006; MEF, 2007; MENR, 2007; Yuksel, 2008c).

	2000	2001	2002	2003	2004	2005
Primary energy production (TTOE)	27,621	26,159	24,884	23,779	24,170	28,020
Primary energy consumption (TTOE)	81,193	75,883	78,322	83,936	87,778	94,300
Consumption per capita (KOE)	1204	1111	1131	1196	1234	1249
Electricity installed capacity (MW)	27,264	28,332	31,846	35,587	36,824	39,596
Thermal (MW)	16,070	16,640	19,586	22,990	24,160	26,481
Hydraulic (MW)	11,194	11,692	12,260	12,597	12,664	13,115
Electricity production (GWh)	124,922	122,725	129,400	140,580	150,698	165,346
Thermal (GWh)	94,011	98,653	95,668	105,190	104,556	124,321
Hydraulic (GWh)	30,912	24,072	33,732	35,390	46,142	41,025
Electricity import (GWh)	3786	4579	3588	1158	464	636
Electricity export (GWh)	413	433	435	587	1144	1812
Total Consumption (GWh)	128,295	126,872	132,553	141,151	150,018	
Consumption per capita (kWh)	1903	1857	1914	2011	2109	2240

Source: Ref. [8] (DPT, 2006; MENR, 2007; Yuksel, 2008c).
Table 3. Developments in production and consumption of energy between 2000–2005 in Turkey.

Conventional electricity supply options include thermal (coal, oil, and gas), nuclear and hydropower. These technologies currently dominate global electricity generation (thermal 60%, hydraulic 20%, nuclear 17% and all others 3%, approximately). Use of cogeneration, particularly geothermal and wind generation, both for isolated supply and small- to medium-scale grid-feeding applications, is small but increasing globally (Altinbilek, 2000; Ceylan and Ozturk, 2004; IHA, 2006; Yuksel and Sandalci, 2009).
The generation of hydropower provides an alternative to burning fossil fuels or nuclear power, which allows for the power demand to be met without producing heated water, air emissions, ash, or radioactive waste. Of the two alternatives to hydropower, in the last decade, much attention has been given to thermal power production because of the adverse effect of CO_2 emissions. With the increasing threat of greenhouse gases originating from

such anthropogenic activities on the climate, it was decided to take action. Thus the Framework Convention on Climate Change was enacted on 21 March 1994 and has been signed by 174 countries to date (Yuksel, 2008c; Yuksel and Sandalci, 2009).
Dams that produce electricity by this most productive renewable clean energy source in the world provide an important contribution to the reduction of air pollution. The result of an investigation held in the USA suggests that the productivity of hydroelectric power-plants is higher than 90% of thermal plants and this figure is twice that of thermal plants. In case of Turkey, the public has been wrongly informed. Some people have claimed that hydro plants do not produce as much energy as planned because of irregular hydrological conditions and rapid sedimentation of reservoirs. It is also claimed that the cost of the removal of dams entirely filled by sediment at the end of their physical lives is not considered in the total project cost, and that there are major problems in recovering the cost of investment and environmental issues (UNDP, 2000; Yuksel, 2008c; Yuksel and Sandalci, 2009).

8. Energy policy in Turkey

The preceding discussion already has laid the foundations for an analytical framework necessary to understand the structural dynamics and political forces at work. The discussion of the determinants of energy intensity and energy sources makes it clear that specific policy outcomes can be understood as a function of two conceptual categories concerning policy-making: regulation and technology. While these two conceptual categories account for most aspects of environmental and energy policy outcomes, a third indicator, political outlook, is required to fully capture the domestic and geopolitical forces at work in Turkey.
The first category, regulation, concerns both the means of devising regulatory frameworks on energy and the overarching goal of such policies. The second category also comprises two variables: the relationship between technology and risk and the nature of technology implementation. Finally, the category of political outlook comprises a discursive alignment and outlook on the nature of international relations (MEF, 2007; MENR, 2007; IEA, 2005; Yuksel 2010).
Using these three categories, it is possible to construct a matrix of the competing energy and environment discourses in Turkey. For the sake of simplicity, this chapter uses only two major orientations, though a variety of combinations are possible. These do not necessarily correspond with real world actors as the matrix is merely intended as a heuristic device to chart the profile of the ongoing policy debates in Turkey. Naturally, the real world of energy politics has various shades of gray, and it is not uncommon for actors to borrow from each camp over time. Nevertheless, these two positions, Greens and Developmentalists, capture the tenor of the ongoing debate in Turkey (Kaygusuz and Arsel, 2005; Yuksel 2010).
Greens believe in extensive environmental regulation. In line with their European and North American counterparts, Greens in Turkey articulate their positions with an implicit critique of markets that question both their desirability as social institutions and effectiveness as regulatory tools. Thus, this position is characterized by calls for the direct involvement of the state in protecting the environment through command-and-control mechanisms. Moreover, Greens privilege ecological protection over continued economic growth. This is not to suggest that this position rejects economic growth entirely, since such deep ecology-inspired movements in Turkey remain relatively rare. The practical upshot of this for their energy policy is built around small-scale and alternative technologies, such as wind farms and solar panels. Finally, in their political outlook, the Greens in Turkey parallel

the 'liberal' school of international relations, constructing their discourse around concepts such as multiculturalism and universal human rights, believing on the one hand that non-state actors are increasingly important in energy politics and on the other interpreting the interstate system as one characterized by win–win cooperation (Kaygusuz and Arsel, 2005; Yuksel 2010).

Air quality standards for four pollutants, namely SO_2, nitrogen dioxide (NO_2), particulate matter (PM) and ozone (O_3) are set under the 1986 Air Quality Protection regulation. The monitoring of ambient air pollution has improved over recent years but remains a problem, particularly with regards to NO_2 and O_3. On the other hand, until recently, the 1986 regulation was also responsible for setting air pollution standards for combustion plants. It was amended in October 2004 by the new Industrial Air Pollution Control Regulation (ESMAP, 2003; MEF, 2007; Yuksel and Sandalci, 2009).

The emissions standards for power plants remain significantly less stringent than those currently in force at the European Union (EU) level as defined by the revised Large Combustion Plants (LCP) Directive. For example, for new solid fuel-fired power plants with a thermal input greater than 300 MW, the NOx emissions limit is set at 200 mg/Nm^3 at the EU level, while the NOx emissions limit is 800 mg/Nm^3 in Turkey. On the other hand, first estimates show that achieving the standards defined under the LCP directive would entail investments of over US$ 1 billion. This would include investments in the retrofitting of installed FGD and ESP equipment and the adoption of advanced and environment-friendly coal technologies. The 2004 Industrial Air Pollution Control Regulation is an important step towards aligning air quality standards with EU regulations, but more efforts will be needed (MEF, 2007; Kaygusuz, 2009; Yuksel and Sandalci, 2009).

Construction of one power plant based on circulating fluidized bed technology has recently been completed. The plant is the first application of advanced coal technology in Turkey and has been designed to use low-quality lignite with high sulfur content. The industry and residential sectors are also responsible for significant air pollution, mainly as a result of lignite consumption. In order to reduce emissions from these sectors, the state-owned Turkish Coal Enterprises (TKI) has developed significant lignite washing capacity. By the end of 2006, total washing capacity was approximately 10.8 Mt, equivalent to current coal demand from both sectors. In addition, the use of high-sulfur coal in residential heating is prohibited. Lastly, the substitution of gas as distribution networks are expanded in urban areas should further contribute to reduce air pollution (TEDAS, 2006; MEF, 2007; MENR, 2007; Yuksel and Sandalci, 2009).

9. Conclusions

The following concluding remarks may be drawn from this study (Yuksel, 2008c):

- There are a number of environmental problems in the country that we face today. These problems span a continuously growing range of pollutants, hazards and ecosystem degradation over the country. So, all government agencies and other non-governmental agencies in the country must work together to utilize their renewable energy and choose the appropriate application in Turkey.
- Take into account the effects of liberalization in the energy forecasts. Continue to revise forecasts regularly to enable the creation of a robust long-term energy policy framework in light of the sharp demand growth and increase focus on the demand side in energy policy planning.

- The technology of hydropower involved has proven itself over a long period of time and is therefore very reliable. So, the government and private sectors should be consider steps to accelerate economic hydropower projects, including refurbishment, consistent with the protection of the environment, to utilize the remaining hydropower potential.
- Fuelwood and modern biomass energy presents a considerable opportunity for Turkey to obtain a significant part of our future energy needs from this sustainable and domestic energy source, since, at present, modern technologies are increasingly being applied to fuelwood development.
- Growing environmental and social concerns, both on the part of decision makers and public opinion, have brought a new perspective to the perception of renewable energy sources as a valid alternative in the long-term, and a useful and practical complement to traditional sources of energy in the short and medium-term. In this respect, geothermal, solar and wind energy sources present a considerable opportunity for our country to obtain a significant part of future energy needs from this sustainable, clean and domestic sources.
- The government should be enact the renewable energy law as envisaged and monitor and evaluate its cost and effectiveness. Share information and experience with other countries introducing quota and certificate-based promotional schemes for renewables.
- Assess the impact on the network reliability and stability resulting from increased penetration of intermittent wind power and explore ways to minimize such an impact. Consider a combination of wind power and pumped storage hydropower for this purpose. Share information and experience with other countries on technical and regulatory approaches to intermittency.

Hydropower represents an alternative to fossil fuel generation, and doesn't contribute to either greenhouse gas emissions or other atmospheric pollutants. However, developing the remaining hydropower potential offers many challenges and pressures from some environmental action groups over its impact has tended to increase over time. Moreover, in the context of the restructuring of the electricity sector, markets may favour more polluting and less costly options. On the other hand, small hydropower's main challenges relate to both economics and ecology. Especially small hydropower can be successfully developed as long as it produces electricity at competitive prices and under conditions that respect the environment. In addition, hydro plants are often superior to other power plants from the standpoint of socio-economic and environmental considerations. The environmental impacts of hydropower plants are at the lowest level compared with the other alternative resources.

10. References

Altinbilek, D. 2000. Hydroelectric development plans in Turkey.

Arikan, Y., Kumbaroğlu, G. 2001. Endogenising emission taxes: a general equilibrium type optimisation model applied for turkey. *Energy Policy* 29: 1045-1056.

Briden, JC., Downing, TE. (Eds). 2002. Managing the Earth, Oxford Univ. Pres, Oxford, UK.

Ceylan, H. and Oztürk, H.K. 2004. Modeling hydraulic and thermal electricity production based on genetic algorithm-time series (GATS). *Int. J. of Green Energy* 1: 393-406.

DIE, State Institute of Statistics. 2006. Statistic yearbook of Turkey in 2005, Prime Ministry, Republic of Turkey, Ankara, 2004.

Downing, TE., Munasinghe, M., Depledge, J. 2003. Special supplement on climate change and sustainable development. *Climate Policy* 3S1: S3-S8.
DPT, State Planning Organization. 2006. Ninth Development plan 2007-2013, Ankara, Turkey.
DSI, State Water Works. 2005. Hydropower potential in Turkey, Ankara, Turkey.
ESMAP, Energy Sector Management Assistance Program (ESMAP). 2003. Turkey-Energy and the Environment Review: Synthesis Report, World Bank.
IEA, International Energy Agency. 2002. World Energy Outlook 2002, OECD/IEA, Paris.
IEA, International Energy Agency. 2005. Energy policies of IEA countries: Turkey review, OECD/IEA, Paris.
IEA, International Energy Agency. 2008. CO_2 emissions from fuel combustion, 2008 Edition, OECD/IEA. Available from www.iea.org.
IHA, International Hydropower Association. 2003. The role of hydropower in sustainable development, IHA White Paper.
IHA, International Hydropower Association. 2006. Hydropower information and country report for Turkey. Available from http://www.hydropower.org.
IPCC, Intergovernmental Panel on Climate Change. 2001. Climate Change 2001: Mitigation, IPCC Third Assessment Report, //www.ipcc-nggip.iges.or.jp.
Kaygusuz, K. 2001. Environmental impacts of energy utilization and renewable energy sources in Turkey. *Energy Exploration and Exploitation*, Vol.19, pp. 497-509.
Kaygusuz, K. 2003a. Energy policy and climate change in Turkey, *Energy Conversion and Management* 44, pp. 1671–1688.
Kaygusuz, K. 2003b. Climate change mitigation in Turkey. *Energy Sources* 26: 563-573.
Kaygusuz, K. and Sarı A. 2003. Renewable energy potential and utilization in Turkey, *Energy Conversion and Management* 44, pp. 459–478.
Kaygusuz. K. 2004a. Energy policies and climate change mitigation in Turkey, *Energy Exploration and Exploitation* 22, pp. 145–160.
Kaygusuz. K. 2004b. Climate change mitigation in Turkey, *Energy Sources* 26, pp. 563–573.
Kaygusuz, K. and Arsel, M. 2005. Energy politics and policy in "Environmentalism in Turkey: between democracy and development". In: F. Adaman and M. Arsel, Editors, Ashgate Publishing Limited, UK, pp. 149–165.
Kaygusuz, K. 2009. Energy and environmental issues relating to greenhouse gas emissions for Sustainable development in Turkey. *Renewable Sustainable Energy Reviews* 13: 253-270.
MEF, Ministry of Environment and Forestry. 2007. First national Communication of Turkey on Climate Change (Eds. Apak, G and Ubay, B): 60-150.
MENR. Ministry of Energy and Natural Resources. 2005. Greenhouse Gas Mitigation in Energy Sector for Turkey, *Working Group Report.*
MENR, Ministry of Energy and Natural Resources. 2007. *Energy report of Turkey*, Ankara, Turkey. Available from http://www.enerji.gov.tr.
Munasinghe, M., Swart, R. 2004. Primer on Climate Change and Sustainable Development, Cambridge University Press, Cambridge UK.
Ogutcu. M. 2002. Turkey's energy policies in the context of Eurasian geopolitics. Event report, Kennedy School of Government, Caspian Studies Program.
Paish, O. 2002. Small hydro power: technology and current status. Renewable and Sustainable *Energy Reviews* 6: 537–556.
PEWCLIMATE. 2002. Climate change mitigation in developing countries: Brazil, China, India, Mexico, South Africa, and Turkey, available from www.pewclimate.org.

Say, NP. 2006. Lignite-fired thermal power plants and SO_2 pollution in Turkey. *Energy Policy* (article in press).
TEDAS, Turkish Electricity Distribution Corporation. 2006. Annual Report, Ankara, Turkey.
TEIAS, Turkish Electricity Transmission Corporation Turkey, 2005. *Annual report*.
UNDP, United Nations Development program. World Energy Assessment Report, 2000. New York: United Nations
UNFCCC, United Nations Framework Convention on Climate Change. 1999. Information Unit for Conventions, UNEP, Geneva.
WEC, World Energy Council, 2001. Survey of Energy Resources, www.worldenergy.org
WEC, World Energy Council. 2004. 19th WEC Energy Congress Conclusions. Sydney, Australia.
WECTNC, World Energy Council Turkish National Committee. 2003. *Energy report of Turkey in 2002*, Ankara, Turkey.
WECTNC, World Energy Council Turkish National Committee. 2004. *Turkey energy report 2003*, Ankara, Turkey.
World Bank, 1992. World Development Report 1992: development and the Environment. Oxford University Press.
World Bank, 2000. Entering the 21st century: World Development Report 1999/2000. Oxford University Press,
Yuksel, I. 2007. Development of hydropower: a case study in developing countries. *Energy Sources* 2: 113-121.
Yuksel, I. 2008a. "Energy Utilization, Renewables and Climate Change Mitigation in Turkey", *Journal of Energy Exploration & Exploitation*, 26 (1), 35-52, 2008.
Yuksel, I. 2008b. Hydropower in Turkey for a clean and sustainable energy future. *Renewable and Sustainable Energy Reviews*, Vol. 12, pp. 1622-1640.
Yuksel, I. 2008c. Yuksel, I., "The Role of Hydropower in Energy Utilization and Environmental Pollution in Turkey", *Journal of Energy Source, Part A*, ID: ESO-07/223, (Article in Press).
Yuksel, I. and Sandalci, M. 2009. "Climate Change, Energy and Environment in Turkey" *Journal of Energy Source, Part A*, ID: 403246 UESO-2009-0105, (Article in Press).
Yuksel, I. 2010. "Energy production and sustainable energy policies in Turkey" *Journal of Renewable Energy*, 35 (2010) 1469–1476.

Role of Nuclear Energy to a Low Carbon Society

Shinzo SAITO[1], Masuro OGAWA and Ryutaro HINO
Japan Atomic Energy Research Institute
(At present : Japan Atomic Energy Agency)
[1]*At present: Radiation Application Development Association*
Japan

1. Introduction

More than 10 billion tons of oil equivalent energy are consumed a year in the world in the present time and over 80 % of it is provided by fossil fuels such as coal, oil and natural gas. Many specialists, institutes, international agencies and organizations have foreseen or estimated an increase of energy consumption in future, remaining fossil fuel resources, and the period of consumption of them.

On the other hand, global warming due to green house gases (GHG) emissions, especially carbon dioxide (CO_2) emitted by burning of fossil fuels has become a serious issue. The IPCC (Inter-governmental Panel on Climate Change) opened their Fourth Assessment Report [1] to the public last year indicating that anthropogenic warming over the last three decades has likely had a discernible influence at the global scale on observed changes in many physical and biological systems. The report also describes that altered frequencies and intensities of extreme weather, together with sea level rise, are expected to have mostly adverse effects on natural and human systems.

Most of the countries in the world confirmed the significance of the Fourth Assessment Report of the IPCC as providing the most comprehensive assessment of the science and encouraged the continuation of the science-based approach that should guide our climate protection efforts. The COP (Conference of the Parties on United Nations Framework Convention on Climate Change) 15 was held in December, 2009, to construct the new protocol on reduction of CO_2 emission following the Kyoto protocol which was valid until 2012.The new protocol is to form agreement of reduction of CO_2 emission by 2020 in each country to avoiding the most serious consequences of climate change and determined to achieve the stabilization of atmospheric concentrations of global greenhouse gases considering and adopting the goal of achieving at least 50 % reduction of global emissions by 2050. Negotiations in the COP continue in 2010.

Various considerations and measures to mitigate climate change are expected in various sectors such as energy supply, transport and its infrastructure, residential and commercial buildings, industry, agriculture, forestry and waste management. Enhancement of energy utilization efficiency is one of the key issues and adoption of renewable energy such as solar and wind energies are progressing in many countries. Among them, nuclear energy is an essential instrument of energy supply to mitigate global warming from the viewpoints of stable energy supply with necessary amounts, harmonization with global environment and economical competitiveness. The present status and perspective of electricity generation by nuclear power are discussed, covering that growing number of countries have recently expressed their interests in nuclear power programs as means to resolve climate change and energy security issues. Furthermore, nuclear energy can also produce high temperature gas to be used as process heat in chemical and petrochemical industries and production of hydrogen which can be used for steel making, fuel cell vehicles and so on. The Japan Atomic Energy Research Institute (JAERI, currently the Japan Atomic Energy Research and Development Agency (JAEA)) developed the HTGR technology capable of producing high temperature gas and succeeded in obtaining helium gas of 950 °C at the reactor outlet in the HTTR through the development of various materials and introduction of new design concepts. On the other hand, the JAEA has took over from the JAERI development of a carbon free hydrogen production process in which the high temperature process heat can be provided by an HTGR. The process is high temperature thermo-chemical water splitting method using iodine and sulfur (IS process). So, nuclear energy can greatly contribute to build a low carbon society by providing electricity as well as process heat in various industries.

2. Present status and perspective of energy consumption and CO_2 emissions

The total amount of energy consumption in the world is 11.4 billion tons of oil equivalents in the present time. The USA's share is 20 %, China's is 15 %, Russia's is 6 %, and India's is 5%, etc. A projection of energy consumption by several regions for longer time span [2] was made by the Institute of International Association on System Analysis, IIASA-WEC as shown in Fig. 1. The total amount of energy consumption in the developing countries will exceed that in the developed countries in 2030, and will continue to increase dramatically. The total amount of energy consumption in 2100 will reach to 6.2 times of that in 2000 in the developing countries. This leads to an obvious question: are there so many energy resources in the earth?

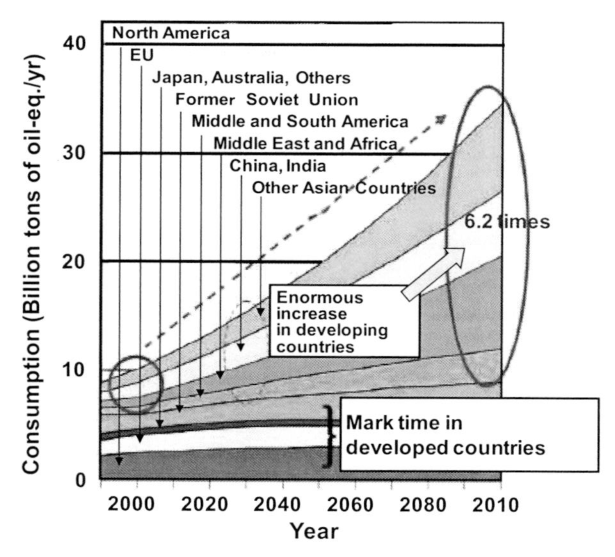

Fig. 1. History and perspective of world energy consumption by region

As concerns share and amount of consumption of each energy resource, the OECD/IEA integrated the past results and projected future consumption of various energy sources from 1970 to 2030 as shown in Fig.2 [3]. The Agency estimated further increase of consumption of fossil fuels and that the total amount of energy consumption in 2030 will become 1.6 times higher than that in the present time. Furthermore, a great attention should be paid to the fact that fossil fuel holds over 80 % of the total energy consumption. Are there inexhaustible fossil fuel resources?

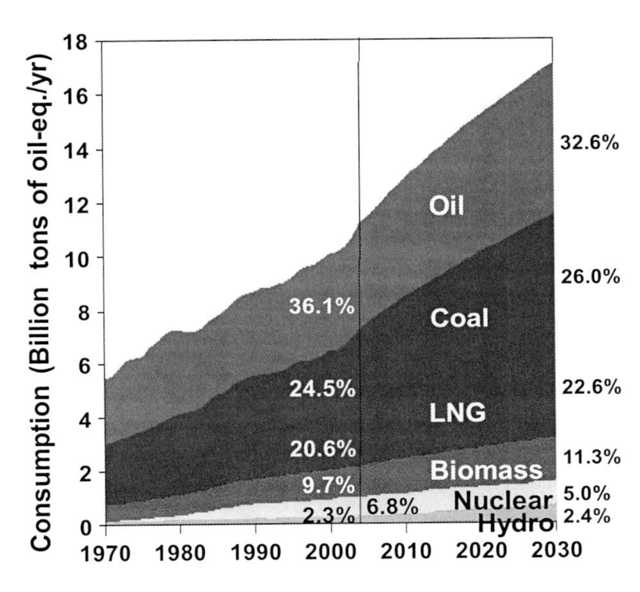

Fig. 2. History and perspective of world energy consumption by energy sources

The British Petroleum evaluated energy resource reserves and reserve-production ratio for fossil fuels [4] and IAEA and OECD/NEA projected them for uranium [5], as shown in Fig. 3. The reserve-production ratios of oil and natural gas are only 40 and 60 years, respectively. The definition of reserve-production ratio, here, is the reserve remaining at the end of year per production in that year. So, as far as new energy resources are not discovered and production is constant, the reserve-production ratio decreases 1 year for each energy source every year. If production in some year increases much more, the reserve-production ratio decreases much rapidly. As concerns uranium resources, the reserve is 5.47 million tons and the reserve-production ratio is more than 100 years. Furthermore, it becomes over 3000 years if a Fast Breeder Reactor (FBR) which produces more new plutonium fuel than spent plutonium becomes commercial. Namely, utilization efficiency of uranium resources reaches about 60 % in the FBR cycle due to breeding plutonium fuel from uranium, recycling plutonium fuel and un-necessity of uranium enrichment with loss of uranium resources although it is about 0.5 % in once-through use of uranium in a light water reactor. The reserve-production ratio sets here conservatively 30 times larger than that of once-through use case considering loss of recycling plutonium and uranium in the processes of re-processing of spent fuels and fuel fabrication.

There is another subject to be discussed. The energy consumption per person in Canada and USA is around 8 tons of oil equivalent energy per year; that is 4.5 times higher than the global average. Most of European countries and Japan consume energy about a half of that of the former two countries per person. On the other hand, China and India consume one third and one eighth, respectively, of the European energy use per capita. It is thus reasonably expected that the developing countries will consume more energy than the present amount to facilitate continuous improvement in the standards of living to levels close to those of the developed countries.

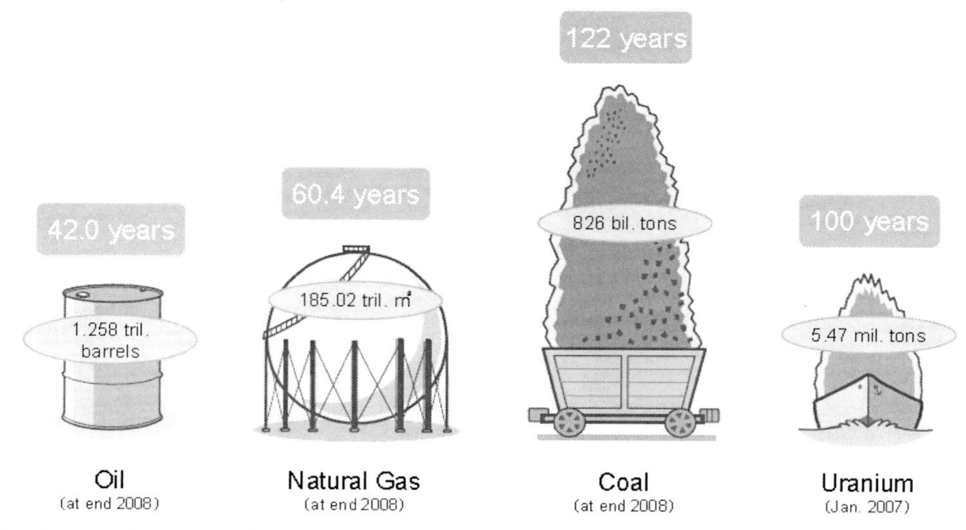

Fig. 3. Proved reserves of energy resources

Global warming due to green house gases, especially carbon dioxide (CO_2) emission has become a serious issue. Carbon dioxide emissions by burning of fossil fuels scarcely occurred before the industrial revolution and atmospheric carbon dioxide concentration was stable at about 280 ppm. CO_2 emissions have increased at first as the amount of coal consumption increased after the revolution, and then again after World War II together with oil consumption with industrial progress and economical expansion in developed countries. Recently, CO_2 emissions due to burning of natural gas have been added. An increase of CO_2 emissions in the last 35~40 years has been substantial and the total amount of CO_2 emissions due to burning of fossil fuels reaches to about 26 billion tons. In accordance to this tendency, CO_2 concentration in the atmosphere has increased to about 380 ppm in the present time.

The IPCC reports that warming of the climate system is unequivocal, as is now evident from observations of increases in global average air and ocean temperatures, widespread melting of snow and ice, and rising global average sea level [1]. Anthropogenic warming over the last three decades has likely had a discernible influence on the global scale on observed changes in many physical and biological systems.

Several international organizations and institutes have projected CO_2 emissions. Figure 4 shows CO_2 emissions per year by countries in 2004 and estimated ones in 2030 by IEA [6]. The total CO_2 emissions in the world per year will increase from 26 billion tons to more than 40 billion tons between 2004 and 2030, 1.6 times higher than the present CO_2 emissions.

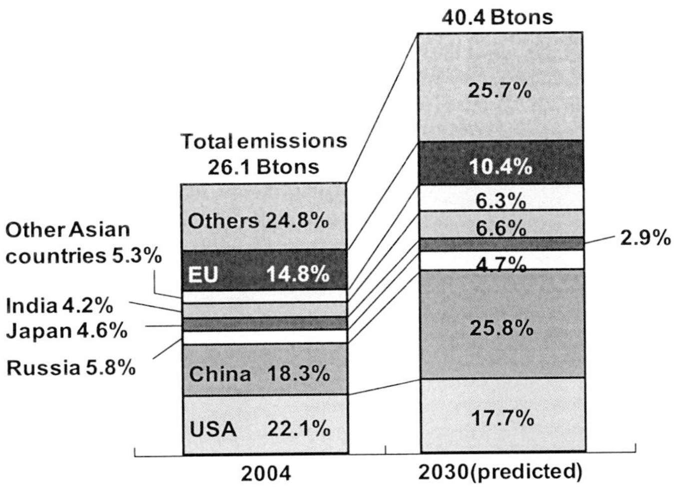

Fig. 4. Present stat1us and outlook of CO_2 emissions/year by countries

Every country and region will emit more amount of CO_2 per year. The IIASA estimated that CO_2 emissions per year in 2100 would reach 3.5 times higher than those in 2000 [2], mostly due to increase of CO_2 emissions in the developing countries as shown in Fig.5.

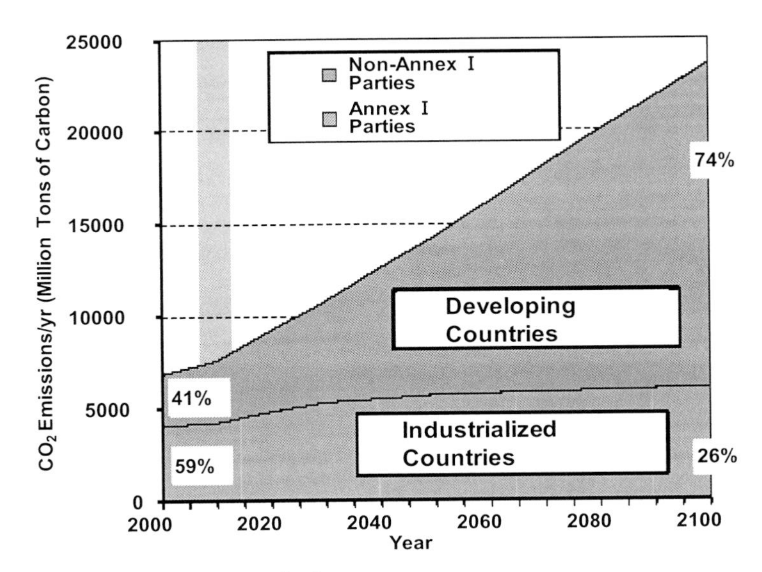

Fig. 5. Long range CO_2 emission outlook

On the other hand, the IPCC suggested to maintain the temperature increase within 2 °C reducing CO_2 emissions in 2050 by 50~85 % of those in 2000 together with establishment of peaking year of CO_2 emissions by 2015 in order to achieve less impact on global physical and biological systems.

3. Countermeasures against global warming and contribution of renewable energy to a low carbon society

It can be recognized that there are several subjects to be resolved in order to construct a low carbon society under the present situation and projection of energy consumption, strong dependence on fossil fuels resulting in increasing emission of CO_2 in future.
Several countermeasures against global warming are considered as follows.

- to increase energy efficiencies in various industries fields, and to save energy consumption, switching off the unnecessary lights and house-hold apparatus, changing the setting temperature of air- conditioners, etc.
- to introduce hybrid cars and electric vehicles instead of gasoline and diesel driven vehicles and to promote modal-shift.
- to introduce renewable energies and nuclear energy instead of fossil fuels.
- to develop and introduce carbon capture and storage system, if it is technically feasible and cost effective.

  And, so on.

The introduction and limits of renewable energy and possibility of introduction of carbon capture and storage system are described in the chapter. The contribution of nuclear energy is analyzed and proposed in the next chapter.
Renewable energy is energy which comes from natural resources such as sunlight, wind, rain, tides, and geothermal heat, which are renewable (naturally replenished). Biomass and biofuels are also generally categorized as renewable energy because plants absorb carbon during growing up although they emit carbon during being used.
Renewable energy accounts for around 13 % of primary energy supply of which 90 % is traditional biomass for cooking and heating in developing countries in 2007 [8]. Biofuels contribute less than 2 % of total transport liquid fuel supply.
Hydropower accounts for 16 % of world electricity, and wind, solar and biomass together account for another 2 % of electricity supply. As concerns hydropower, large scale hydroelectricity systems have been already mostly developed, therefore, only a small hydro system is discussed to be as new renewable energy.
A massive investment of over 100 billion US$ has been made for development of technologies and installation of various renewable energies together with large subsidy to install them by the governments in the world. As the result, wind power is growing at the rate of 30 % annually, with a worldwide installed capacity of 121 GW, solar photovoltaic power reaches 13 GW in 2009 as shown in Table 1. Figure 6 shows installed capacities of solar photovoltaic power (PV) and wind power by countries as of March, 2009. As concerns PV, Germany, Spain and Japan are big three countries, and as for wind power USA, Germany and Spain are top three countries. Amounts of introduction of the above-mentioned power quite depend on various political decisions by the government such as subsidy for installation and purchase of generated electricity by them in every country. A share of the total renewable energy power capacity becomes 6 % of the total electricity power capacity from Table 1, however, it should pay attention that contribution of renewable energy to total electricity generation is only a few percent because capacity factors of wind power, PV, etc. are 10 to 20 %, although these are 80 to 90 % in fossil fueled power and nuclear power, in general.
The utilization of renewable energy should be promoted together with technological innovation to bear a part of construction of a low carbon society from view points of not

only reduction of CO_2 emitted by burning of fossil fuels but also fear of shortage of fossil fuel resources. Table 2 summarizes general evaluation result of various energy resources.

Technology	Electric Power Capacity (GW)
Wind power	121
Small hydropower	85
Biomass power	52
Solar photovoltaic power	13
Geothermal power	10
Solar termal power	0,5
Tidel power	0,3
Total renewable power	280
Total electric power capacity	4,700

Table 1. Renewable electric power capacity

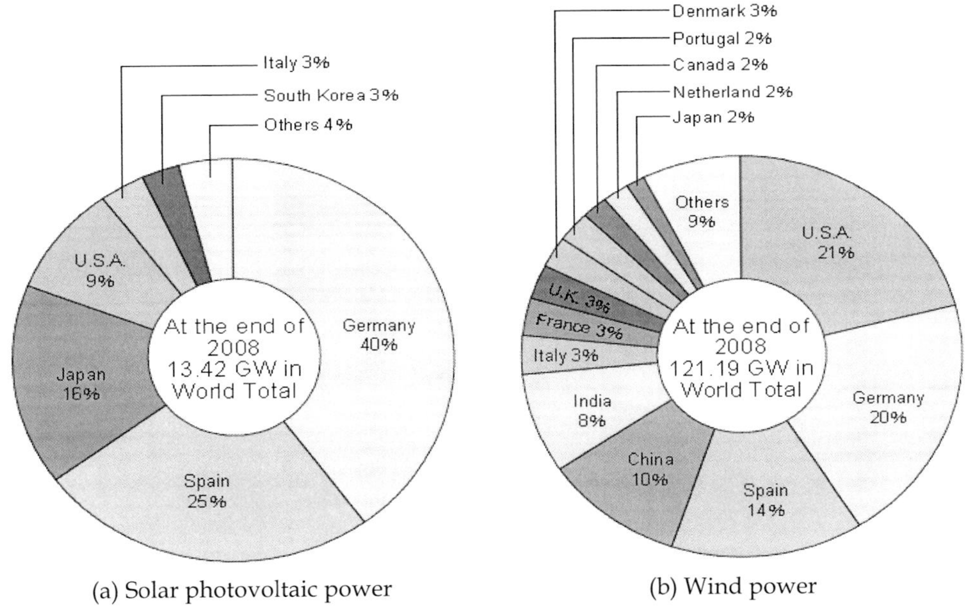

(a) Solar photovoltaic power (b) Wind power

Fig. 6. Photovoltaic power and wind power generation capacities in the world

Many countries have introduced wind power and solar energy, however, amounts of electricity generation by them is small in general and unstable. Furthermore, energy intensity of them is very low, then, huge space is needed to achieve some amounts of electricity generation by them. Therefore, electricity generation cost is very high, especially in PV, then, the governments have offered large amounts of subsidy for installation of them which comes from tax paid by people. Smart grid which connects PV and/or wind power with battery, in some case battery installed in electric vehicles is discussed and developing currently. It might be an idea to improve to use wind power and solar energy effectively and more cost-efficiently. On the other hand, there is some optimistic estimation that the long-term technical potential of wind energy will be five times total current global energy production, or 40 times current electricity demand. This could require large amounts of land to be used for wind turbines, particularly in areas of higher wind resources. Offshore resources experience mean wind speeds of ~90 % greater than that of land, so offshore resources could contribute substantially more energy although it is not applicable to every country. As concerns PV, building-integrated photovoltaics or "onsite" PV systems have the advantage of being matched to end use energy needs in terms of scale. So the energy is supplied close to where it is needed.

	Wind power	Solar photovoltaic	Geothermal energy	Biomass
Resource (or scale)	△	△	△	△
Cost	△	×	△	△
No CO_2 emission	◎	◎	◎	◎
Public acceptance	◎	◎	△	◎
Subjects to be solved or difficulties	Cost and limitation of introduction	Cost and limitation of introduction	Limitation of resource	Limitation of resource
Solution	Dispersal use, smart grid	Innovative technology, dispersal use, smart grid	Innovative technology	Innovative technology
	Biofuel	Oil	Coal	Nuclear
Resource (or scale)	△	△	○	○
Cost	△	△	○	◎
No CO_2 emission	◎	×	×	◎
Public acceptance	○	○	○	△
Subjects to be solved or difficulties	Production from other plants than sugar cane, corn	Limitation of resource	Gasification technology, Carbon capture and storage technology	Public acceptance, radioactive waste disposal
Solution	Innovative technology	Increase utilization efficiency	Innovative technology	Communication with public

Table 2. General evaluation result of various energy resources

According to the BLUE Map scenario by IEA, in which CO_2 emissions are halved by 2050, biomass would become by far the most important renewable energy source. Its use would increase nearly four-fold by 2050, accounting for around 23 % of total world primary energy. Such a level of use would require approximately 15,000 Mt of biomass to be delivered to processing plants annually. Around half of this would come from crop and forest residues, with the remainder from purpose-grown energy crops. The scenario seems to be very hardly possible.

Another recent attention and controversy have focused on biofuels, which have been growing at a rapid rate. Some of the current "first generation" biofuels (derived from grains and oil-seed crops) raise questions of sustainability, as they compete with food production and contribute to environmental degradation, with dubious CO_2 benefits. However, introduction of "second generation" biofuels, e.g. from grasses, trees and biomass wastes, should help overcome most problems and provide sustainable fuels with large GHG reductions. Major deployment of second generation biofuels should be replaced with first generation biofuels.

Apart renewable energies, carbon capture and storage (CCS) is a means of mitigating CO_2 emission based on capturing CO_2 from large point sources such as fossil fuel power plants, and storing it away from the atmosphere by different means. CCS will bring great contribution to reduction of CO_2 emission to the atmosphere, if it becomes technically and economically feasible. However, there are many technical subjects to be solved in the process of capturing CO_2, transportation of CO_2 by pipe line, injection of CO_2 into storage site together with its safety and public acceptance. As concerns CO_2 capture from the point source, broadly, three different types of technologies exist: post-combustion, pre-combustion, and oxyfuel combustion. In the post-combustion capture, the technology is well understood and is currently used in other industrial applications, although not at the same scale as might be required in a commercial scale power station. A few engineering proposals have been made for the more difficult task of capturing CO_2 directly from the air, but work in this area is still in its infancy.

Storage of the CO_2 is envisaged either in deep geological formations, in deep ocean masses, or in the form of mineral carbonates [9]. In the case of deep ocean storage, there is a risk of greatly increasing the problem of ocean acidification, a problem that also stems from the excess of carbon dioxide already in the atmosphere and oceans. Geological formations are currently considered the most promising sequestration sites although there are not so many appropriate sites. Purpose-built plants near a storage location are recommended and applying the technology to preexisting plants or plants far from a storage location will be more expensive. Safety issue of CCS is leakage of CO_2 from transportation piping system and storage location. In fact, a large leakage of naturally sequestered carbon dioxide rose from Lake Nyos in Cameroon and asphyxiated 1,700 people in 1986.

CCS applied to a modern conventional power plant could reduce CO_2 emissions to the atmosphere by approximately 80~90 % compared to a plant without CCS. The IPCC estimates that the economic potential of CCS could be between 10 % and 55 % of the total carbon mitigation effort until year 2100, considering Capturing and compressing CO_2 requires much energy and would increase the fuel needs of a coal-fired plant with CCS by 25 %~40 %.

Micro hydro systems are hydroelectric power installations that typically produce up to 100 kW of power. They are often used in water rich areas as a remote-area power supply. There are many of these installations around the world, which are also renewable energy.

4. Current and future role of nuclear energy

4.1 Electricity generation

Although nuclear energy has a misfortune and tragic history to be used first as nuclear bomb, peaceful use of nuclear energy was initiated and has been promoted based on the speech of "Atoms for Peace" by USA President Eisenhower at United Nations in 1953. Many developed countries started and promoted the construction of nuclear power plants mostly due to oil crises and energy security. However, the pace of construction of nuclear power plants became stagnant in several countries after Three Mile Island (TMI) and Chernobyl accidents. Currently, 432 nuclear power plants are operating world-wide, producing 16 % of the total electricity generation, or 6 % of all primary energy production with total plant capacity of 390 GWe [10] as shown in Fig.7. USA has a quarter of the total producing 20 % of the total electricity generation in the country, nuclear power produces about 80 % of the total electricity generation which reaches to truly 43 % of primary energy production in France and one third of the total, or 14 % of all primary energy production in Japan.

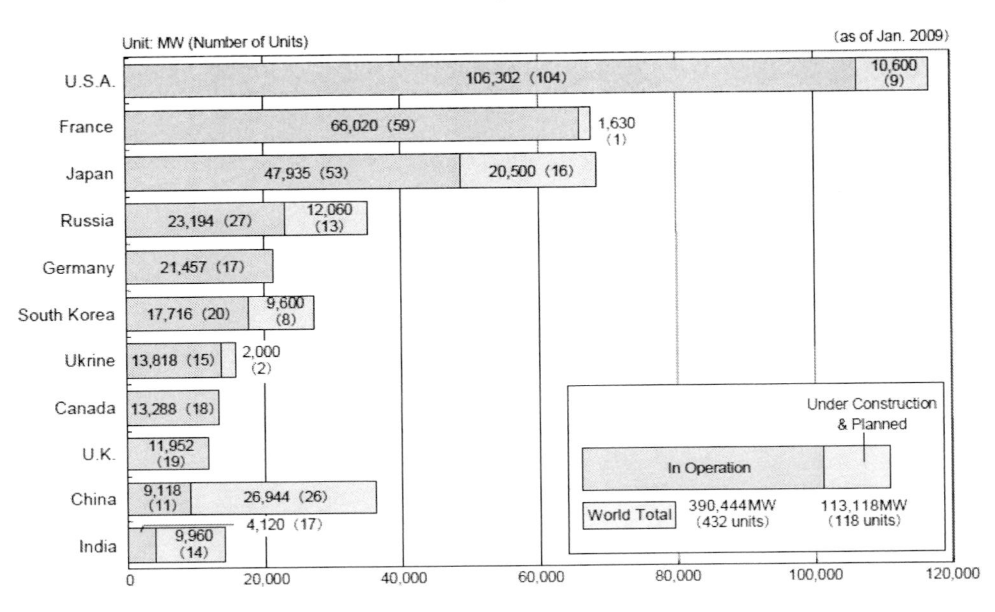

Fig. 7. Generated capacity of nuclear power plants in major countries

As described in the G8 Summit leaders declaration, a growing number of countries currently regard nuclear power as an essential instrument in reducing dependence on fossil fuels, and hence greenhouse gas emissions. Fig.8 shows amount of CO_2 emissions through life cycle of each electricity energy source in unit of g-CO_2 per kWeh [11]. Clearly, fossil fuel fired power plants emit enormous amounts of CO_2 from about 500 g~1 kg/kWeh compared with renewable energies and nuclear power which emit CO_2 only from 10 to 50 g/kWeh. In fact, amount of CO_2 emission by nuclear power is 1/25~1/45 of that by fossil fuel. If the existing nuclear power plants are replaced with oil and coal fired power plants, for example, amount of CO_2 emissions would increase by 230 million tons, which is equivalent to about 20 % of the total CO_2 emissions in Japan. Furthermore, nuclear power is the cheapest electricity source at least in Japan and in a similar situation internationally as shown in Fig.9. A number of countries have recently expressed their interests in nuclear power programs as means to addressing climate change and energy security concerns based on the situation described above, so it is said that we are entering a "Nuclear Renaissance". In fact, USA is going to re-start construction of new nuclear power plants after the TMI accident, France and Japan are steadily constructing new nuclear plants. Russia, China and India have big plans to build 13~26 new nuclear plants by 2020 or 2030, and several plants are being constructed already as added in Fig.7. A plant unit capacity of them is 1000~1600MWe mostly. Many other countries in Asia, Middle East, Africa and South America are considering introduction of nuclear power. According to the latest data as of March, 2010 [12], 55 nuclear power plants are under construction in the world with an installed capacity of 51 GWe, equivalent to 14 % of present capacity, in 15 countries including 21 plants in China, 8 in Russia, 6 in Republic of Korea, and 5 in India. In addition, WNA (World Nuclear Association) reported in April, 2010 that 195 new nuclear power plants will be constructed by 2020 and another 344 nuclear plants will follow by 2025 in the world, including 156 plants in China, 54 plants in India and 46 plants in Russia [13].

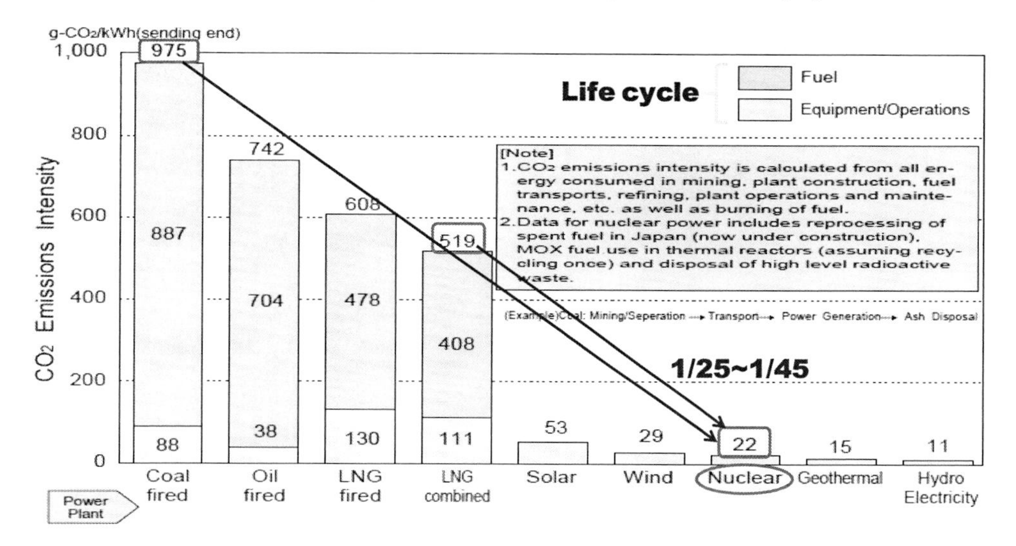

Fig. 8. CO_2 Emissions Intensity by electric source

On the other hand, an increase of world-wide energy consumption in 2030 is projected to be 60 % over the present level. In order to maintain the current level contribution of nuclear power of 16 % to the total electricity generation in the world, another 250 GWe nuclear power is needed by 2030 under the assumption of the same ratio of electric power contribution to the total energy consumption, besides replacing retired nuclear plants with new ones meantime. The current contribution of nuclear power to the reduction of CO_2 emissions is about 9 % in the world. If we wish to raise this figure to 20 % in 2030, new nuclear power plants with about 700 GWe are needed by 2030, that is, construction of 700 nuclear power plants with a capacity of 1000 MWe in the world. These are summarized in Table 3.

4.2 Nuclear heat utilization in various industries

Another type of nuclear energy system has a great possibility to contribute to create a low carbon future society together with current nuclear power system. That is a high temperature gas-cooled reactor, HTGR, which can produce helium gas of about 1000 °Cat the reactor outlet. If so high temperature gas can be obtained, fields of nuclear energy utilization are surely widen in not only electricity generation but also hydrogen production, direct steel making by deoxidization of iron ore, process heat in various chemical and petrochemical industries, and so on, as shown in Fig.10. That means also to contribute as countermeasure against shortage of oil, coal and natural gas. Currently, only two HTGR test reactors, namely, HTTR in Japan and HTR-10 in China, are operating in the world. The HTR-10 is a very small reactor and helium gas temperature at the reactor outlet is 700°C. Furthermore, the technology of high temperature thermo-chemical decomposition of water utilizing iodine and sulfur has most progressed in the JAEA in the world. Therefore, the most advanced technologies in these fields in the JAEA are described here.

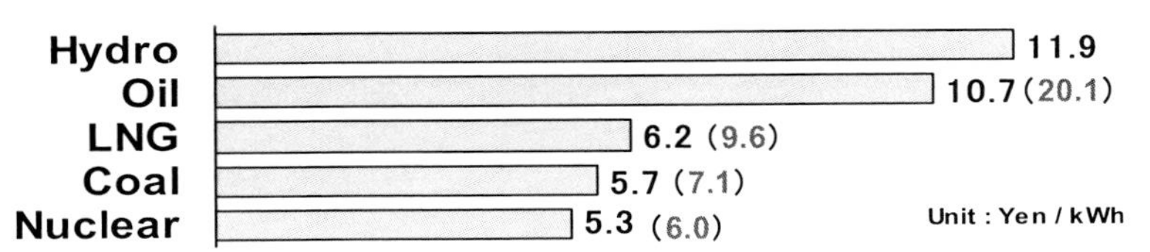

Japanese Case (black letter : based on fuel price on 2002, by METI ; red letter : based on fuel price on Feb., 2008, by FEPC)

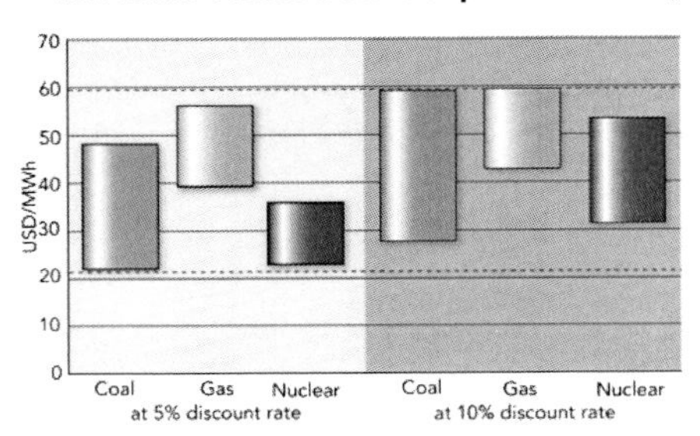

Range of Levelised Costs (OECD/NEA Nuclear Energy Outlook 2008)

Fig. 9. Comparison of electricity generation cost

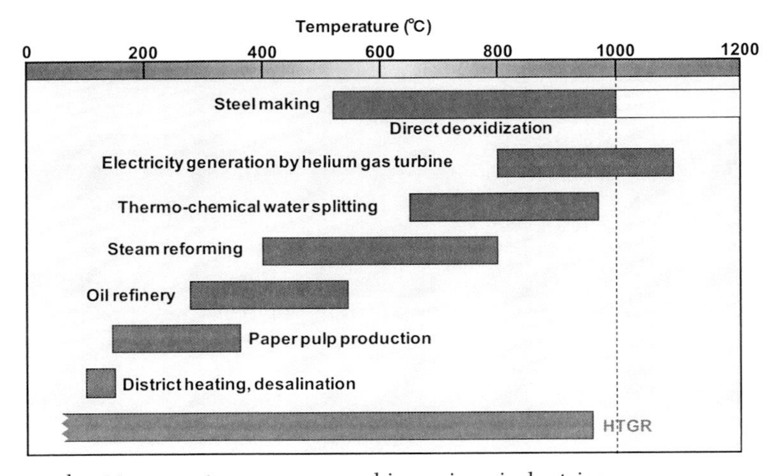

Fig. 10. Process heat temperature ranges used in various industries

The JAERI developed the HTTR [14], a 30 MWt HTGR test reactor, and succeeded in getting helium gas of 950°C at reactor outlet of the HTTR in 2004 for the first time in the world. Several key technologies are described below. One of big differences between an LWR and an HTGR is that no metal is used in the reactor core of the HTGR. The fuel element of HTTR, for example, is quite different from that of LWR as shown in Fig.11. In the

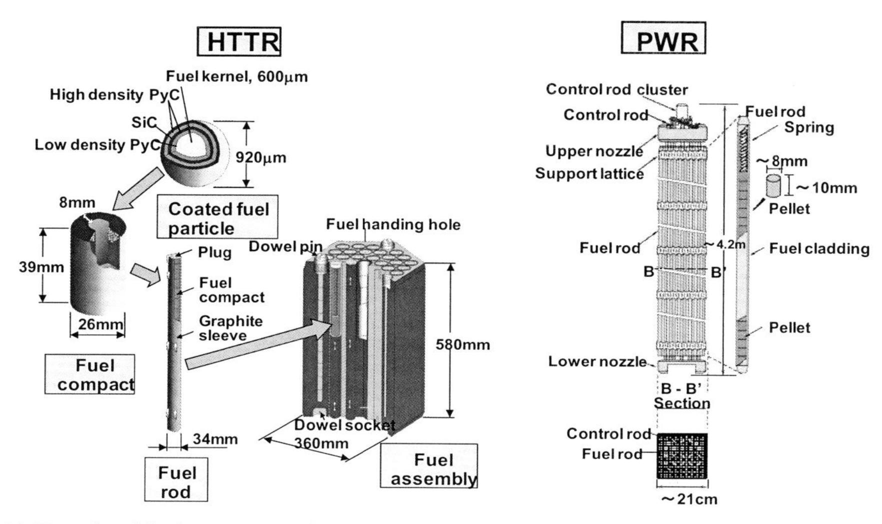

Fig. 11. Details of fuel structure of HTTR and LWR

HTTR, coated fuel particles consisted of low enriched UO_2 kernel with TRISO coating are combined with graphite powder to form a fuel compact which is equivalent to UO_2 pellet in LWR. A fuel rod is composed of graphite sleeve in which fuel compacts are contained. A fuel assembly is a pin-in-block type hexagonal fuel element, that is, helium gas flows through the gap between a vertical hole and a fuel rod to remove the heat produced by fission and gamma heating. Excellent graphite for core and its surrounding components which has less dimensional change due to neutron irradiation, large tensile strength and high corrosion resistance is needed. The JAERI succeeded in development of IG-110 which satisfies the above-mentioned requirements as shown in Fig. 12. As concerns the coated fuel particle, great efforts had been made to improve fabrication technologies having made neutron irradiation tests resulting in production of very high quality one. As for heat resistant alloy for piping systems, Ni-base Hastelloy XR with very high corrosion resistance had been finally developed.

Also, the JAEA has been developing operation technologies of HTGR by using the HTTR so as to supply high temperature heat stably to heat utilization systems, and succeeded in continuous operation for 50-days at high-temperature of about 950 °C in 2010, which was the first demonstration making stable nuclear heat supply possible. Due to the successful long-term operation, nuclear heat utilization with the HTGR became realistic. One of the promising nuclear heat utilization is a large amount of hydrogen production aiming for reduction of CO_2 emission, because hydrogen is said to be a most promising energy carrier for low carbon society. However, if it is produced by utilizing fossil fuels as it was, such as in steam reforming process with CO_2 emissions, hydrogen is not really clean energy.

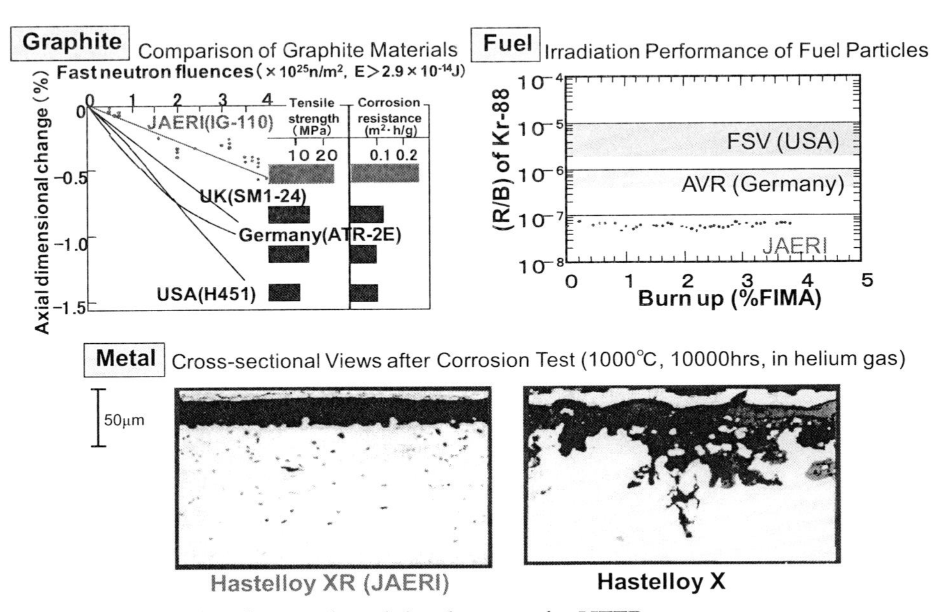

Fig. 12. Several results of research and development for HTTR

Therefore, the JAEA has devoted substantial resources to develop a high temperature thermo-chemical decomposition of water utilizing iodine (I) and sulfur (S), the IS process as shown in Fig. 13 and successfully achieved continuous hydrogen production [15] and [16]. In this process, high temperature process heat is used in sulfuric acid and iodine hydride decomposition reactions. Iodine and sulfur are used cyclically, water is alone the feedstock to produce hydrogen and oxygen. The IS process coupled with HTGR (HTGR-IS), is a really clean hydrogen production system and economically competitive to those of steam reforming of methane and coal and superior to that of water electrolysis [17]. In fact, Ewan and Allen evaluated hydrogen cost for various production routes considered [18].

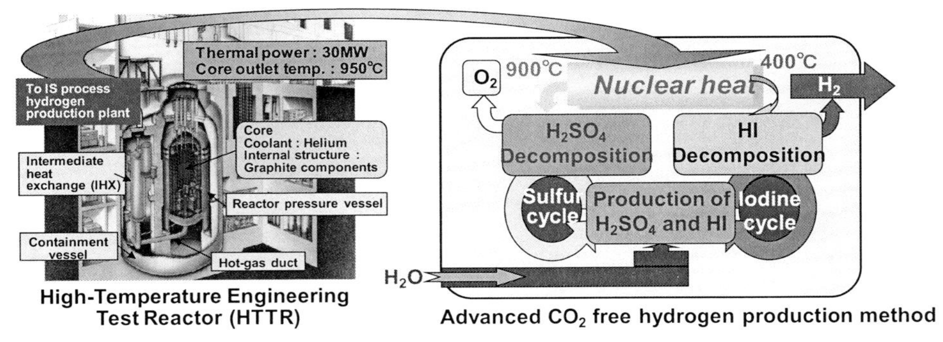

Fig. 13. Nuclear heat application from HTGR to IS-hydrogen production

According to their report, hydrogen production costs per ton are 982 US$ for steam reforming of methane (SMR), 1575 US$ for SMR + carbon capture, 1270 US$ for nuclear/thermocycle, 1621 US$ for coal, 3114 US$ for coal + carbon capture, 4725 US$ for hydroelectric, 14,950 US$ for solar PV, etc. On the other hand, several methods are recently proposed to produce hydrogen utilizing an HTGR and other types of reactors [19], [20], [21] and [22], however, IS method is considered to be the most progressed, promising and good cost performance one without emission of CO_2 among them.

To apply nuclear energy by HTGR to extensive non-electricity fields, the JAEA proposed the original HTGR system, GTHTR300C as shown in Fig.14 [23]. The GTHTR300C is the first commercial-scale HTGR cogeneration plant with 600MWt combining electricity generation by a direct cycle gas turbine and hydrogen production by the thermochemical IS process.

The direct cycle gas turbine of a recuperated Brayton cycle generates electricity and circulates reactor coolant, performing both tasks most efficiently relative to all other forms of process arrangement. Hydrogen cogeneration is enabled by adding an intermediate heat exchanger (IHX) in serial between the reactor and the gas turbine. A secondary loop delivers hot helium gas from the IHX to the IS process hydrogen plant over a sufficient distance that together with the isolation valves located in the secondary loop circuits provides safe and environmental separation between the nuclear plant and the conventional-grade hydrogen plant [24].

Additionally, the seawater desalination plant can be provided readily as a cooling system of removing the sensible waste heat of the Brayton cycle gas turbine power conversion and without an efficiency penalty to either the power generation or high temperature process heat utilization. Providing a seawater desalination plant making freshwater as shown in Fig.14, the HTGR system has an exceedingly high thermal efficiency up to 80 %, which is called an "HTGR cascade energy plant" utilizing heat in a cascade manner from high

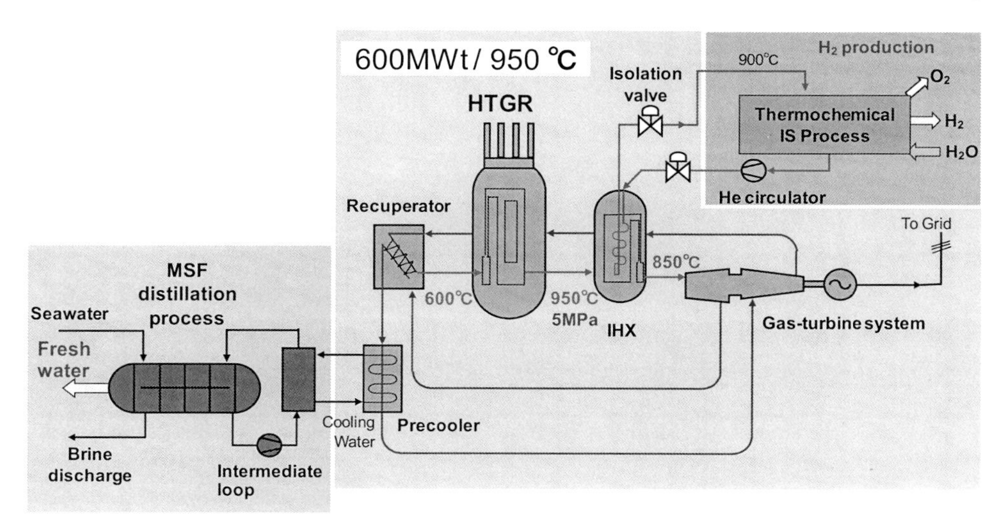

Fig. 14. HTGR cascade energy plant for 80 % efficient production of hydrogen, electricity and freshwater

temperature to low temperature, for example, although thermal efficiency of a current light water reactor (LWR) is 34 %. Due to this high thermal efficiency, the HTGR system can operate by using a small cooling system, whose cooling water consumption is reduced to less than one-third of what the existing LWR needs. For the same reason that the desalination can be completely driven by the high temperature (160ºC) exhaust heat of the gas turbine. On the other hand, the HTGR power and heat cogeneration plant can be operated by using economical air cooler, excluding any need for a cooling water source near the plant site due to exhaust heat is rather small compared to conventional power plants and light water reactors. These environmentally friendly characteristics make the HTGR uniquely suited to barren inland provinces, and other regions, where cooling water resource is scarce.

The HTGR of 600MWt can produce a maximum of 300MWe electricity, 650 tonnes/hour of quality steam at 500ºC, and 85,000 m^3/hour of hydrogen, or it can simultaneously co-generate fractions of all these products by the HTGR cascade plant said above and with the addition of a steam boiler in parallel with or in place of the hydrogen production plant.

Many industrial and market applications are possible for the energy and feedstock obtained from the HTGR. The steam and hydrogen products can be used to refine and hydrogenate profitable clean petroleum products from the crude oils. The steam can be used to reform coal to produce synthetic gas and transportation liquid fuel. The hydrogen produced from a 600MWt HTGR is sufficient to provide fuel to more than half a million of fuel cell vehicles and eliminate 1.45 million tonnes of CO_2 emission by replacing the same number of gasoline cars. New and environmentally friendly industries can be created. As an example, in the current steel making process, huge amount of coke produced from coal is used for the reduction of iron ore with a significant CO_2 emission ($Fe_2O_3 + CO \rightarrow 2Fe + 3CO_2$). In order to reduce the CO_2 emission, the substitution of coke by hydrogen in the steel making is being studied in the Japanese 'Course 50' plan. The direct steel making using hydrogen ($Fe_2O_3 + 3H_2 \rightarrow 2Fe+3H_2O$) by a 600MWt-HTGR for hydrogen supply can produce over half a million tonnes of steels while reducing CO_2 emission by 1.24 million tonnes per year, compared with the current steel making process using coke.

A preliminary evaluation on the reduction of CO_2 emissions is made for the case in Japan [25]. A reduction of CO_2 of 170 million tons (13 %) could be realized through the replacement of 50 million automobiles (2/3 of all cars in Japan) with fuel cell vehicles, 100 million tons (8 %) by the adoption of direct steel making utilizing hydrogen and 30 million tons (2.3 %) in the chemical and petrochemical industrial complexes by the adoption of process heat and electricity produced by the HTGR system, respectively. Namely, a total amount of CO_2 reduction reaches to 23 % of the total emission of 1.3 billion tons in Japan.

As for spent fuel treatment and disposal, coated particle fuels are very convenient to direct disposal because fuel kernel is coated by ceramics triply. Re-processing of spent fuels is also possible by the current Purex method. Technologies of the pretreatment consisting of, in the case of prismatic fuel elements, separation of fuel particles from fuel compact and the following extraction of fuel kernel from a coated fuel particle by crashing have already been performed for HTTR fuels in a laboratory scale [15]. Concerning the chemical waste of the HTGR+IS, it will not bring a special issue to be considered since the IS process constitutes a closed cycle in terms of the sulfur- and iodine-compounds, in principle.

Commercialization of HTGR and HTGR-IS system could be attained through demonstration of nuclear hydrogen production by the IS process connected with HTTR (HTTR-IS system) shown in Fig.15, and operation of a demonstration HTGR with about 30 MWt. Since the utilization system of high temperature heat obtained by an HTGR can be flexibly designed based on user's needs, HTGR technology can widely applied to the non-electricity fields, so that, it would be expected to dramatically reduce global CO_2 emissions.

Primary energy of about 60 % is consumed in non-electricity fields in the world. Hence, the worldwide deployment of the HTGR system, i.e., clean and high efficiency nuclear energy, in the near future is expected to reduce huge amount of the CO_2 emission, which can contribute to build a low carbon society.

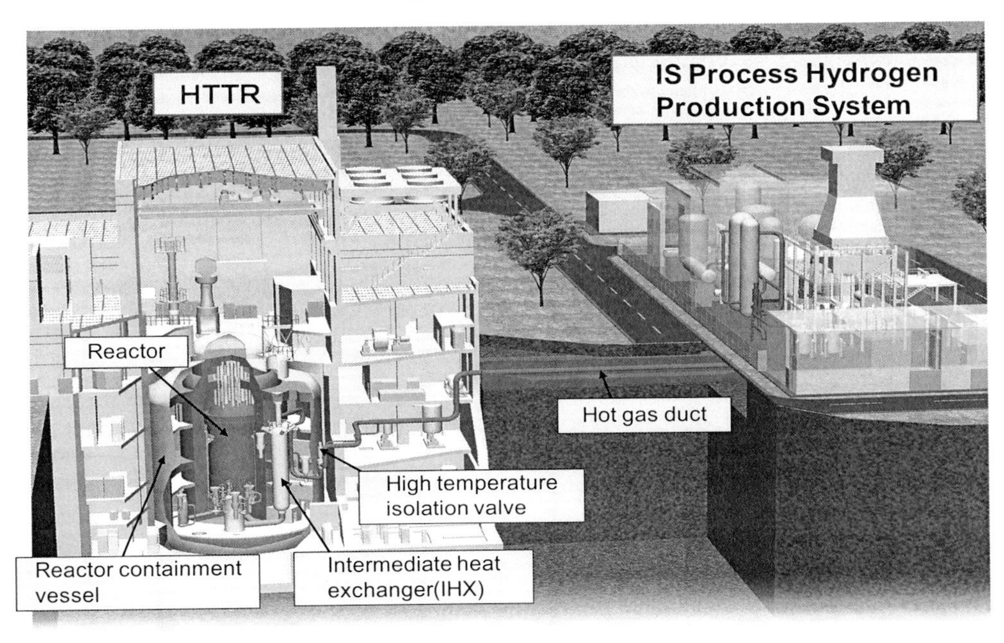

Fig. 15. Demonstration of nuclear hydrogen production by the IS process connected with HTTR

5. Conclusions

1. More than 10 billion tons of oil equivalent energy are consumed a year in the world in the present time, in which over 80 % is provided by fossil fuels. Energy consumption is projected to increase by 60 % in 2030 and by 240 % in 2100, mostly in the developing countries despite a protected shortage of fossil fuels, especially oil and natural gas, within a few decades. On the other hand, consumption of large amounts of fossil fuels may have influenced global climate change. We will face the most serious consequences of climate change unless we stabilize the atmospheric concentrations of global greenhouse gases (GHG) considering and adopting the goal of achieving at least 50 % reduction of GHG emissions to the present figure by 2050.
2. Nuclear energy must play an essential role in reducing the dependence on fossil fuels and hence CO_2 emissions, together with recognition of importance of renewable energy. Therefore, a growing number of countries have recently expressed their interests in nuclear power programs, so it is said that time is "Nuclear Renaissance". Nuclear energy can contribute as means to energy security and reduction of CO_2 emissions not only through electricity generation but also by heat application in various industries such as steel making,

chemical and petrochemical industries, together with hydrogen production for transportation, for example. Commercialization of High Temperature Gas-cooled Reactor (HTGR) that can produce very high temperature heat of about 1000 °C based on the existing technologies will be vital to the realization of these goals, because HTGR is characterized by its flexibility of system design enable to meet heat application demands in various industries of non-electricity fields. We should expand utilization of nuclear energy to non-electricity fields which holds about 60 % of total energy consumption.

6. References

[1] IPCC Fourth Assessment Report, November, 2007.
[2] World Population Prospects, 2006 Revision (UN).
[3] Energy Balances of OECD Countries and Energy Balances of Non-OECD Countries 2005–2006.
[4] BP Statistical Preview of World Energy, June, 2009.
[5] OECD/NEA and IAEA, Uranium, 2007.
[6] OECD/IEA, World Energy Outlook, 2006.
[7] REN21(2009) Global Status Report 2009 Update
[8] IEA, Agency for Natural Resources and Energy 2009
[9] IPCC "Special Report on Carbon Capture and Storage, 2010.
[10] Japan Atomic Industrial Forum, Inc., World Nuclear Power Plants, 2006.
[11] Central Research Institute of Electric Power Industry Report.
[12] http://www.eurnuclear.org/info/npp-ww.htm.
[13] http://www.world-nuclear.com/info/default.aspx?id=27636&terms=World+Nuclear.
[14] S. Saito et al., JAERI 1332, September, 1994.
[15] S. Saito, Report IAEA-TECDOC-761, 1994.
[16] K. Onuki *et al., Energy Environ. Sci.* 2 (2009).
[17] T. Inoue et al., Genshiryoku Eye 53 (4) (2007) (in Japanese).
[18] B.C.R. Ewan and R.W.K. Allen, *Int. J. Hydrogen Energy* 30 (2005).
[19] H.J. Hamel et al., Proc. of the ICONE14, Paper No. 89035 (2006).
[20] C.O. Bolthrunis et al., Proc. of the HTR2006, Paper No. I00000118 (2006).
[21] M.G. McKellar et al., Proc. of the ICONE14, Paper No. 89694 (2006).
[22] W.S. Summers et al., Proc. of the ICAPP'06, Paper No. 6107 (2006).
[23] X. Yan. et al., Proc. of the OECD/NEA 3rd Information Exchange Meeting on the Nuclear Production of Hydrogen, OECD/NEA, 121 (2005).
[24] T. Nishihara et al., AESJ Transaction 3 (4) (2004).
[25] S. Saito, *J. Atom. Energy Soc. Jpn.* 51 (2) (2009) (in Japanese).

10

Global Warming

John O'M. Bockris
Texas A&M University, Retired
Gainesville, Florida,
USA

1. Introduction

The first person to write a paper on the possibility of Global Warming by a mechanism he outlined was Svante Arrhenius (1859-1927) {National Research Council, 2004} [1], a renowned Swedish physical chemist who was known particularly by his early ideas on electrolytes and their conductivity.
His idea about Global Warming depended upon the reflected light from the sun that he deduced would be likely to be absorbed by CO_2.
The date that this paper was first written indicates that it hardly caused a flutter on future ideas about the methods of obtaining energy.[1]

1.1 Global warming due to CO_2

The stress upon our dealing with Global Warming, predicted by Arrhenius has been thrust upon the CO_2 in the atmosphere that clearly depends on the amount of fossil fuels burned per unit time and therefore reflects the degree by which we use carbon-containing fuels to run our civilization.
Now, one has to understand first of all, the radiation from the sun comes into the earth's atmosphere at wavelengths which correspond to the temperature of the surface of the sun, the emitter, 6 million degrees and the wavelength of the irradiated light from a body of that temperature would be far from that which would get absorbed by the earth's atmosphere. After it has struck the earth, the earth itself absorbs about half of it whilst about half of it is reradiated into space, (Figure 1 {Robert A. Rohde, 1997}) from published data and is part of the Global Warming Art project) and is that part of the solar radiation that is partly absorbed by the CO_2.
However, this second half of the reradiated light comes at wavelengths that correspond to the temperature of the radiating body, i.e. our earth, so that the reflected light is in a wavelength corresponding to light coming from a body with at temperature of around 300° K.

[1] **Friedrich Wilhelm Ostwald** (September 1853 - 4 April 1932) [2], a renowned German chemist of the early part of the 20th century, wrote a paper which can be looked at, as parallel to that of Arrhenius. Ostwald was a savvy physical chemist and he saw something else which was parallel to the observations Arrhenius had made somewhat earlier. Ostwald spoke before the German society of scientists pointing out that if we went on burning the fossil fuels we would gradually evolve so much heat that the atmosphere itself would warm.

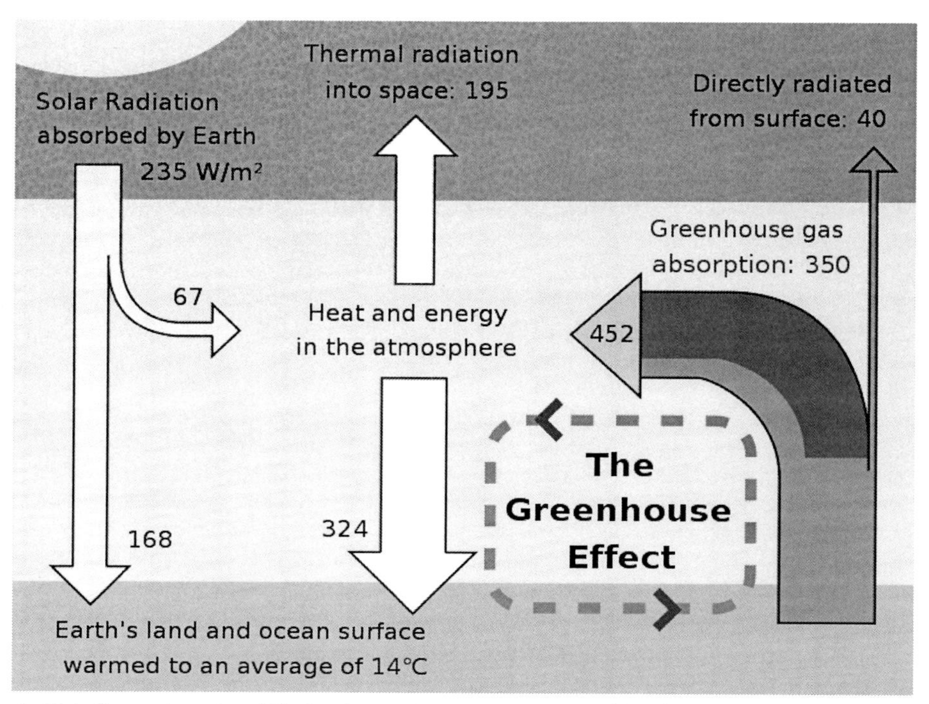

Fig. 1. This figure is a simplified, schematic representation of the flows of energy between space, the atmosphere, and the Earth's surface, and shows how these flows combine to trap heat near the surface and create the greenhouse effect. Energy exchanges are expressed in watts per square meter (W/m2) and derived from Kiehl & Trenberth (1997).The sun is ultimately responsible for virtually all energy that reaches the Earth's surface. Direct overhead sunlight at the top of the atmosphere provides 1366 W/m2; however, geometric effects and reflective surfaces limit the light which is absorbed at the typical location to an annual average of ~235 W/m2. If this were the total heat received at the surface, then, neglecting changes in albedo, the Earth's surface would be expected to have an average temperature of -18 °C (Lashof 1989). Instead, the Earth's atmosphere recycles heat coming from the surface and delivers an additional 324 W/m2, which results in an average surface temperature of roughly +14 °C.Of the surface heat captured by the atmosphere, more than 75% can be attributed to the action of greenhouse gases that absorb thermal radiation emitted by the Earth's surface. The atmosphere in turn transfers the energy it receives both into space (38%) and back to the Earth's surface (62%), where the amount transferred in each direction depends on the thermal and density structure of the atmosphere.This process by which energy is recycled in the atmosphere to warm the Earth's surface is known as the greenhouse effect and is an essential piece of Earth's climate. Under stable conditions, the total amount of energy entering the system from solar radiation will exactly balance the amount being radiated into space, thus allowing the Earth to maintain a constant average temperature over time. However, recent measurements indicate that the Earth is presently absorbing 0.85 ± 0.15 W/m2 more than it emits into space (Hansen et al. 2005). An overwhelming majority of climate scientists believe that this asymmetry in the flow of energy has been significantly increased by human emissions of greenhouse gases.

Now, the shape of the solar spectrum (see Figure 1) i.e. the plot of intensity against wavelength depends sharply upon the temperature of the emitter. The solar light incoming, as we have said, does not overlap the absorption bands of the CO_2 in the atmosphere. Conversely however, the radiation coming from the 300-degree emitter, our earth does indeed contain bands that correspond to those in which CO_2 absorbs. (Figure 2 {Robert A. Rohde, 2008}); Figure 3 {Tapan Bose & Pierre Malbrunot, 2006}).

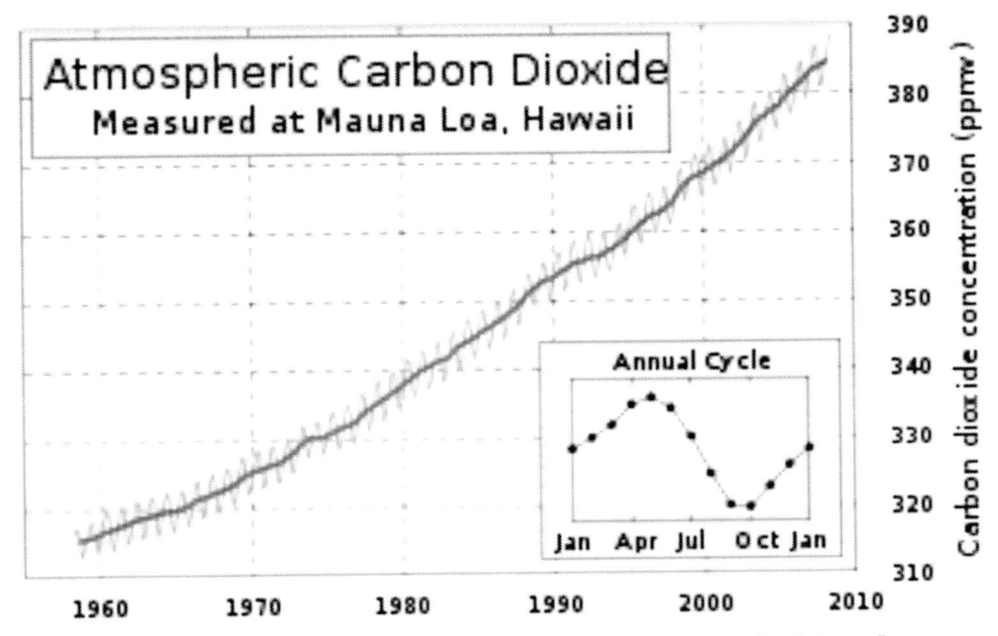

Fig. 2. The Keeling Curve of atmospheric CO_2 concentrations measured at Mauna Loa Observatory.This figure shows the history of atmospheric carbon dioxide concentrations as directly measured at Mauna Loa, Hawaii. This curve is known as the Keeling curve, and is an essential piece of evidence of the man-made increases in greenhouse gases that are believed to be the cause of global warming. The longest such record exists at Mauna Loa, but these measurements have been independently confirmed at many other sites around the world. The annual fluctuation in carbon dioxide is caused by seasonal variations in carbon dioxide uptake by land plants. Since many more forests are concentrated in the Northern Hemisphere, more carbon dioxide is removed from the atmosphere during Northern Hemisphere summer than Southern Hemisphere summer. This annual cycle is shown in the inset figure by taking the average concentration for each month across all measured years. Own work, from Image:Mauna Loa Carbon Dioxide.png, uploaded in Commons by Nils Simon under licence GFDL & CC-NC-SA ; itself created by Robert A. Rohde (2008) from NOAA published data and is incorporated into the Global Warming Art project. *Permission is granted to copy, distribute and/or modify this document under the terms of the* ***GNU Free Documentation License****, Version 1.2 or any later version published by the Free software Foundation; with no Invariant Sections, no Front-Cover Texts, and no Back-Cover Texts. A copy of the license is included in the section entitled "GNU Free Documentation license"*

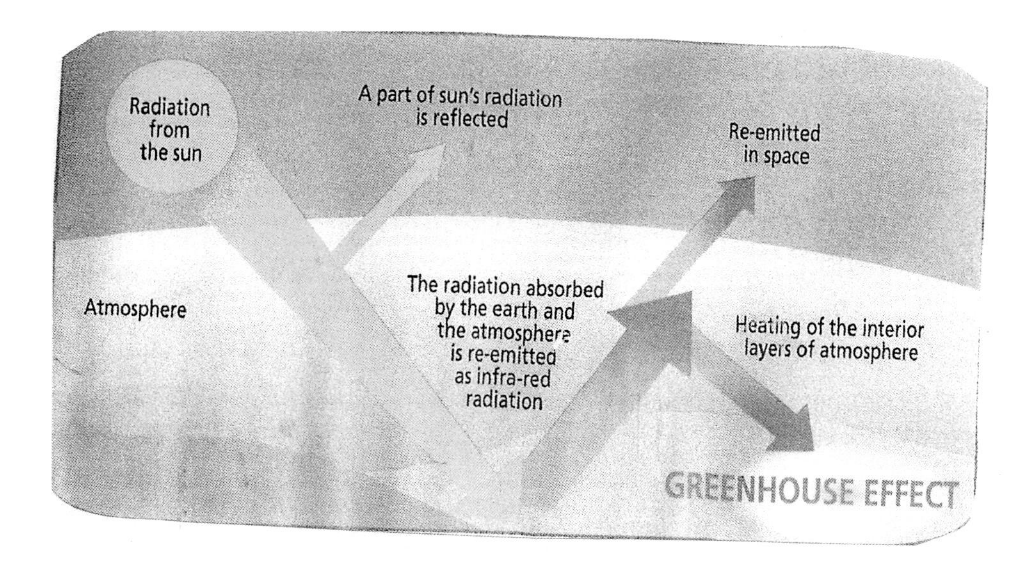

Fig. 3. From Tapan Bose and Pierre Malbrunot, et al, Hydrogen: Facing the Energy Challenge of the 21st Century, John Libby Eurotext, UK, December 2006, page 17.

It is possible to look at Global Warming in a mathematical way and that is exactly what the Turkish-American scientist, Veziroglu {Veziroglu, Gurkin, and Padki, 1989} with colleagues did in a paper to which we shall refer later on when considering contributions which could be made for the earth's temperature by other gases, e.g. methane [3].

Figure 2 shows the temperature rise in the atmosphere and it can be seen that the increase of the CO_2 with time has been of an exponential character.

The anxiety that has been produced in some citizens, who conclude that the earth will become too hot to sustain human life, can now be looked at with the facts. The first reaction is perhaps a sigh of relief. It's not going to happen at once but there are societies that would be sensitive in respect to the maintenance of life, and even due to a further rise of, say, 5 ^{o}C. (See section on methane.)

Such a country is Saudi Arabia, and also the surrounding countries in the Middle East. The government of Saudi Arabia has made a law there that should the surrounding temperature increase got to more than 50 ^{o}C (122 ^{o} F), then as far as is possible: no traffic, no machines operating, which produce significant heat. Heat bursts at 40 ^{o}C were experienced in France in 2007 and more than 1000 did not survive, but these people were above 75 years in age.

Looking then at Figure 4 {Jones, P.D. and Moberg, A., 2003}, it is seen that we have, at 2010, that the increase has already exceeded 1.4 ^{o} F.[2]

---

[2] The actual mechanism of the heat rise of the atmosphere comes through an intermediate stage when the excited CO2 molecules, absorbing the reflected light, collide with very many surrounding nitrogen and oxygen molecules of the air and transfer some of the excited energy in the vibrational bands to the translational energy of the air molecules. This means that they in turn travel faster, i.e. their molecular energy is increased and that in turn is the essence of Global Warming.

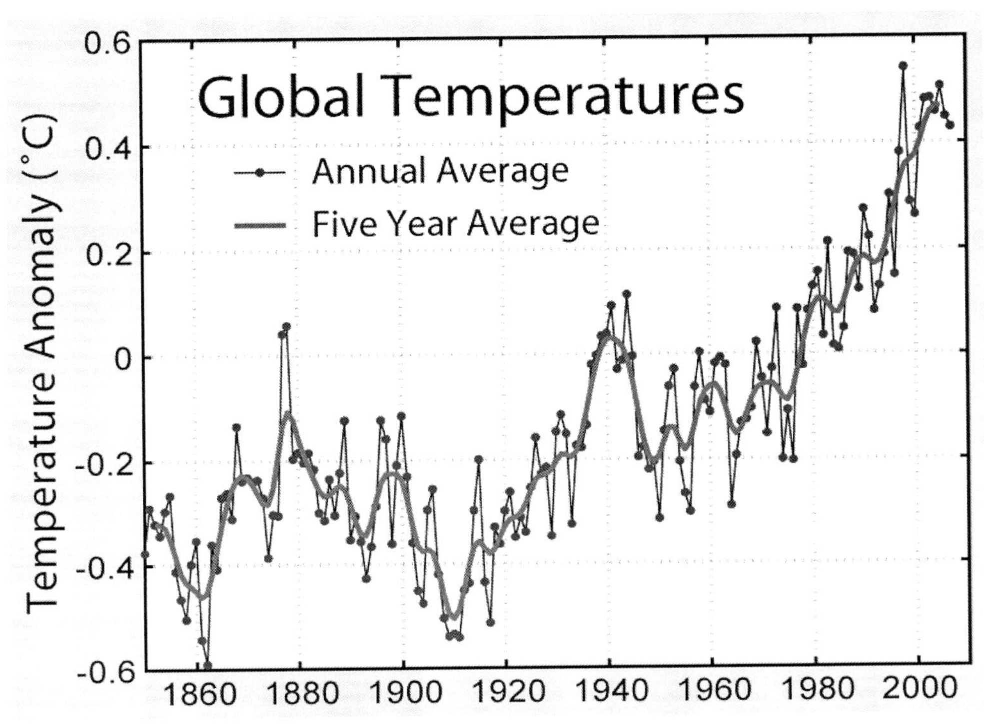

Fig. 4. This figure shows the instrumental record of global average temperatures as compiled by the Climatic Research Unit of the University of East Anglia and the Hadley Centre of the0 UK Meteorological Office. Data set TaveGL2v was used. The most recent documentation for this data set is Jones, P.D. and Moberg, A. (2003) "Hemispheric and large-scale surface air temperature variations: An extensive revision and an update to 2001". Journal of Climate, 16, 206-223.

Many interested in this area of Global Warming would like to know how many years do we have before an unattended problem becomes too much for us [3]? Now, the answer to such a question depends upon how citizens react to very high atmospheric temperatures. 50°C, the Saudi limit, is 123 ° F and that is not an unknown temperature in the United States, in such places as Death Valley in California. However, the prospect of living under such temperatures seems to be out of the question.

Now, to answer the question, when will it get too hot, is difficult for two reasons. First of all (and this is easily understood) the answer can only be given for a given region of earth, or at least a section of a large country such as the USA. Indeed, if one moves a thousand miles north into arctic Canada, one can see some years of happiness there, occurring during the later stages of Global Warming because Canada, too, would be a gigantic country were it not for the fact that most of it is at present frozen.[3]

[3] It is possible to treat the degree of curvature in Figure 2 and we would do better with an equation for a relation which has curvature in it were we to have a few more points.

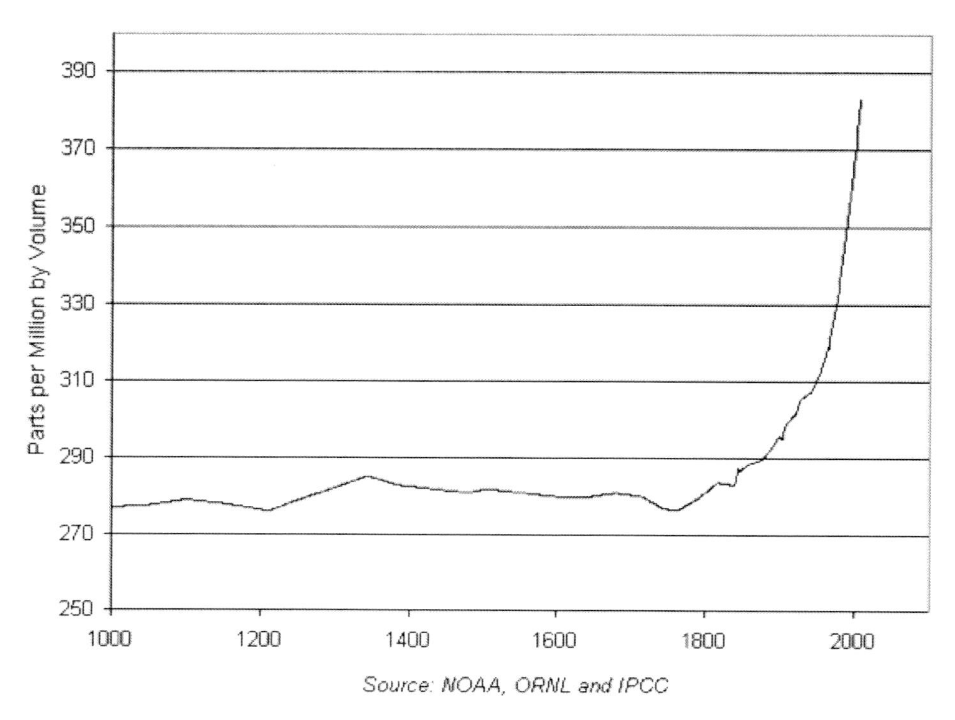

Fig. 5. CO_2 over 1000 years. The Hydrogen Economy. Opportunities, Costs, Barriers and R&D Needs. National Research Council and National Academy of Engineering, National Academies Press, Washington DC, 2004 [4].

1.2 Global warming due to the presence of methane in the atmosphere?

In most articles on Global Warming, the entire problem is put on CO_2, but this may be too optimistic because there is another gas that is gradually increasing in our atmosphere and it is the simple molecule methane, CH_4.

Now, at present, 2010, there is a contribution of methane to the temperature of the atmosphere, which at first seems quite low, 8%.

However, in considering this figure, one has to understand something after which methane can be looked at differently {H. Blake, 2010} [5]. Thus, the individual methane molecule absorbs 23 times more of the reflected energy from the sun than the CO_2 molecule when both, in our atmosphere, get reflected light upon our surface.

In other words, methane, CH_4, is a more dangerous molecule than CO_2 and the only reason why there has been so much discussion of CO_2 and almost no public discussion about methane is that hitherto the concentration of methane in the atmosphere has been small.

Now, there is a reason why we might have to be more concerned with methane for not only its absorptive power, 23 times greater than that of CO_2, but also there is a reason whereby methane could significantly increase its concentration in our atmosphere.

Estimates have been made of the total amount of methane that may be in fact hidden from us at the moment because it is largely in the tundra in the northern climes of the world {National Oceanic and Atmospheric Administration, 2007; and H. Blake, 2010} [4,5].

This tundra is dark-colored vegetation that is met in the far north and it is inside this that the methane at present is largely hidden. This area of the world is still frozen and the methane is in the frozen tundra {University of Toronto, Chemistry Department, 2008} [6].

Predictions have been made (but I must caution they are not reliable) about the total amount of methane that may be hidden in the tundra {BBC News, 2006; N. Shakhova & I. Semiletov, 2007; University of Cambridge Press, 2001; and Walter et al., 2006} [7, 8, 9, 10, 11]. The figure I have obtained is 380 billion tons and were this huge amount of methane to be released, the question is what would happen to it?

One way of looking at this is to observe that methane is lighter per molecule than oxygen, nitrogen or CO_2 and therefore, according to the Archimedean principle, it should rise and eventually escape our atmosphere into space {http://globalwarmingcycles.info/, 2010} [12].

This is comforting but then we come across a disagreeable fact. CO_2 is heavier than the other molecules in the atmosphere and if Archimedean principles were the only thing to consider, CO_2 would sink among the other constituents in the atmosphere until it blanketed the earth down low on us. This would not be good at all. Luckily, our measurements show that CO_2 is evenly distributed for at least 10 miles up.

Thus, we cannot complacently expect the methane to escape upwards. What is it that makes the CO_2 be uniformly distributed?

The answer the climatologists give us is that as one goes upwards from the earth, there is increasing turbulence. The temperature gets colder and the winds greater, so the CO_2, jostled around in its collisions with the other molecules until the affect of the Archimedean drop becomes negligible. Indeed the CO_2 has been there for much of the earth's life, because the green plants and their growth depend directly upon it.

The principal thing that I tried to draw out of DOE was the rate of the movement of the ice line towards the north. It's clear that it's retreating, but what is the rate of that retreat for it will eventually melt the frozen tundra?

Some discussions I had with a senior expert from the Washington DOE {Private communications, 2009} [14], who warned me that I should be cautious in stirring anxiety. I decided that the only thing I could do was to assume that eventually, be it in one year or ten, that the tundra were going to melt and I wanted to know what would happen then {Private communications, 2009} [13].

Thus, to assume the entire 380 billion tons would all go to the atmosphere was an extreme but unlikely assumption. The tundra is not growing on the surface of the earth but deep inside it as well.

Further, to get the 380 billion tons estimated was to assume that the whole tundra was inundated with methane now whereas the creation of methane is a biological reaction going on at a speed of which we know little.

It is not that the 380 billion tons that may be there right now might hit us immediately. The question is how much methane is being created inside the tundra and what will be the rate of that growth compared with the time at which the tundra will melt.

The truth is the methane in the tundra is a possible threat {D. Roberts et al., 2007} [15]. We should be aware of it and look at calculations with certain assumptions. Certainly the maximum likely effect is dire, but its severity is unlikely to be realized.

1.3 Attempted calculation of the maximum effect of methane on the world's temperature

I made a number of positive assumptions in order to get the worst that the assumptions predict. The first assumption is that the 380 billion tons of methane is a number that may become reality in our time.

A second assumption is: will the distribution of methane, were it to mix with air, be uniform and how long would it take to become so? At first I assumed that the methane would spread along the near earth surface and then diffuse upwards. The figure I got was four years, for the methane to diffuse up 10 miles that is around about the extent of 90% of our atmosphere. (Some information on the albedo can help in estimating a uniformity of the mixture of gases (Figure 6) {Dar A. Roberts a, Eliza S. Bradley a, Ross Cheung b, Ira Leifer c, Philip E. Dennison d, Jack S. Margolis, 2006}.)

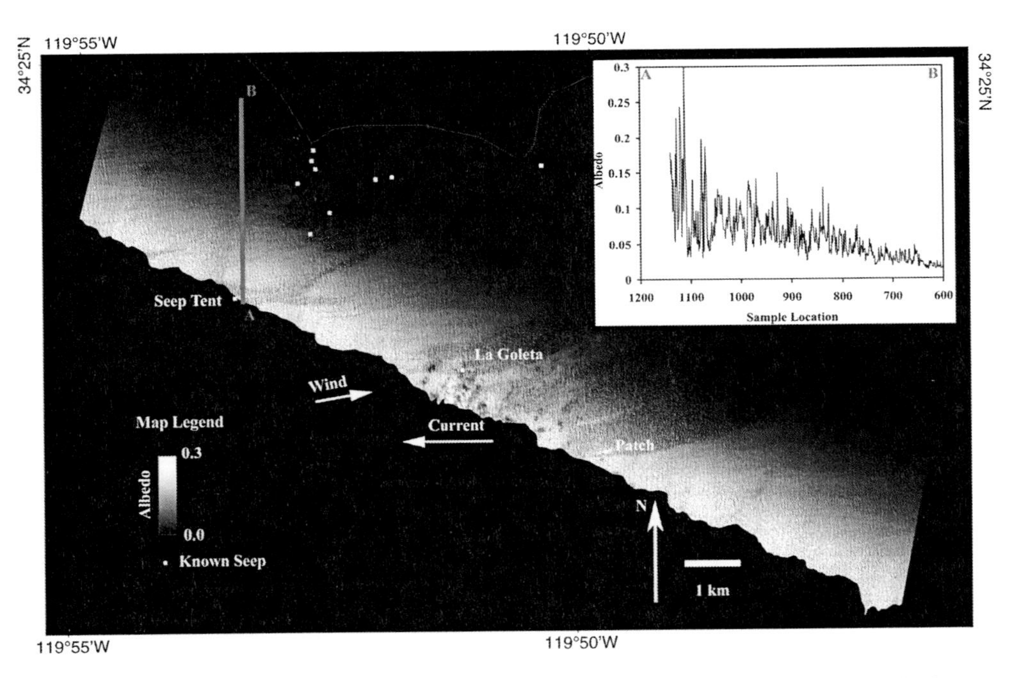

Fig. 6. Estimated albedo for 6 August 2007 Run R04. The location of the coast is marked in very faint green. Wind direction, from a coastal weather station (www.geog.ucsb.edu/ideas) and codar-derived currents, measured by the Interdisciplinary Oceanography Group (http://www.icess.ucsb.edu/iog/archive/25) are marked. Inset shows north-south albedo transect (red line) that includes the Seep Tent area. Some named seeps are marked by white squares [15]

However, I abandoned this approach because, of disturbances which interfere grossly with the condition diffusion requires. It's going to spread further and faster than that, egged on by the Archimedean thrust to rise but mixed up with wind and temperature changes it will meet.
I therefore assumed uniformity and of course it's a simple calculation to find out the concentration per liter of methane if the whole 380 billions tons were uniformly distributed in the 10 miles (upward in our atmosphere).
With these limiting assumptions then, I turned to the mathematics which Veziroglu {Veziroglu et al., 1989} and his associates produced and fitted my assumptions into his calculations [3]. What the Veziroglu paper actually calculates is the temperature change in the atmosphere and so far as the CO_2 changes its concentration, climbing slowly as we show in Figures 2 and 4. So I assumed one could equate a single methane molecule to 23 CO_2 molecules. Of course this simplifying assumption made it easy to get results from the Veziroglu theoretical formulations on CO_2 and the result I got, with all the positive assumptions I had made, was 6 ^{o}C in ten years {Veziroglu, Gurkin, and Padki, 1989} [16].
I asked myself then when it would begin a decline in our atmosphere and was there any end to it, and here I took to a Professor in Meteorology at the University of Florida, who seemed knowledgeable in discussions of methane and the dynamics of its presence in the atmosphere.
Qualitatively, his view was that there was a conflict between the Archimedean rise idea and the wind and temperature disturbance idea. He brushed aside the CO_2 and the fact it has remained stable and uniform for millennia. He said he had made a calculation which suggested that the best model would be to assume a quick distribution of the methane after the tundra had melted and then he thought that ten years would be about the time at which the tendency of the light methane molecule would escape into space.
For a moment, let us consider that my 6-degree calculation from Veziroglu's theory has value.
One can see at once there were some places on earth that would be stricken. Imagine what it would be like in Saudi Arabia at 123 ^{o} F. Now, add to that, 6 ^{o}C or c. 12° F, and you will see that the inhabitants of Saudi Arabia could be really threatened if the temperature rose as I think is possible.
Of course it wouldn't be only Saudi Arabia but their surrounding countries, too. This is something that they have to confront (and they have the money to launch a more accurate investigation than the rough one I did in using what DOE would give, together with the calculations of Veziroglu et al {Veziroglu, Gurkin, and Padki, 1989} [17].

1.4 Disagreement as to the cause of global warming

Among those who have studied the CO_2 theory of Global Warming, may be somewhat surprised to know that there is a group of people (are they scientists?) in our community who disagree that CO_2 is the main cause {Edward Townes, 2007} [18].
This has always been the case from the beginning of concern about Global Warming way back in the 1970's.
The argument of the anti- CO_2 group begins by pointing out that ice cores taken deep into the earth show that the temperature of the earth has varied greatly over thousands of years. The opponents of this theory point to much greater variations in the earth's temperature

than we see at the moment. Some anti-reactions will occur on earth that will compensate the temperature rise we are now seeing and it's better to find out the true cause of the present rise before we put too much money into fighting it {B. Pelham, 2009} [19].
Another part of the strength of the anti- CO_2 group is largely from the public itself. The distressing truth is that the majority does not believe in Global Warming and that naturally this affects the vote in congress when it comes to research and money spent in that direction. The answer is that the change is very slow but indeed it is faster than the changes in the past (the really big changes) to which people refer. The idea that there is "no change really"

2. Sources unencumbered by CO_2

The general presentation of this treatment of Global Warming is to point out that there are a total of six different sources of energy, some of which we could develop and rely upon. They're inexhaustible and clean, and it's easy to profit from them, compared to gasoline that comes from oil buried in the earth and has to be processed, but also damages the environment.
The first thing then is to present clean sources of energy. They are mainly wind {J. Usaola, E. Castronuovo, 2009; C. Osphey, 2009; H. Green, 2008} [20, 21, 22], solar, and enhanced geothermal.
Then having given the stated main sources on each of them, I go on to treat several others {J. Bockris, 2009} [23], for example, the enhanced geothermal energy ("Hot Rock Geothermal"), which could be a major source of energy, together with the less realized ones, the massive development of tidal energies and et cetera {C. Osphey, 2009; H. Green, 2008} [21, 22].
Later on in the article you will find there is a discussion of the mediums because each of these main energy sources {J. Bockris, 2009} [23] must have a partner which is in a form of energy which can be spread and be introduced into households and factories {J. Bockris, 2009} [23].
Among the discussion of these mediums there is an introduction to a concept, the power relay satellite. German inventions of World War II but never developed. It's development concerns diurnal difficulties of solar light and it would be possible, if we had a sufficient collection of solar energy, - and the Australian Continent is such {B. Roberts et al, 2007}[24], - to spread this solar energy and operate not only within a few tens of miles of the original source, but to anywhere in the world and therefore as the times of darkness are different in different parts of the world, but varying the opposite direction to the periods of light, it should be possible in principle to bring solar energy {J. Bockris, 2009} [25] to anywhere in the earth and thus counteract its principal hazard {J. Bockris, 1975} [26].

2.1 General philosophy of dealing with global warming

The general philosophy in this article in dealing with Global Warming is to take the attitude that the principal cause of Global Warming; the influx of CO_2 into the atmosphere, must be reduced towards zero. This therefore is only a scientific matter in respect to what comes after {N. Muradov, N. Veziroglu, 2009} [27]; because of course there is no point in shutting off the gasoline unless we replace it. The task is large so that is seems reasonable that there should be a central authority for the development of replacement energy systems for the fossil fuels.
As to the fossil fuels, - coal, oil, and natural gas, - I believe that what has to be done with them, - a very political matter, - is arranged between the government and their very wealthy owners, for the government has the right to tax their products.

Thus, in the following pages we are going to review our energy future in two ways {J. Bockris, 2009} [28]. Firstly, we are going to think that discretion is the better part of valor in respect to dealing with the oil companies. It is a matter that the government has to do and the president of our country has to be careful to be sure that special interests do not have any part in the decision as to when and how the fossil fuels will be made too expensive.
It will be necessary to allow time to build across the country the replacement energy systems of wind, solar, and hot rock geothermal.
There are various estimates on how quickly the change can be made. The Chinese government has made public their plan to change their transportation system in eleven years.
Let us adopt a pathway that is a little less demanding and decide that we are going to change over in twenty years with the extension to thirty years being acceptable, but not joyfully.
We will begin then by illuminating here first wind energy because it is the lowest cost. Then after we have the best source for our part of the world, other matters such as the transfer of energy over long distances, - will come in.

2.2 Wind:

Many who are told that wind may be part of our future energy supply find it hard to believe because wind is sporadic, and cannot be relied upon at any particular time or place.
Hence, it is important to understand the concept of averages when applied to wind energy. The usual thing is to look at the average or the cubes of the reported wind velocity taken daily. This gives the effective wind speed for the year, and the cube of this is the usual quoted figure. It's important not to take the cube of the average of the wind energies, but rather the average of the cubes. (See Equation 1 below.)
Another important preliminary to discussion of wind energy is wind belts. Of course, there are minor variations from year to year of the wind velocities in a given location, but on the whole if the average of the cubes is taken every year for a number of years, and the average of this figure is used in planning, such results will be effective.
In the USA, the part of the country for wind belt location is in Middle USA., north to south. The Wind Energy Association publishes maps of wind belts (DOE does the same). To show the sensitivity of a wind generator to values of v, the wind speed, one can take the example of going from 15mph to 18mph (apparently a small difference), but when one takes the cubes, it turns out that 18 mph is some 75 percent over 15 mph as the rates at which energy can be gathered.

2.3 Wind to electricity

The transfer of wind energy to electricity is carried out by using the combination of the energy of a rotating series of blades in the path of the wind, coupled with an electricity generator built into the apparatus. The axle of a rotor may weigh many tons {J. Usaola, E. Castronuovo, 2009} [20].
If untreated the supply of electrical energy from a wind generator would vary with the cube of the speed of the wind, and the occasional wind gusts. In order to avoid irregularity of supply, most wind generators are fitted with electronic devices that smooth out the supply in terms of volts. Powerful wind gusts, however, are a different matter and there is research to be done on how to capture the considerable energy that does come in gusts where the v may go to six to ten times the average velocity {J. Usaola, E. Castronuovo, 2009; C. Osphey, 2009} [20,21].

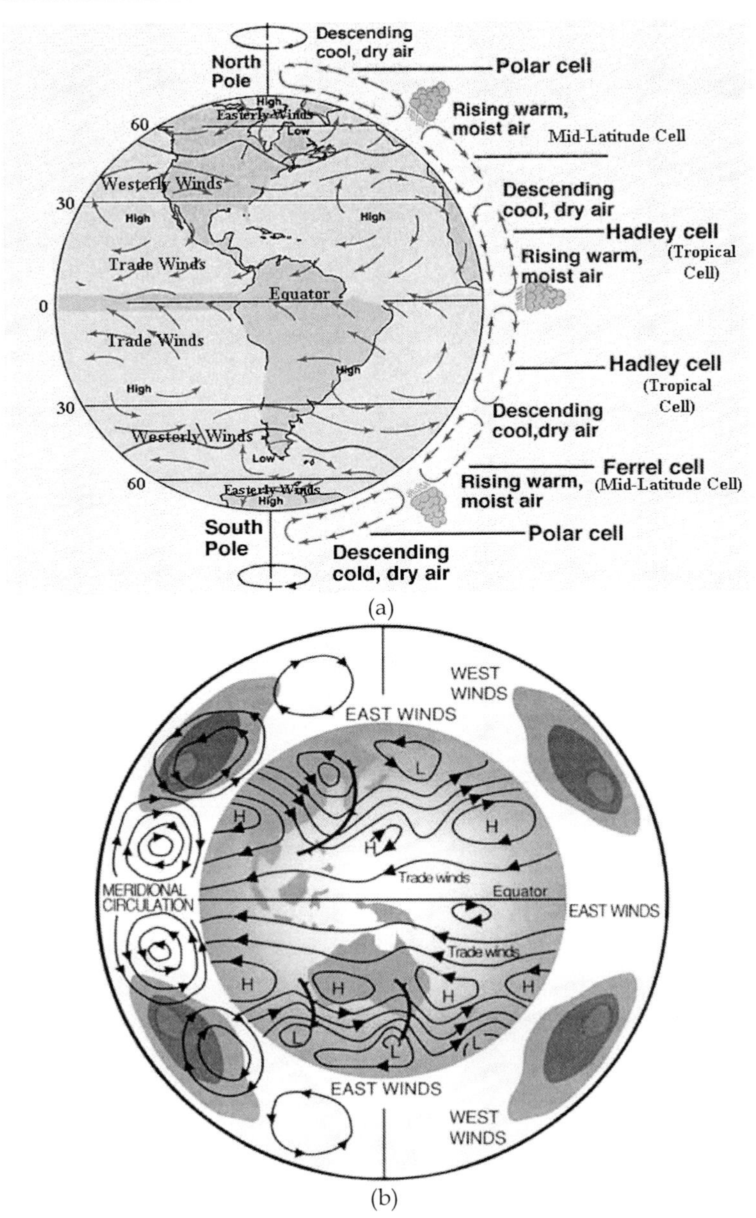

Fig. 7. a. Wind map of the USA
http://www.cnsm.csulb.edu/departments/geology/people/bperry/geology303/_derived/geol303text.html_txt_atmoscell_big.gif
b. Wind maps of northern regions.
http://mabryonline.org/blogs/woolsey/images/global%20winds%202-1.jpg

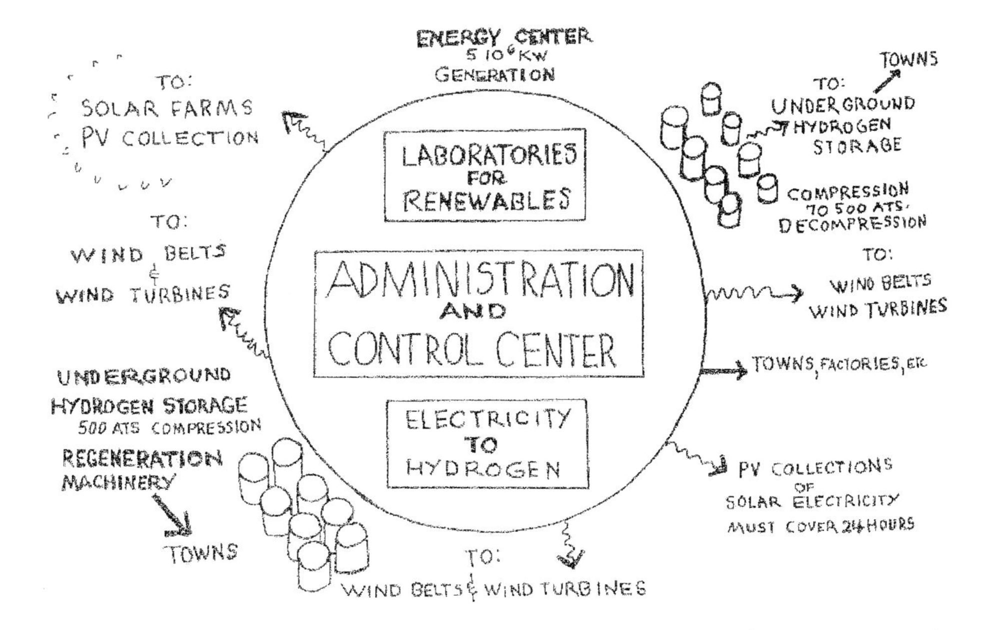

Fig. 8. Energy Center, J. O'M. Bockris Original, 2009

Many of the earlier wind generators often broke down in gusts, having been built to sustain only the average wind energy in a given location.

An energy center (See Figure 8 above) has to be made if wind is to be used on a massive scale for the supply of towns. The idea here is to place the wind generators in a circle surrounding the energy center with no greater distance than 50 miles between generator and center.

A possible energy center is shown in the Figure 8 above.

The Center contains apparatus for mixing various incoming electrical energies from the wind generators. These are then divided into supply lines that go out from the wind (or solar) center to surrounding towns. Details of arrangements will depend upon the population density of the area, however, the center may supply only large towns of say 1 million in population or larger.

Then, these supply towns would act as sub centers for other smaller towns. So, a one million people town may branch out to supply, say, ten smaller towns, down to the supply of villages from nearby larger house groups.

In large cities such as New York, several centers would have to be used.

After much research the optimal shape of wind generators has been reduced to two, {H. Green, 2008} [22] (see 9A and B). The main one is that well known one, horizontal propeller and such wind generators are found to last about fifty to 100 years. However, there is another type of wind generator as shown in the Figure 9B which is called a vertical axis generator, and it can be seen that the wind is gathered in the cusp type shape of half the blades, and these then rotate around the vertical shaft, bringing in to face the wind, a sloping area of the other half type cusps so that when this swings around to face the wind, the pull on it is much less than when the wind is being collected in the cusp type part of the generator.

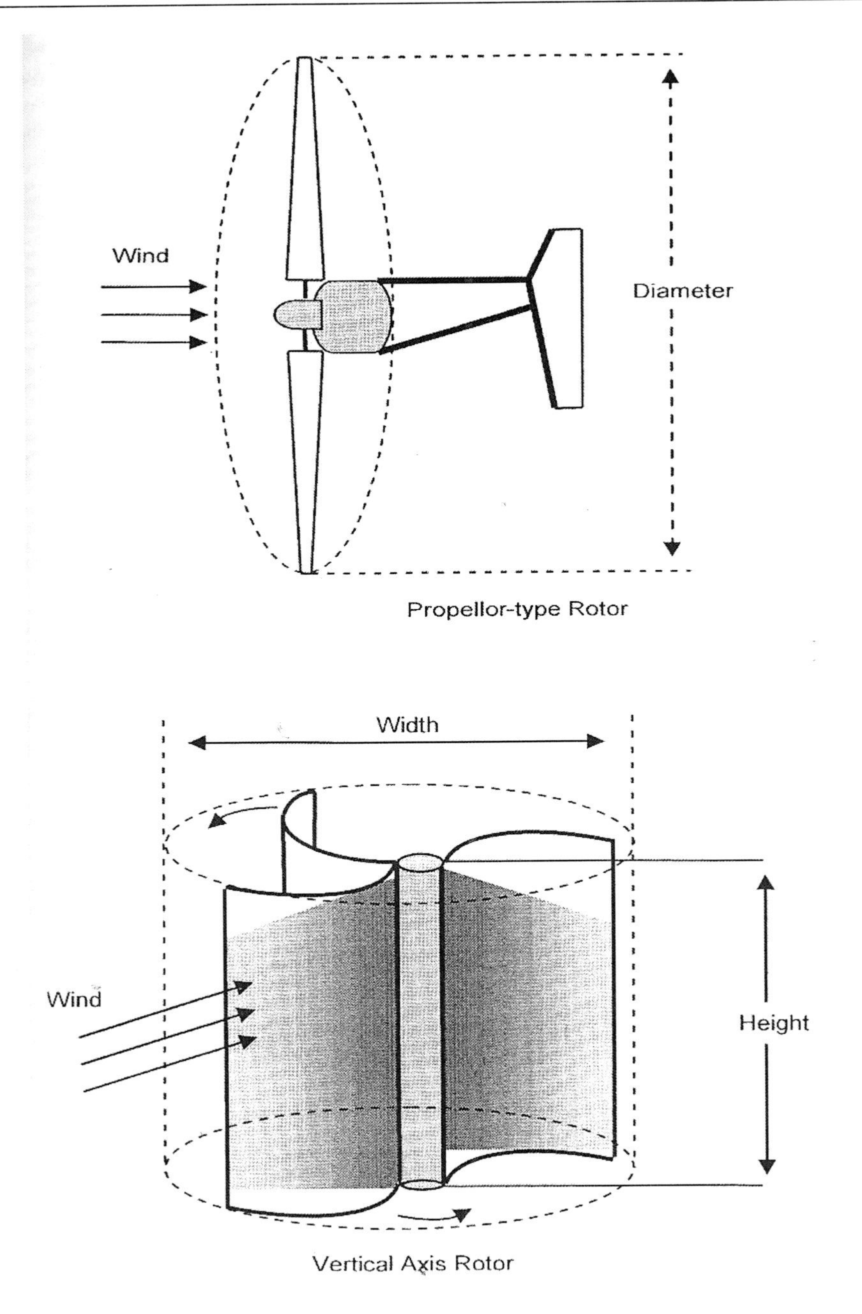

Fig. 9. A & B: {Iowa Energy Center, 2006}

It might be thought that four blades would increase the use of a single shaft but the manufacturers tell us that the material and machinery for accommodating multi-blade generators do not pay.

Wind generators can also be set up to work at sea. At first sight, there is much advantage in this because winds at sea tend to be greater and even up to twice times the winds on land. The reason is the lack of obstructions to the wind that occur on the ground.

However, there are compensating factors that make the positioning of the generators at sea, a questionable matter. Firstly, the construction of the actual generator has to be strengthened because of the higher intensity of the winds. This strengthening must include balancing weights underwater as shown in Figures 10 and 11.

Another negative feature of the wind borne generator is the cost of delivering the energy back to land. This can be done by cable but in extreme cases, ships collect the product.

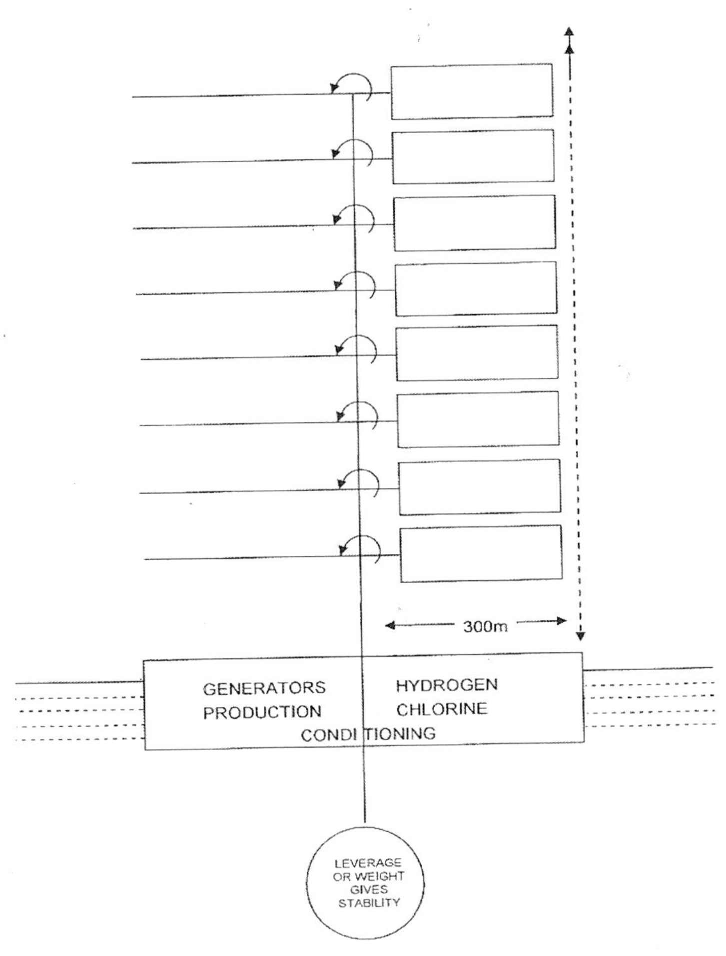

Fig. 10. A possible arrangement for a sea-borne generator. {J.Bockris, 1975}

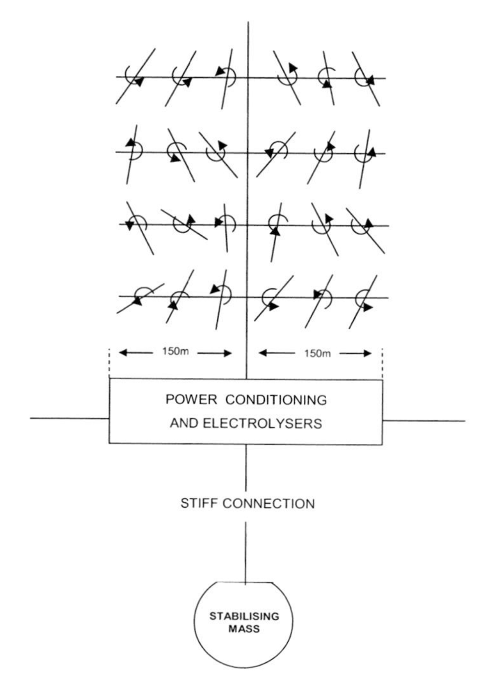

Fig. 11. An alternative arrangement for a sea-borne generator. {J.Bockris, 1975}

One of the newer concepts that have been introduced into wind generator construction is the magna lev concept, i.e. the shaft of the generator that of course normally is fitted into a socket that causes friction but is lifted from the socket contact by electro-magnetism. This concept is not commercial, but the designers say the lessening of the cost of the wind is up to 10x, and if this can be verified in practice, it is obvious that it will be introduced into newer generators which make wind even lower cost.
One may be forced to go to sea, where there is always plenty of room, - and more wind.[4]

2.4 A theory of wind generation of energy at speeds of up to 20 mph

Wind generators have not been considered on a massive scale such as that which will be needed for the supply of towns. However the economic attraction of the wind generator is great on an economic basis because the owner who receives his generator can start using it to produce a profit within weeks of delivery.
With several renewable energies, there may be preliminary building to be made that could delay the receipt of profit by the owner for years.
Of course, a study has to be made firstly about the detailed conditions of wind in the place considered, and this must include not only the minimal economic velocity of the wind

[4] It may be important to lower the cost of wind generators, which at the moment on land, produce energy as low as $.03c/kWh. Wind as a main source of energy in the future must face the hot rock geothermal situation and therefore lowering it would be needed.

average, about 12mph, but also the question of wind gusts and whether they would be a threat to the stability of the wind generators {H. Green, 2008; J. Bockris, 2009} [22, 23].
A primary engineering objective therefore is the mechanical engineering one of producing wind generators, should always take into account the question of whether the generator can withstand gusts {B. Roberts et al, 2007} [24].
Now, a simple theory of the energy obtained from a wind generator starts by recalling that the kinetic energy of a moving mass is given by ½ mv ^{2}. The hydrodynamics of the actual transfer of the wind energy due to the rotational action of the blades is the kinetic energy multiplied by the factor 16/27 {J. Bockris, 2009} [28].
Thus, the energy of the generator, taken in this ideal picture is $\frac{1}{2}\rho v^3 \frac{16}{27}$ where Δ is the density of the air and v the average energy of the wind (over one year).
This is the simplest basic expression possible for a wind generator. However, it is still insufficient and has to be aided by an experimentally added factor, which for most generators is about ½ the ideal value {AWEA, 2009} [30].
It's important to realize that even this simple equation only applies in the lower regions of wind speeds. The important information that the energy of a wind generator depends on the 3rd power of v, the average wind speed for the year, makes it important to ascertain when the equation begins to break down as the average speed is increased past 20 mph.
Thus, does it apply where much higher average wind speeds than that typical of North America (15-20mph) are available {J. Bockris, 2009} [25}? In Patagonia at the tip of South America, there are regions where the average wind speed for nine months of the year, is 40mph.

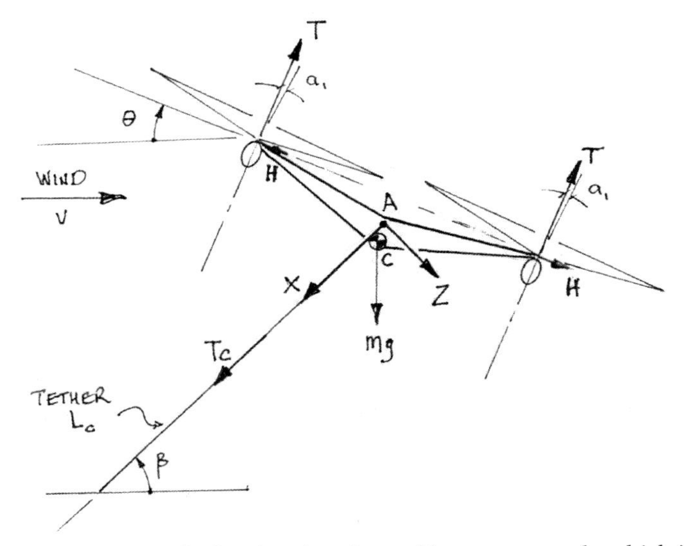

Fig. 12. Diagram of the FEG in flight, showing the craft's nose-up angle which is identical to the control axis, as no cyclic pitch use is planned. The rotor's fore and aft flapping angle, a1, is shown as the angle between the normal to the tip-path plane and the control axis. The total rotor thrust component along the control axis is T, and normal to this axis is the component force H. If T and H forces are combined vectorally the total rotor force is almost normal to the tip-path plane {B. Roberts et al., 2007: [29].

Fig. 13. Rendering of Sky Wind Power Corp's planned 240 kW, four-rotor demonstration craft {B. Roberts, D. Shepherd, et al, 2007: [29].

The difficulty of putting such great winds to a useful purpose is not only engineering generators that will withstand the gusts from such winds (>120mph?) but also the fact that we do not know the upper limit of the equation derived above. There are qualitative indications, however, that the equation begins to breakdown at about 25mph and this makes it difficult for in engineering research that might meet the problem of the stability of generators in very high winds {AWEA, 2009} [30].
P is the kinetic energy of the wind per unit volume in time and c is a hydrodynamic factor for the extraction of energy. However, the equation neglects the effect of rotor-air resistance. The basic empirical equation (Equation 1) is:

$$\text{Power} = \frac{16}{27} c \frac{1}{2} \rho v^3 \quad (1)$$

where c is a parameter, generally taken as 1/2 but falling with an increase in wind velocity. The cube law dependence of power on wind velocity v^3 is noteworthy.
The wind equation requires the *mean of the cubes* of the instantaneous wind velocities over the year. If the mean of the velocities is cubed, results are 2 to 3 times too small {Bockris, 1975} [31]."

2.5 Wind belts

It is important to locate the rotors in areas ("wind belts") in which the average wind speed is maximal. Due to the rotation of the earth, gravity forces air raised by heat over the equator to drop, colder air on the earth beneath. (Figure 14) [32].

Two main systems are shown in Figure 14. The southern pink winds, "trade winds," were vital to sailing ships en route from England to Australia. The ships traveled south of the Cape of Good Hope to reach the West to East wind that blew them eastward to Western Australia and onwards, at about 14 knots. Clearly, the lower the velocity at which winds could be useful, the better. Experience, however, shows that wind speeds below 12 mph are no longer economically attractive. As to the higher speed limit and its practicality, that is not sharply defined. Great advantage is offered by higher winds.

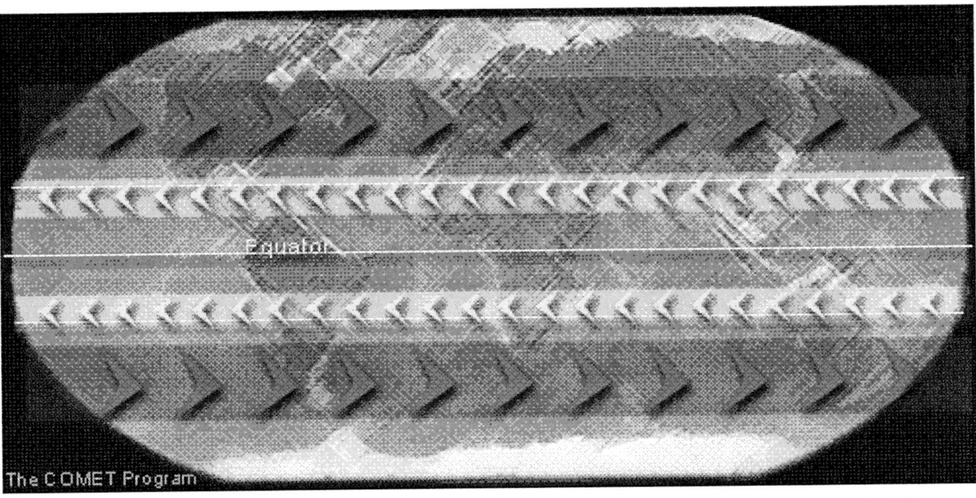

Fig. 14. "World map showing two mid latitude westerly wind belts (shown in pink). The northern belt blows from west to east across North America, the North Atlantic Ocean, Europe, and Asia. The southern belt blows from west to east across the South Pacific Ocean, Chile, Argentina, the South Atlantic Ocean, South Africa, the South Indian Ocean, Southern Australia, and New Zealand. The yellow arrows in the picture also show two tropical easterly wind belts blowing from east to west on either side of the equator. The northern tropical easterly belt blows across the Pacific Ocean, Southeast Asia, India, the North Indian Ocean, the Arabian Peninsula, Saharan Africa, the Atlantic Ocean, the Caribbean Sea, Southern Mexico, and Central America. The southern belt blows from east to west across Northern Australia, the Indian Ocean, Southern Africa, the South Atlantic Ocean, the middle of South America, and the South Pacific Ocean." [32]

2.6 The distribution of winds

A picture of the wind belts of the world has been given (Figure 14). However, it is of interest to identify places where the big winds blow. Both the Department of Energy and the Wind Energy Association publish maps of yearly average wind speeds in most parts of the world and particularly those in North America. The following quotations are from documents published by these organizations. [33] The terminology is explained in Table 1.

"Areas that are potentially suitable for wind energy applications (wind power class 3 and above) are dispersed throughout much of the United States. Areas which have useful wind energy resources include: the Great Plains from northwestern Texas and eastern New

Class 3	(Marginal)	12 mph year average
Class 4	(Satisfactory)	13 mph year average
Class 5	(Good)	14 mph year average
Class 6	(Excellent)	15 and above mph year average
Class 7	(Outstanding)	16 and above mph year average[5]

Table 1. Wind classes and wind speed
"Alaska is Class 7.
Great Plains (North Dakota) area is Class 5
Montana hilltops and uplands are Class 4
Hawaii area has areas of Class 6 but includes Oahu with Class 7 winds."

Mexico northward to Montana, North Dakota, and western Minnesota; the Atlantic coast from North Carolina to Maine; the Pacific coast from Point Conception, California, to Washington; the Texas Gulf coast; the Great Lakes; portions of Alaska, Hawaii, Puerto Rico, the Virgin Islands, and the Pacific Islands; exposed ridge crests and mountain summits throughout the Appalachians and the western United States; and specific wind corridors throughout the mountainous western states."

"Exposed coastal areas in the Northeast from Maine to New Jersey and in the Northwest southward to northern California indicate class 4 or higher wind resource. Class 4 or higher wind resources also occur over much of the Great Lakes and coastal areas where prevailing winds (from the strong southwest to northwest sector) have a long, open-water stretch. The Texas coast and Cape Cod in Massachusetts are the seats of coastal wind resources which extend inland a considerable distance." [33]

"Offshore data from Middleton Island indicate class 7 wind power. Shore data such as Cape Spencer, Cape Decision, Cape Hinchinbrook, and North Dutch Islands reflect class 5 or higher power." "Most of the coastlines associated with these areas are heavily wooded, so wind power estimates are very site-specific."

"Interactions between prevailing trade winds and island topography determine the distribution of wind power. On all major islands, trades accelerate over coastal regions, especially at the corners. The best examples are regions of class 6 or higher wind power on Oahu, Kauai, Molokai, and Hawaii. The rampart-like mountain crests of Oahu enhance prevailing winds to class 6. On other islands, circular mountain shapes and extreme elevations prevent the type of wind acceleration observed, e.g., on the Oahu ranges."

"On Oahu (Honolulu County), the long Koolau mountain rampart and shorter Waianae Range enhance trades to class 6, although the rugged topography, watershed value, and turbulent air flows over these ranges make practical application more difficult. The northeastern (Kahuku) and southeastern (Koko-head) tips of Oahu have areas of class 7 and broad areas of class 3 or higher. A class 3 and 4 area exists at Kaena Point on the island's northwestern tip, and class 3 areas exist along the southern coast west of Honolulu and southeastern coast north of Makapuu Point." [33]

---

[5] The v^3 law makes the difference in wind energy of the outstanding winds as giving more than an twice times termed increase in energy from "outstanding" sources 16 mph) compared with those term satisfactory 12 mph). [18, 18a]

2.7 Storage of wind energy [34]

The peak in the world oil production (apart from tar sands) is likely to come before 2060. Unfortunately, any new method of obtaining energy, - and wind is the cheapest and the simplest to build, - is going to take more than ten years to build throughout the country, - and suggested resources after the peak are the tar sands, coal, or the use of solar energy to grow plants. (1 percent efficiency and using more energy to make the alcohol than can be got from it).

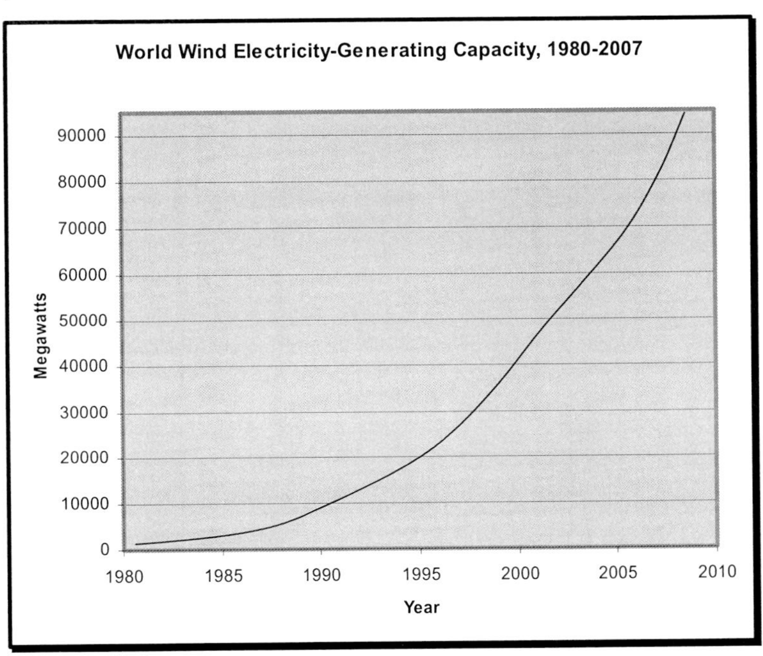

Source: GWEC, Worldwatch.

Fig. 15. From Wind Energy Fact Sheet, American Wind Energy Association, 2004. {AWEA, 2001} [35].

There is a new and attractive method for storing electrolytic hydrogen: combine it with atmosphere origined CO_2 to form CH_3OH, a liquid. If the CO_2 is extracted from the atmosphere, then, when burned, it simply replaces the CO_2, - it would be a CO_2 neutral process. The combination of H_2 and CO_2 needs a special catalyst but there is much evidence that it occurs.

The handling, storage, and transportation of methanol will be similar to that of gasoline.

2.8 The US position in the development of wind technology

Since the middle 80's the US began to lag behind European Nations (particularly Denmark, Holland and Germany) in the development of wind power (Figure 15) [35]. However, in 2005, the USA installed more new energy capacity than that of any other country (2,431 megawatts). The total cultivated wind energy in this country (2007) is equivalent to only about ten nuclear plants.

In cases in which yearly average winds above 15 mph are available, the upper limit of wind velocity that can be used in practice depends on engineering resources. Disasters that have befallen wind generators in the past have been brought about by storm-borne gusts of an intensity unallowed for in the design.
Reports of an extension of the capacity of modern wind generators so that they can operate in winds of 50 mph are available {R. Heinberg, 2007} [36].

2.9 Effect of height

Wind increases with height and there has to be a trade off between extra costs of building above the ground and the gain in average wind speed. 30 feet above the ground is used in measuring average yearly speeds. The use of mountain regions looks attractive but wind farms are difficult to build there.
Sometimes natural geographic arrangements give helpful situations such as that in which an approaching wind is increased in velocity by being compressed in a geographically natural funnel.

2.10 Wind belts at 15,000 feet

In recent times, a new technology has been born {Roberts et al, 2007} [24]. It has been made clear above that it's desirable to stick to wind belts on the ground, but a discovery was made in 2008 {J. Bockris, 2009} [25] that similar wind belts exist at heights of 15,000 feet. They found that they are stable and usable for three-fourths of the year. (cf storage devices below).
The Australian American team has been concerned with the collection of wind energy at 15,000 feet. They have used a helicopter modified to contain four rotors. As no forward motion is required of them on these helicopters they do not contain a forward thruster but are tethered to the ground. Tests have been made above the Mojave Desert and also above the Australian outback.
The electricity developed by means of the helicopter rotors is taken down the tether to an area containing water electrolysis plants, which electrolyze the water to produce hydrogen that can be stored at several hundred-atmosphere pressures.
This initiative is, of course, preliminary to any commercialization. The principal doubt is the lastingness of the rotors.
However, the paper published by Roberts et al, (2007)[29] seems to be a fundamental one for the future and if in 2008 it's possible to show that the 15,000 feet winds are usable then it might well be possible within, say, 25 years to look into the jet stream, 40 thousand feet, with speeds of more than 100 mph.
It is necessary to look at least thirty years into the future as we build massive low cost supplies to replace the fossil fuels {K. Deffeyes, 2003; R. Heinberg, 2007; M. Simmons, 2005} [37, 38, 39}.

2.11 Could wind energies be transferred over long distances?

It has been suggested by Muradov and Veziroglu (2008) [40] that the massive winds available at the tip of South America (Patagonia and southward to and in the Antarctic) could be used as massive energy generation areas.
Of course the first problem to solve is the equation that tells us what the energy received would be under extremely high average winds of 40mph year average.

However, there are two possibilities for transferring the large energy amounts that could be made in these artic areas. On the one hand, we can assume that, because of the wind speeds available, the cost of electrical energy is reduced below 1cent per kWh. If this were so, then it would be feasible to think of liquefying hydrogen produced by the electrolysis of seawater. The remote location signifies that care in avoiding transfer to the atmosphere of chlorine is not needed (if it were it can be pumped into the sea).
The circumstances portrayed would justify building modified tankers to take liquefied hydrogen to the northerly parts of the world needing energy.
However, there is another concept which has been documented and which may turn out to be cheaper than the transfer of hydrogen in the liquid form {K. Deffeyes, 2003} [37].

2.12 Potential transfer of energy in a power relay satellite

Kraft-Ehricke (1973), one of the German rocket team left behind some interesting calculations and diagrams of his concept of transferring large amounts of energy thorough a power relay satellite {Kraft-Ehricke, 1973} [41].
In Kraft-Ehricke's concepts the heavy parts of the system are retained on the ground, and the light parts would be put into orbit and be a satellite which is to be hung over the equator. The cost of such a system is largely the cost of putting the satellite into orbit {Kraft-Ehricke, 1973} [42].
The satellite should respond to energies on the ground between 30° north and 30° south of the equator. Once the beamed energy at microwave frequencies reaches the satellite, it can be directed more or less anywhere in the world and beamed to receiving stations on the ground. This has the possibility of transferring energy virtually anywhere, because, once the

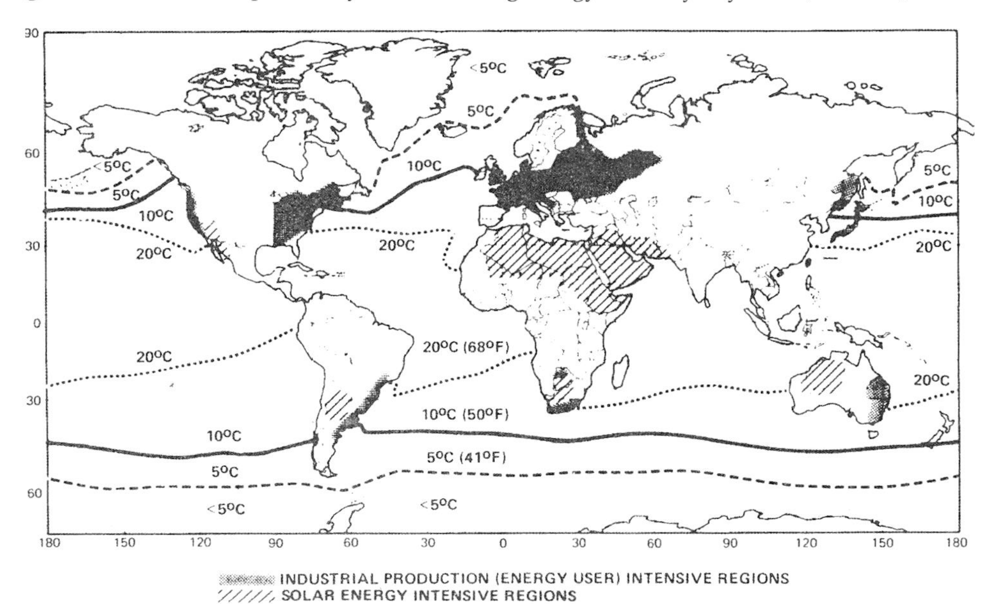

Fig. 16. Figure shows the great distances between areas of high insolation; and those of high concentration of affluent groups with manufacture {Kraft-Ehricke, 1973} [41].

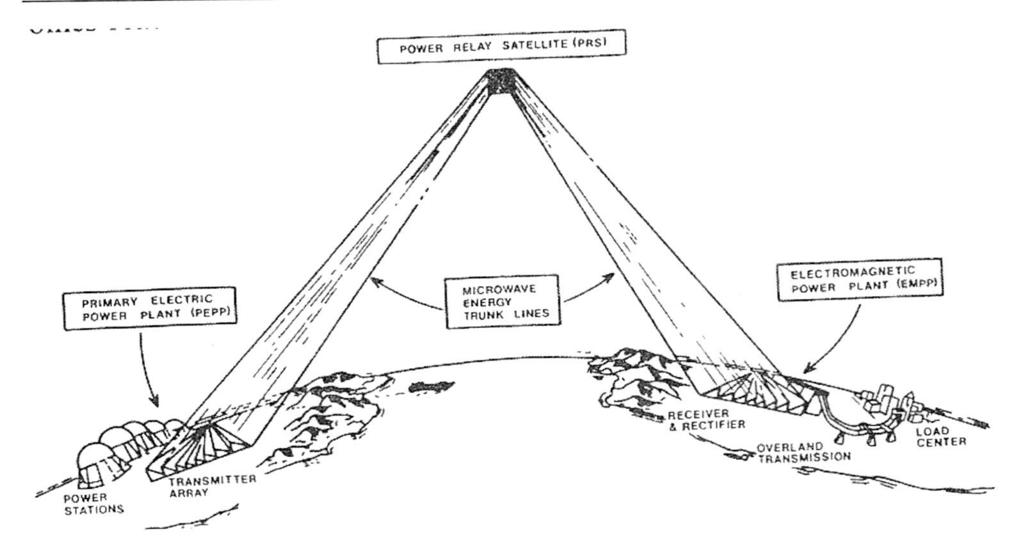

Fig. 17. Power relay satellite concept {Kraft-Ehricke, 1973} [44].

POWER RELAY SATELLITE CONCEPT - PERFORMANCE AND COST DATA

Power generated	13,800,000 kwe	Maximum values.
Power delivered	9,000,000 kwe*	Beam power can be
Energy delivered over 30 years	2.1 trillion kwhe*	less.
Transmitter array:		
Construction cost	$3.1 billion	$40/kwe*
Maintenance cost (30 years)	$12 billion	
Overall cost	$15.1 billion	7 mils/kwhe*
Power Relay Satellite		
(1 kW^2; 300 tons)		
Construction/delivery/erection	$0.75 billion	$84/kwe*
Maintenance cost (30 years)	$1.5 billion	
Overall cost	$2.25 billion	1 mil/kwhe*
Electromagnetic power plant (EMPP)		
Construction cost	$1.5 billion	$166/kwe*
Maintenance cost (30 years)	$0.95 billion	
Overall cost	$2.45 billion	1.15 mils/kwe
Grand total		
Construction cost	$5.3 billion	$590/kwe*
Maintenance cost (30 years)	$14.45 billion	
Overall energy transmission cost	$20 billion	9.3 mils/kwhe* †

* Power delivered.
† According to a newer analysis, taking into account the lessened maintenance cost caused by the development of platinum-coated amplitrons – this cost goes down to about 5.2 mils/kWH^{-1} (but this does not include the capital cost, which puts the figure up to 10 mils/kWH^{-1}).[27]

Table 2. {Kraft-Ehricke, 1973} [44].

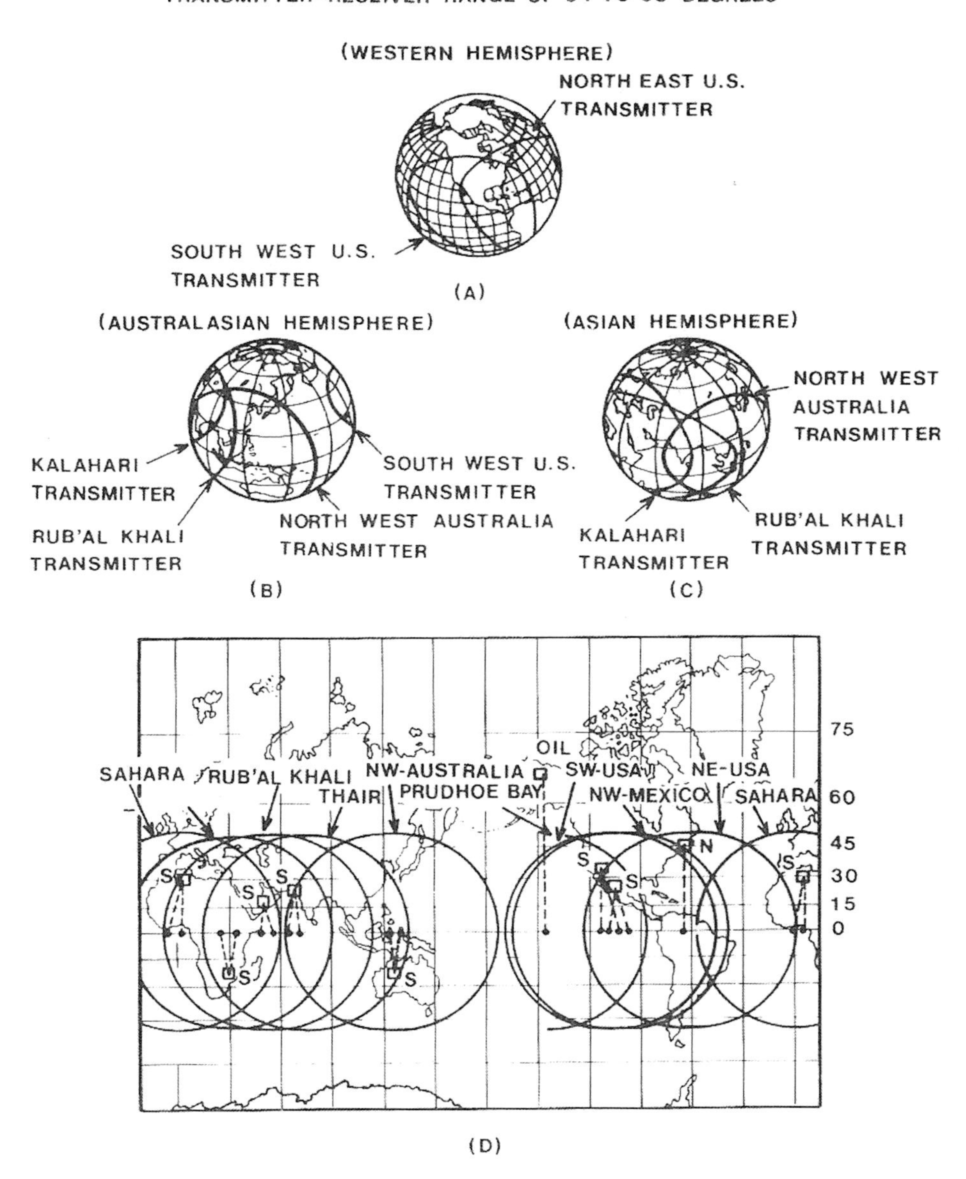

Fig. 18. Range of a number of Primary Energy Power Plant Systems {Kraft-Ehricke, 1973} [44].

CHARACTERISTICS OF MICROWAVE-PRS ENERGY SYSTEM

Microwave Energy (point-to-point transmission via power relay satellite).

NOTE:
MW = microwave
PRS = power relay satellite
kwb = kilowatt beam power; as used in (C), kwb refers to the beam power at transmitter aperture
kwe* = dc electric kilowatt as delivered at the bus bar of the receiver power plant.

A *General*
Avoids questions of engineering overseas pipeline; building tanker fleet. Because of high cost of the transmitter, is economically important to use energy sources lasting at least 30 years, i.e. largely for transmission solar energy from high insolation areas or exporting fusion energy.

B *Processing*
Conversion to MW energy at high efficiency ($\gtrsim$90%) in microwave power generators at the expected state of the art at mid-1980s. Transmission efficiency over several thousand miles 50-60%.

C *Loading*
As phase-controlled MW beam in transmitter antenna. Antenna size large, but not prohibitive ($\gtrsim$ 138,000 kwe*/km^2). Relatively high construction cost, due to stringent engineering requirements (~\$3/ft^2; ~\$31M/km^2; ~\$295/kwb; ~\$340/kwe*).

D *Shipping*
Power Relay Satellite (PRS) in geosynchronous orbit needed for beam redirection. Stringent PRS engineering requirements. But moderate construction cost (~\$84/kwe*) due to small size (~1 km^2) and to shuttle which keeps maintenance low. Environment effects of beam on atmosphere are small.

E *Unloading*
Receiver-rectifier system (rectenna) of 85% conversion efficiency to dc current. Therefore, little thermal load. Environmental burden? Receiver size is larger, but not prohibitive ($\gtrsim$90,000 kwe*/km^2). Receiver power plant construction cost is small (~\$166/kwe*). Operating costs low. Operation is simple, making system suitable for export and use in developing countries.

F *Storage*
Has no storage capacity, would need to convert to hydrogen, anyway.

G *Conversion to electricity*
In rectenna, conversion to ac after electric transmission close to load centre.

H *Transmission to user*
Receiver power plant can be located near load centres, due to its low socio-environmental burden quotient. Dc-power transmission yields superior efficiency to ac-power transmission.

* Table originates in ref. 5, but modified.

Table 3. Systems Kraft-Ehricke, 1973} [44].

energy has left the ground in microwave beam form, its transfer is more or less equal in cost if it's transferred 1000 miles or 5000 miles, it depends upon the orientation given in the satellite {Kraft-Ehricke, 1973} [43].

Thus, solar energy from the ground could be converted to electricity and eventually beamed at microwave frequencies to strike the satellite, which then orients it toward any desired location. Australia, North Africa, Saudi Arabia, would be places from which solar energy in massive amounts could be beamed.

Transmission of energy by microwave beams must have a load reception center at the end, where a country needing energy receives the beam. For example, 59% of the entire Australian continent is open for solar energy exploitation{Kraft-Ehricke, 1973} [44].

In Ehricke's plan, (1973) {45], transmitting and receiving antennae would consist of very many individual elements {J. Bockris, 1975} [46]. He suggests a helix antenna 1.4" in diameter, 14" in length.

Receiving areas depend on many things, such as Osaka, Japan, or London, England, are places where large amounts of energy are needed and Australia is a place from which very large amounts of solar energy can be created {Kraft-Ehricke, 1973} [43].

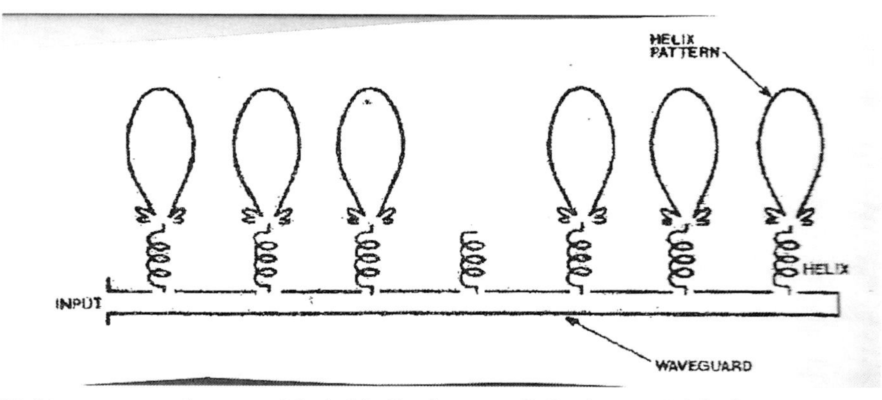

Fig. 19. Linear array of waveguide-fed helix elements {J. Bockris, 1975} [46].

2.13 Wind and Sun

There are many considerations that could influence a community that would have to decide if it wanted wind and sun as the origins of its energy supply {R. Heinberg, 2007} [38].

Of course there is a need for a detailed study of the average available solar or wind before a decision. Solar intensity is optimal roughly 3000 miles on each side of the equator. Wind tends to be the superior source outside this area but one other aspect of the matter is that the solar source is available for only six to eight hours per day. (Except for OTEC.)

Wind energies in general are available for 24 hours per day but whereas the solar energy can be reliable knowing the history of the location, wind energy is more subject to sporadicity. At the present time, around 2010, North Africa is the place where the commercial development of the solar source is making progress {M. Simmons, 2005} [39] particularly important as it is with an exhausting oil supply {A. Cristian, 2008} [47].

On the other hand, Europe is the place where there is a major development of wind energy (particularly in Denmark and North Germany)

2.14 Cost of wind energy

Discussions of wind energy in the 2008 literature are often aimed at small-scale wind farms or even individual users. The problem with them is that they mix up the (large) amortization costs of construction with the (small) cost of operating and servicing the equipment. The amortization costs are spread out over the expected life of the plant (twenty to thirty years) so that the low costs of wind energy, free of repayment for the costs of construction, are seldom brought out [48, 49]. (2008 forecasts of wind energy by 2010 are quoted at 3.5 cents per kWh - well below the corresponding prices of commercial electricity in the USA at that time) [50].

2.15 Range of practical wind energies

With wind turbine technology, commercially available in the U.S. in 2010, the acceptable wind velocities ranges are from 12-15 mph, this is the practical range of wind energy for use under 2008 conditions and acceptable to the US Department of Energy in that year {N. Muradov and N. Veziroglu, 2005} [51].

This small range of practical wind speeds explains why the costs of wind energy are often stated without defining the wind speed. In 2006, the range of total costs (construction and operating) quoted by DOE, are 4-6 cents per kWh, but the National American Wind Energy predicts 3 and even 2 cents per kWh within a decade from 2007. No other source, except paid off hydro could compare with these costs, half the costs of polluting fossil fuel based electricity.

Among published costs of recent times are those of some wind farms of 0.51 MW. The dependence of cost on wind speed, experimentally established is as follow:

$$16\text{mph} = 4.8 \text{ cents kwh}^{-1}$$

$$18\text{mph} = 3.6 \text{ cents kwh}^{-1}$$

$$21\text{mph} = 2.6 \text{ cents kwh}^{-1}$$

and thus show a sizable effect of wind velocity in present practice. Reports from non-governmental sources in the USA extend acceptable wind speeds to higher values and lower costs {DOE, 2010} [52].

One tends to look back to Churchill's description of the defeat of the Nazi Air Force by the Royal Air Force, in the Battle of Britain in World War II (1941). "Never has so much been owed by so many to so few". Applied to the present situation of development of clean energy in the USA, one might write "Never has so much been left unused by so few, when needed by so many" {DOE, 2010} [52].

2.16 Summary of wind energy

The main advantage of wind energy is low cost. The only cost lower than that obtainable from winds, is that of paid off hydroelectric plants, massively developed in Canada.

One of the advantages of wind energy is that the developer can receive a profit from his purchase within days of the machinery being delivered to him whereas with some other developments of renewable energies, extensive building may have to be done.

On the other hand, wind is challenged by the Enhanced Geothermal source. It is too early, - only two plants in hot rock geothermal have been built, - to make a well-informed comparison as to cost. Present production of futuristic schemes for wind might be thought to out range those for the hot rock costs.

3. The Earth's temperature

The amount of energy, e.g. from the sun, varies over the long term, and for many centuries there has been a slow but small decrease. Then, there is the question of heat from the earth, which contains heat-emitting radionuclides.

There may be other causes for the variation of the earth's temperature. The reason why these changes are little discussed in dealing with Global Warming is that they are much slower in respect to rate of change than those we are seeing. (This warming correlates with the increase in the use of carbon-containing fuels).

3.1 Attitude of the oil companies to global warming

Although the general talk among citizens has been for many years that oil is exhausting, the oil companies have often denied this. On the other hand, books are now being written about Saudi Arabia in particular and what we take from them is that the main well (huge in extent) in that country is no longer a sure supply for the future. There have been many values put forward for the Hubbert peak (Hubbert made the first scientific estimate of the amount of remaining oil) {K. Deffeyes, 2003} [53].

It is a matter of good business that oil companies will continue to sell oil (and damage the atmosphere) whilst it is still a desired product, i.e., until either there is a cheaper fuel (from wind) or our government has the votes to introduce a carbon tax to make alternative fuels relatively cheaper.

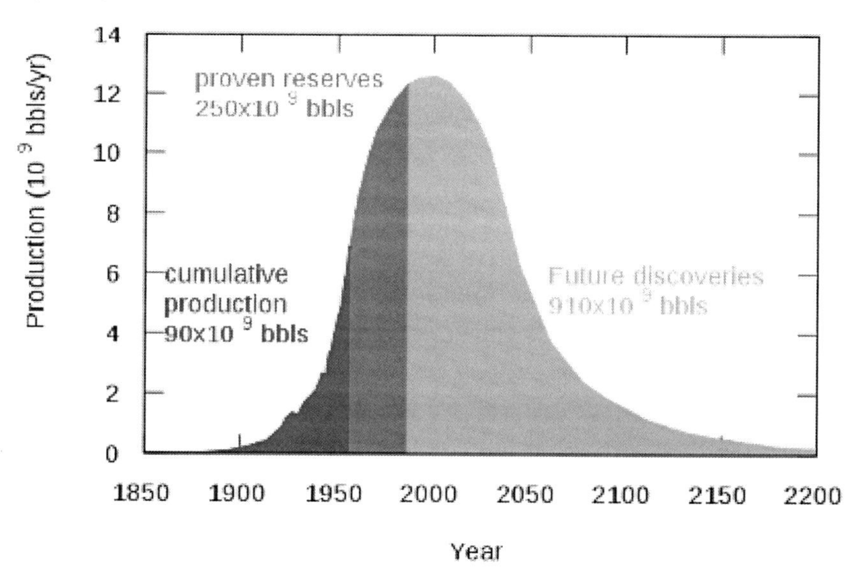

Fig. 20. New presentation of data in figure 20 of http://www.hubbertpeak.com/hubbert/1956/1956.pdf. Meant as replacement for non-free en::Image:Hubbert-fig-20.png 2007-03-04 (original upload date) Transferred from en.wikipedia; transferred to Commons by User:Pline using CommonsHelper Original uploader was Hankwang at en.wikipedia CC-BY-2.5; Released under the GNU Free Documentation License

Some analogy may be drawn between the damage scientifically proven to those smoking tobacco and the present population, damaged in health by inhaling polluted air from certain CO_2 -producing fuels.
During the last ten years we have identified three successive peaks. The one Hubbert put forward came at the year 2000, but after that there have been successive predictions by seriously minded experts on oil supplies. Every time a later one has followed the prediction of a peak, and the cause of these changes is that from time to time, even now, discoveries of new oil are being made.
Now, these discoveries are not always of oil, but rather in getting access to it. There are still sources of oil within the United States that have not been tapped. The reason why they are not usually counted is that they are often covered with thick layers of rock that, in the past, have been thought of as impenetrable, hence useless.
On the other hand, progress is being made in drilling which can indeed penetrate thick rock layers. For example, quite recently, a major find became operable near the Montana, North Dakota, and Saskatchewan border {A. Cristian, 2008} [54].
What we hear is that this deposit should provide us with more oil than we expect to get from Saudi Arabia. Consequently, the greatest burden on the budget is our armed forces may be resolved. Looking, then, to a fifty year future, our greatest danger is not exhaustion of oil, - but the temperatures of the future atmosphere.

3.2 Solutions to global warming

General

Discussions of Global Warming are often obscured by the fact that people who make proposals are often interested in short-term gains whereas anything we do to eliminate the negative effects of the warming climate would have to last at least thirty years in which time we expect still to be using some oil.
A good example of this is the activity of Virgin Airlines companies {2008} [55,56, 57] that offer a multi million-dollar prize to anyone who could solve the problem of Global Warming {K. Deffeyes, 2003; R. Heinberg, 2007; M. Simmons, 2005} [53, 58, 59]. However, it became clear that the winner would be he who found how to eliminate CO_2 whilst still burning the fossil fuels.
There are numerous ideas about what is called "sequestering"
The CO_2 is to be removed from plants producing electricity by burning coal and of course from the automobile. The difficulty off this approach involves catching the CO_2 in some kind of cheap compound, for example, lime, CaO. CO_2 easily combines with lime and therefore devices, which will be attached to cars producing large amounts of CO_2, might be followed with machinery to remove calcium carbonate. (Bury it?)
However, the problem here is that the amounts of the carbonate produced per day would be huge, and the problem then would become where to put it and the cost of getting it there.
Another kind of solution to sequestration is to bury the CO_2 in the sea but at deep levels, more than 3,000 feet when CO_2 becomes a hydrate and sinks.
One other partial solution to Global Warming would be to adopt a reaction first studied by Muradov (2005) [60]. The latter found that natural gas, passed through a zone at about 950° C, containing low cost catalysts, methane becomes carbon and hydrogen. The carbon can be dealt with, e.g. by burial. Pure hydrogen is liberated.

The problem becomes the limitation to the available natural gas and the problem of where to put the carbon and the cost of transporting it there. This is hardly a permanent solution and would require moving to a Hydrogen Economy.

3.3 Solar energy as a replacement for that from fossil fuels

Solar energy is undoubtedly the public's view of a future without fossil fuels or nuclear energy. Its antipathy towards the latter arises because of Chernobyl and other nuclear accidents that have killed thousands of people. U.S. workers now claim to pack the nuclear material in such a way that a meltdown is difficult to imagine.

The sun's light can be turned into electricity in a number of ways.

The easiest one to describe, and also at present the cheapest, is called the "solar thermal" method [61].

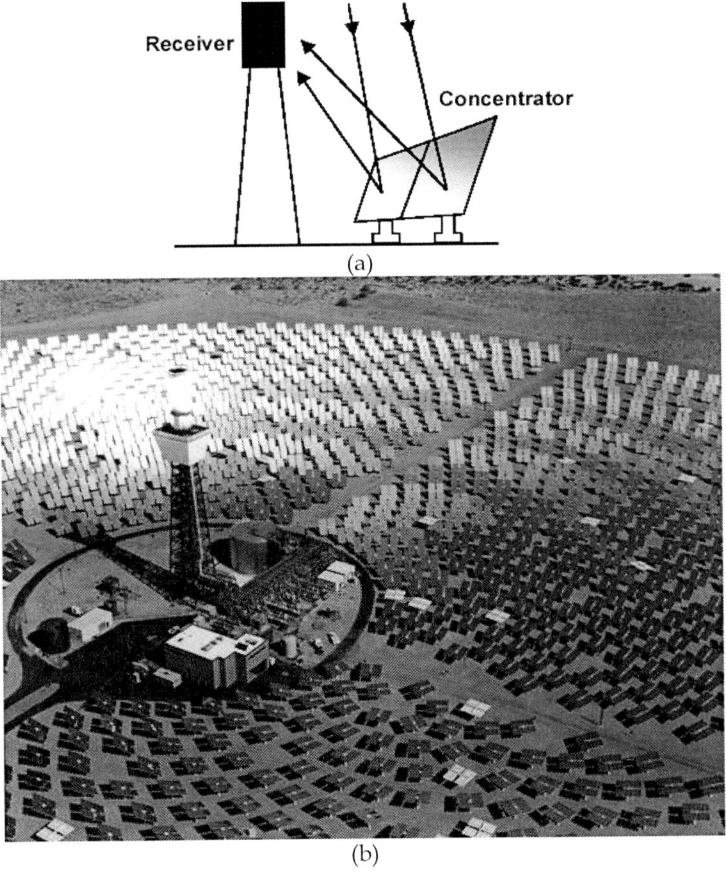

Fig. 21. (a) Schematic of a power tower. Image adapted from Energy Efficiency Renewable Energy Network {J. Tidwell, 2005} [61];
(b): Solar Two, power tower. Image courtesy of NREL's Photographic Information Exchange [62].

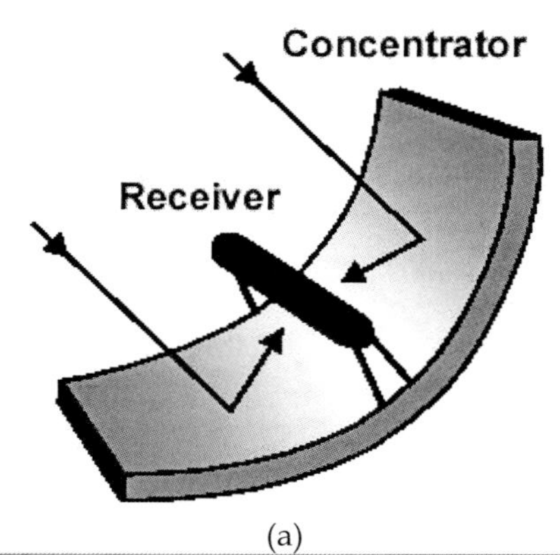

(a)

(b)

Fig. 22. (a) Schematic of a parabolic trough concentrator. Image adapted from Energy Efficiency Renewable Energy Network {Council of Australian Governments, 2006} [63]. (b) Trough concentrator system at the Australian National University, which is designed to incorporate photovoltaic power generation or water heating and steam production. (Image courtesy of the Centre for Sustainable Energy systems, Australian National University, {Wyld Group, 2009} [64].

It is remarkably simple and consists of many mirrors that are oriented towards the sun so that they can all focus the reflected beams on something that exists at the top of a tower. Usually this latter is a boiler containing water, which boils as a result of the sun's light, the steam being led to a conventional steam turbine. The electricity producing machinery is held underneath the tower.

At present, 2010, about ½ of the practical solar energy in use (largely in North Africa) uses this solar thermal method.

3.4 Photovoltaics

This is the second most well known method for converting solar light to electricity; the heart of the method is two slabs of (often) silicon. One is called the p type and the other the n type.
The electron concentrations in the n type are high and the p type low. What fills the p type is called holes, a puzzling name but what it means is sites where there are no electrons.
When a beam of light falls upon these couples as they are called, a potential difference is created between the two sides, the n and the p. It is a low potential, about 0.6 volts at open circuit.
A great number of solar cells have to be connected in series to give power to a hypothetical grid that has been tested out by some trials in California but has not yet been commercialized.
The efficiency of the collection of light is only 16% for relatively big cells (the biggest cell is only about eight inches in diameter) but efficiencies can be obtained with small cells created under careful and clean laboratory conditions, together with some kind of roughening technique so that the incoming light is absorbed and reflected light absorbed again. Efficiencies may be increased.
There are big ideas for developing photovoltaics, thus there has been published a plan where the company claimed to be able to paint a photovoltaic onto a big surface, for example a sheet of aluminum foil. A relatively gigantic sheet of this foil, duly covered with a photovoltaic was supposed to hang on the side of buildings in towns {Popular Science, 2007} [65].
The efficiencies of collection from such devices using photovoltaic materials which avoid silicon was found to raise the efficiency of collection, although the report which described this did not publish an efficiency figure for the conversion of light to electricity.
The commercial income from such a production would be so great that the realization of many of the things which had been promised is almost certain to be realized in the next few years. In fact, the idea of "painting" a photovoltaic onto the aluminum foil was suggested in the 1980s.
The commercializations of Nano-Solar's new techniques have been slow and some German academics have denounced Nano-Solar propositions as impractical.

3.5 OTEC (Ocean Thermal Energy Conversion)

The basic idea of this method, which although not yet built on a large scale, is often referred to as the French physicist Arsene d'Arsonoval suggested the most impressive of the solar energy converters at the end of the 19th century.
It depends on the temperature of the surface of the tropical sea that is usually more than 2 5° C and the temperature at the bottom of the ocean that is generally about 4-5° C. It follows that we have around 20° C difference to work with and the physicist's idea was that this was enough to run a heat engine with a low boiling point working fluids (liquid ammonia). A long tube lowered from a floating platform on the sea surface and a pump is used to draw up the very cold water from the sea bottom {A. Aponte} [66]. On the surface of the platform, one would have hot (25-30 °) and cold (4-5°) water. A heat engine can thus be run.

The efficiency of energy conversion is very low, about 4 percent, but this is not important for we have an almost infinite amount of heat from the surface of the sea, and cold from the sea bottom. (Figure 23.)
There is another version of OTEC and this does not use the working fluid ammonia, but simply evaporates and condenses water by means of pressure changes. Water is made to boil by lowering the pressure.

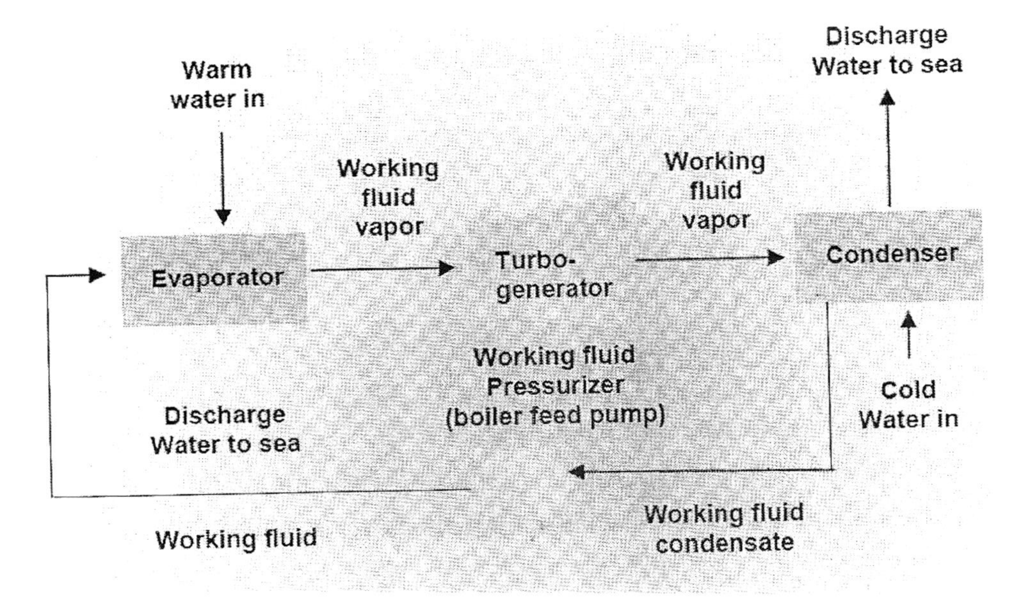

Fig. 23. Schematic of Closed-cycle OTEC system. Closed cycle OTEC Schematic-Ocean Thermal Energy Conversion: Possible application in Certain Pacific Island Nations, a{Alicia Altagracia Aponte} [66].

This second version has the advantage of producing fresh water from seawater.
The potential production of fresh water from seawater on a large scale is financially important. The OTEC machinery needs help when it comes to financing because it is about twice more than other methods (per kw).
However, OTEC would work 24 hours per day because the sea surface is hardly affected by night.

3.6 A new approach to the conversion of light to electricity

It is admitted by all workers in the field the chances that photovoltaics could be really much reduced in cost and increased in efficiency were not very great. It is true that if one wants to go to very tiny cells, then the efficiency is greatly increased past 16 percent right up to more than 32 percent, but the cells were too small to be of commercial value.
In 2006 a new star appeared on the horizon. The name of the company is Nano-Solar, and from its announcement and from the records we have of people who have visited the company, and seen what they do, there's hope that they are making a revolution in photovoltaics seems quite right.

What I am able to tell you here is only from material I have from the year 2007, which is when they admitted a visitor from outside to their headquarters, which was then in the Silicon Valley, and he told the following:
One enormous change which Nano-Solar seems to have been able to engineer has come from a suggestion that was made long ago by John Appleby. John, - and indeed perhaps others, - have always thought about for a long time about the possibility that they could "paint" the photovoltaic onto some other fabric or metal sheet such as aluminum foil, and stretch this foil out having great dimensions so that all talk of centimeters and small numbers would be abandoned and great areas were to be covered with photovoltaic which would then be able to be exposed, e.g., to the sides of houses or factories in towns.
Nano-Solar claimed that they had done this essentially but they also brought to light a new photovoltaic concept by mixing four different photovoltaics together as part of that material which they paint on to the aluminum foil. Obviously this achievement makes a revolutionary difference to the possibilities of using photovoltaics, and instead of having tiny cells, which many thousands had to be used to speak of powering a town, there's now every possibility that photovoltaics could be used in the future developments for such things as powering towns and cities, - so long of course refinement was growing at that time or for that application.
Now, the development of this company has been delayed and this comes out of two causes.
First of all, the company in the fact that it has very large private backing. So there is no problem in respect to monetary support.
Secondly, however, a number of academics, particularly in Germany, have made criticisms of Nano-Solar saying that their technology can't work and it will be unsuccessful when developed to full scale.
One of the problems of reporting, as I am trying to do, progress with Nano-Solar is that they are loath to answer correspondence. They say that they will only answer correspondence, "to their advantage," which I suppose means that when they think that the correspondence will lead to financial gain and advantage for the company, they answer, otherwise they ignore.
Of course, for a person who wants to report the latest progress about the company is very difficult because direct questions such as what is the efficiency of the new photovoltaic are not directly answered, or the price of a kilowatt-hour of electricity in a certain location are not answered.
Therefore, at this time, I can tell you only that the company has already spread out and have headquarters now in Switzerland and having offices in Germany and the United States.
My own perception of Nano-Solar is that it is far the most idea-based, fruitful, and the most ideal development in photovoltaics, and my opinion is that the progress that will be made in the next ten years, at a practical level, from Nano-Solar is very probable and very great.

4. Geothermal: developed to give large amounts of clean energy?

There is an aspect of the earth's heat energy and this might make an important contribution to the clean heat energy we need. As "hot rock geothermal" the idea has been known since the early sixties, but no large-scale plant has yet been built, although recently small exploratory plants have been built in France and Germany and a large plant, worth several nuclear reactor is being built in Australia.
Basically, one focuses attention on flat parts of the earth and builds therein bores into the earth to depths of a few kilometers until one meets heat enough to boil water.

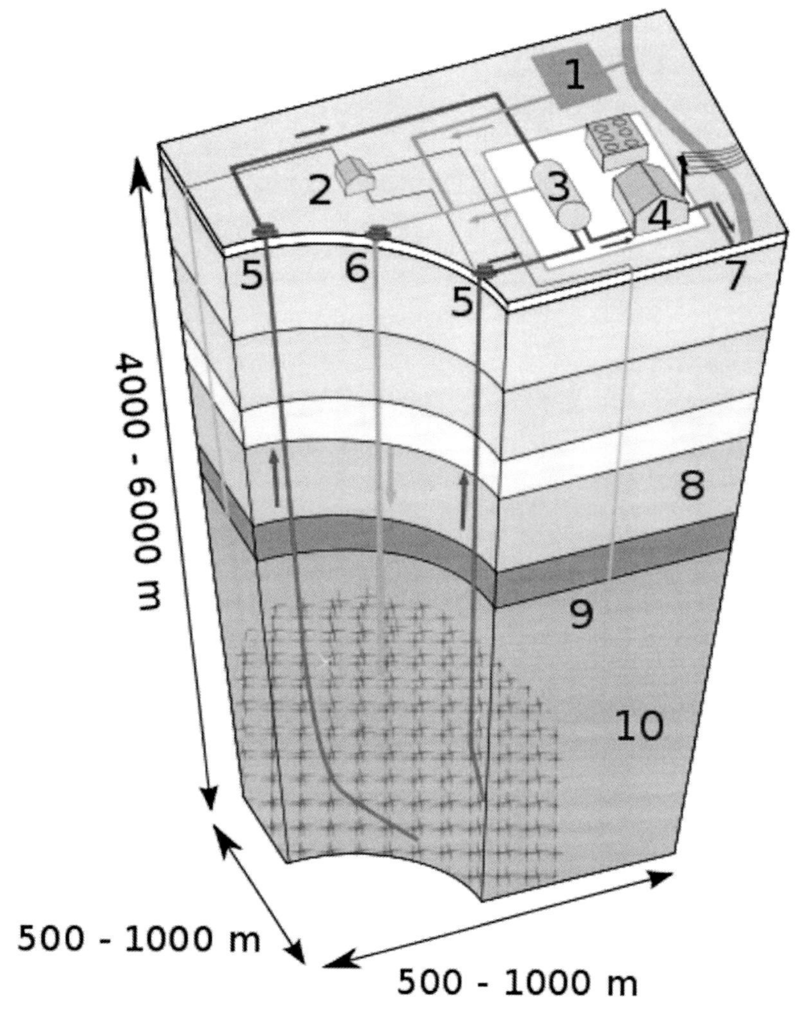

Fig. 24. Diagram of EGS with numeric labels. 1:Reservoir 2:Pump house 3:Heat exchanger 4:Turbine hall 5:Production well 6:Injection well 7:Hot water to district heating 8:Porous rock 9:Well 10:Solid bedrock. 2009-10-24 13:49 (UTC) Geothermie_Prinzip01.jpg Geothermie_Prinzip.svg: Geothermie_Prinzip01.jpg: "Siemens Pressebild" http://www.siemens.com derivative work: FischX (talk) Geothermie_Prinzip01.jpg: "Siemens Pressebild" http://www.siemens.com derivative work: Ytrottier This file is licensed under the Creative Commons Attribution-Share Alike 3.0 Unported license. You are free: **to share** – to copy, distribute and transmit the work; **to remix** – to adapt the work Under the following conditions: **attribution** – You must attribute the work in the manner specified by the author or licensor (but not in any way that suggests that they endorse you or your use of the work). **share alike** – If you alter, transform, or build upon this work, you may distribute the resulting work only under the same or similar license to this one [67].

Then, one injects cold water down this bore, one can expect to receive electricity-generating steam.
Does the bore cool down until eventually one has to rest it? Some designs suggest this, and then work with a twin bore that takes over for a few years. Eventually, this second bore will cool down and heat from the first bore (duly hot again) is used for a cycle.
A recent report has been made on Enhanced Geothermal at MIT.
The advanced Swiss plans only have one bore, and circulate a pipe leading from it inside the hot zone. This brings the entering water up to the requisite steam in this pipe alone, steam is led back and up to the surface and on to the electricity producing machinery, later as "cold" water, it goes back into a bore.
Literature on these geothermal methods describes the fact that when the cold water strikes the bottom of the bore, it may split the rock, expose a further hot region.

Current EGS projects							
Project	**Type**	**Country**	**Size (MW)**	**Plant Type**	**Depth (km)**	**Developer**	**Status**
Soultz	R&D	France (EU)	1.5	Binary	4.2	ENGINE	Operational
Desert Peak	R&D	United States	11–50	Binary		DOE, Ormat, GeothermEx	Development
Landau	Commercial	Germany (EU)	3	Binary	3.3	?	Operational
Paralana (Phase 1)	Commercial	Australia	7–30	Binary	4.1	Petratherm	Drilling
Cooper Basin	Commercial	Australia	250–500	Kalina	4.3	Geodynamics	Drilling
The Geysers	Demonstration	United States	(Unknown)	Flash	3.5 – 3.8	AltaRock Energy, NCPA	Fundraising (Mar 2010) [97]
Bend, Oregon	Demonstration	United States	(Unknown)			AltaRock Energy, Davenport Power	Permitting (Mar 2010) [98]
Ogachi	R&D	Japan	(Unknown)		1.0 – 1.1		CO_2 experiments [99]
United Downs, Redruth	Commercial	United Kingdom	10 MW	Binary	4.5	Geothermal Engineering Ltd	Fundraising [100]
Eden Project	Commercial	United Kingdom	3 MW	Binary	3–4	EGS Energy Ltd.	Fundraising [101]

Table 4. Current enhanced geothermal projects {2009} [68]

4.1 Hydro and tidal

Hydro resources are well known and already widely developed. One thinks of Niagara Falls. There are many falls of this kind around the world, but some are too far from cities where the energy is needed.

Country	Location	Tidal height (m)	Power (thousands of MW'S)
Argentina	San Jose	5.9	6.8
Australia	Cobequin	12.4	5.3
Canada	Cumberland Shepody	10.9 10.0	1.4 1.8
India	Cambey	6.8	7.0
UK	Severn	15.0	8.6
US	Knich Arm Turnequin Mezen	7.5 7.5 9.1	2.9 6.5 15.
Russia	Tuger Penzhmskaya Bay	- -	7.0 50.0

http://en.wikipedia.org/wiki/Tidal_power, October 2007. [69]

Table 5. Some larger tidal power schemes under consideration around the world

Another water resource is the tides. In Table 5 above, is a list of 12 places around the world that had been judged particularly suitable for tidal technology. (See Table 5.)

The key quantity which tidal technology depends upon is the height of the tide. Four meters is minimal for a tide to have an economically worthwhile character to put in the necessary engineering work.

It must not be thought that we are limited to places in the world that are ideal for tides. A suitable spot only has to have a tide of more than 4 meters. An appropriate inlet and a place behind the inlet which can be made easily into a basin-like receptacle for the incoming water, to be held there until the basin is full at high tide, and then released after the tide has gone out, whereupon the energy is converted to electricity from the turbines which are being activated by the flowing water. Incoming water during rising tide can also be used but the usable power of such tides is less than those available during the tidal outflow.

4.2 Would nuclear energy save us from global warming?

4.2.1 Fission reactors

The domes which one sees around the country and the cooling towers comprise nuclear reactors and they work upon a process called fission. There is much drama behind the origin or the process of fission, a nuclear reaction of the type unknown before 1939. The fact that the discovery was made just before the beginning of WWII, made it all the more important, although everything to do with it (as I well remember being in England at the time), was deeply secret.

However, it's indisputable that the discovery of nuclear reactions was made by three Germans, Hahn, Meitner and Strassman. Moreover, these three, who were working in Berlin, were not seeking anything to do with nuclear energy. Uranium was a very heavy metal and the idea which Hahn et al had was that they would like to see if by bombarding it with neutrons, they could make a still heavier element, i.e. add something to the atomic weight.

Indeed, they did bombard uranium with neutrons but they found a result totally unexpected, in those days very peculiar, and the reaction that has altered the world and still does.

What they found was that the result of bombarding the uranium was to produce two other atoms, one called barium and the other a rare gas, called krypton. But, another thing was discovered, and turned out when the use was made militarily of the discovery, to be the key point: although they bombarded one neutron per uranium, they got back the strangely broken up atoms, and three neutrons.

Now the fission reaction which they found and which is now the basis of those towers that we see can be written in a chemical way as follows:

$$^{235}U_{92} + {}_0n \rightarrow {}_{139}Ba + {}^{94}Kr + 3{}_0n \quad (2)$$

One of the most interesting things about the reaction was indeed the three neutrons, because, as those interested in an explosion rather than the peaceful use to provide heat only, saw it, the three neutrons from one could give rise to a spreading reaction of great force because after hitting a uranium atom with one neutron it could then strike 3, 6, 9, etc, and all this would happen in a very short time, providing a super great amount of heat.

But, every one of these reactions of the neutron with uranium was found to be remarkable in another way, instead of producing the heavier element that they sought, the researchers found that they had got two elements instead of one and this was indeed, therefore, a nuclear reaction.[6]

$^{235}U_{92}$ an isotope of uranium was found in the mines when the researchers sought the kind of uranium they wanted. It turned out that there were two kinds of uranium. One had the atomic weight of 235 and this was rare, but the majority of the uranium atoms were of the isotope with the atomic weight of 238 and this was found to be non-fissile, i.e. it would not take part in the nuclear reaction which was found possible with U235.

As most people know, uranium is the heart of the atom bomb and of the peaceful use of nuclear reactions. In this short account, I'm going to neglect the military side completely (it is written up in dozens of books).

Great excitement attended the realization that we now had the ability of generating in a single nuclear reaction using the U_{235} an amount of heat which was of the order of magnitude 1 million times more than we were used to observing in a chemical, i.e. non-nuclear reaction.

For a few years (as I remember) there was a kind of feeling that a great climax had occurred and that all the future energies of the world was taken care of. People used to say "don't bother to turn off the lights in the future, energy will be so cheap" Where the doubts began is when calculations came out about how much uranium was in the ground, and could be

[6] Rutherford at Cambridge in England in 1919 had claimed that he and his coworkers had "split the atom" an early name for a nuclear reaction. His achievement was recognized at the time in these terms but the great excitement which came with the Hahn work in Germany was the three neutrons from one and the realization that it could give rise to an explosion of previously unrecognized force; and also (but latterly) it might, if tamed, be a very convenient source of a great deal of peaceful energy.

recovered and the 235 kind of uranium extracted and how much would have to be rejected, namely the other isotope, the U_{235} kind. It seemed that counting only the USA, the supplies might last 100 years. As time has gone on, and other nations have found out how to do these reactions, and their enormous value, it's quite clear that the 100 years is by no means enough for jubilation when you understand that Russia, at least, and even India and China with their enormous populations, will all want to use the rather rare U_{235} which is discovered to be active and take part in the basic reaction stated above.

Soon after the realization that there was not enough uranium to supply the world for a significant time (several hundred years) an idea was put forward which might solve the problem. Nuclear reactors (which have now been manufactured in numerous countries) take about twelve years each to build. If this astonishes the reader, it must be recalled that a nuclear reactor is not like some normal piece of machinery you could house in a factory. When one visits a nuclear reactor one sees a large spread of land covered with buildings, ending up with the famous dome and the cooling towers.

Apart from what you see, there are of course, in the building stage, many things done to protect the workers from radiation and to minimize the possibilities of an accident which could even cause an explosion.[7]

So let us leave the fission aspect of nuclear reactions here and now. We can say that it has been a success (the only part of nuclear science so far which has been completely successful) but it simply will not do for the further future.[8]

4.2.2 Breeding

It has been clear from what is stated above that when the U_{235} is extracted from the natural uranium found in the ground, a great deal of uranium remains over. Of course this is the non explosive U238 and it didn't take long for nuclear physicists to see that there might be a way of converting the inactive U238 to undergo conversion which could lead to a series of nuclear reactions, and finally to a stable (explosive) isotope of great amount.

The suggested process of breeding was to start with plentiful U238 and from it to form U239.

The U239 then decays spontaneously to Np239 + ß

After this, there is still a third reaction in the sequence in which Np239 becomes Pu239, namely plutonium, and this is fissile.

Thus, when these thoughts were published or became well known, it was thought the problem of our energy needs had been solved. The U238 was in abundance, in wastes that came about when the U235 was extracted, so that there was now no problem about having plenty of uranium. The above reaction to form plutonium had indeed to be carried out, but there was no more problem with the amount of Uranium and therefore for a brief time,

[7] Until after 2000, it was always feared that a meltdown would occur and that a peaceful heat producing reaction will become a menace and perhaps even kill large numbers of people. We only know of one such incident, that at Chernobyl but the nuclear fire and partial explosion which occurred there has caused thousands of death in Russia itself, and damage to people in most parts of the northern hemisphere.

[8] For example in order to supply the USA alone with nuclear reactors, we would have to build 1,800 of them at 12 years each. Of course I understand several can be built simultaneously but to supply our own country with them whilst other countries are also building them, and planning to use the same uranium, simply makes the idea of a nuclear-fission pathway impractical.

there was jubilation again, and again thought that the great problem of the future of energy was a battle won.
Unfortunately however, the question of the efficiency of these reactions of the breeding process had not been taken into account. The early, happy workers thought that these reactions should be carried out "round about" 100% efficient. So, if you knew how much U238 there was in the world, it was simple to calculate how much plutonium could be made, and consequently, how much nuclear energy we would have for our energy supply.
The history of the breeding process has been a series of disappointments. At first the efficiency was about 1%, and over the years with French leadership, the amount of efficiency has grown to 5%. But it gives us hope (that this could be improved to 10-20%, and if worse comes to worse, we will have nuclear energy for at least 200 years more).

4.2.3 Fusion

When it was known that the efficiency of the breeding process was so low, nuclear scientists turned towards another idea that, as we shall see, demanded of remarkable new technology. Thus, what we have been discussing is nuclear fission energy, and what I am now going, very briefly, to survey is fusion, namely the idea that two atoms when brought together with sufficient force, can fuse together instead of breaking up, and if this is done in a nuclear, rather than a chemical way, then great energies could also be released and finally might be tamed in such a way that they could be used in peaceful ways.
The first basic idea here was to use two of the second isotope of hydrogen, mainly deuterium. Calculations showed that if it were possible to raise the temperature of deuterium atoms to some previously unrealized temperature, namely that of the sun (!), then this tremendous heat would cause the two deuterium atoms to fuse together and form the rare gas helium.
Again, there was jubilation, although it was realized even at the beginning that two tasks faced the engineers. One was to obtain the colossal heat and temperature, but the 2nd was the greater difficulty, how could you contain it?
It seemed ludicrous to suspect you could contain something inside another something that was so hot that it would be a greater temperature than the sun itself.
Nuclear scientists are both brave, and daring. They did indeed advance an idea that seemed, at first, to be the solution. Yes, it was not possible for anything to touch the reacting particles, but how about a magnetic field?
Magnetic fields don't break things down but with the atoms in the reaction being in the form of ions, a strong magnetic field engineered to have the shape of a bottle, will theoretically, contain the deuterium ions and when they struck each other, in collisions, at the enormous temperatures we are speaking about (10^{8} ^{0} K) then surely there would be helium produced and a great deal of heat.
You probably understand by now that I'm going to tell you that it didn't work, and with the space available I can only tell you that it was never successful to keep the deuterium inside the "bottle", it always leaked out.
Russians, the US, the British all tried for years to make this work, and the Russians even gave it a name: TOKAMAK, which is the Russian name for bottle, reversed.
Undaunted, the nuclear scientists came out with another idea, maybe even more fantastic than the first. This time they mixed together small amounts of deuterium and tritium and kept them at very low temperatures so that they remained frozen.

Imagine now, a small tower on top of which sits the tiny sphere of mixed deuterium and tritium. The objective is to make the deuterium and tritium fuse together and produce a new particle.
The theoretical concept was to do this with a laser, and a laser of tremendous power, far greater than anything formerly engineered.[9]
Of course the idea was that when the laser stuck the particles, the force of the collision would make the deuterium and tritium fuse together. It was found, however, that the laser that had been built of sensational power was still not enough to fuse the two atoms.
Another difficulty was found but not solved. We're talking about small spheres of deuterium and tritium and thinking they might power the world??
But the physicists had an idea about this, too. The small sphere was simply a working model and if the laser had been powerful enough to make the atoms fuse in it, the engineers were going to drop particle after particle down from the tower and as each particle dropped the laser would strike again and so if you made 100 drops per minute you would get a very considerable amount of energy (remember we're talking nuclear energy, about 1 million times more than chemical energy)
It's not only we in the USA who are held up and very frustrated by the failure of these attempts to supplement and supply something which could last in nuclear energy for many years.
The real thing which bothers the legislators is the amount of money this is all costing. I think that the sum is not often spoken about but it leaks out to us that if you add the Russian expenditures to our own, more than one billion dollars per year has been expended on research into fusion and most of the people outside the USA believe the best thing to do is to stop wasting money.
Hence I think that I can answer the question with which I began this section of this article, and that is to say that I think that nuclear reactions in the future is a very dicey and unsuccessful way to obtain energy without pollution and most of the thoughts of others is the best thing to do is to stop and think before another billion is spent.

4.3 Mediums of energy

Apart from having a certain energy source (wind, solar, and hot rock geothermal being prominent as the first replacements of fossil fuels) all the renewable energy sources, e.g. wind, may need a medium which can couple with the source to produce energy in a form suitable for households, factories and military.

4.4 Electricity: the principal energy medium

This is the obvious medium for new, clean energy and will serve in most situations. We are all familiar with what is called "the grid" which consists of cables carrying energy from sources where the electricity is produced, e.g. solar, wind, geothermal, and bringing it to places where it is needed.
At present, cables carrying 100,000 volts are used. It is important to realize that the potential that is used in the AC transmission of electricity over long distances in cables decides the length of the cable that must not manifest too much IR drop, - wasted energy.

---

[9] By happenstance, I visited the site where the laser was housed just after being shown to be a failure again. It was a large building, especially built to contain the laser.

Until the 1960's long distance cable lines were run at 30,000 volts, but this has been changed to 100,000 volts. The difference is easily calculated because the heating effects, i.e. energy lost in passing an electric current through a cable is given by the equation:

$$\text{Lost heat} = I^2R \tag{3}$$

Hence, if one increased the E, one decreases the I, for the same power, EI and therefore the energy lost in heat is decreased.
There is a limit to how far this raising of the volts can go, because the AC nature of the electricity means that cables radiate and could provide a health hazard to those sufficiently near them. At present, the limit is 100,000 volts.

4.5 Is room temperature super conductivity a possibility?

Were we to have virtually no resistance in cables, we should be able to send energy unlimited distances without loss of energy.
How far along are we with superconductivity research?
The answer is that unexpected strides have been made in this area, and that coming from a situation in which superconductivity was to be observed only near to the absolute zero of temperature. Superconductivity has become something that is still far from large-scale practical application but there are now situations where the working temperature is above that of liquid nitrogen and might be (economically) usable.
In Table 6 a number of superconductors are portrayed as of the present time, 2010, and it's visible that the substances that has been found to have superconducting properties and to allow the temperature to rise as high as 134 K, are complicated substances.
The one that has the highest temperature, which performs as a superconductor there is:

$$HgBa_2Ca_2Cu_3O_8. \tag{4}$$

There are other fundamental problems in realizing practical super conductivity: thus, if the current passing exceeds a carbon value, the phenomenon appears to fade off.
So, there is a long way to go, but the goal here is so important that we can expect a good deal of National Science Foundation funding.

4.6 Hydrogen: could it be a clean replacement for co_2-producing gasoline?

The clean hydrogen could be a medium of energy was proposed in 1971 {John O'M. Bockris} [71]. At this time it was feared that smog could develop over cities with insufficient winds to clear it. So, one of the solutions suggested was that the medium by which we drive our cars should be changed from gasoline to hydrogen, so automotive exhausts would be changed from the material causing smog to pure water vapor. Further, the use of hydrogen would make fuel cells an immediate source of electricity as fuel cells convert chemical to electrical energy at twice the efficiency of batteries.
Since the early seventies there have been changes that affect the need for hydrogen as a medium. The main one has already been mentioned: the potential in the cables for long distance transmission of electricity has been raised, thus extending the practical use of the cables by lessening the energy lost in heat.
The need for storage of large amounts of electricity increases when we think of supplying cities with, say, solar energy with its six to eight hours availability.

Formula	Notation	T_c (K)	No. of Cu-O planes in unit cell	Crystal structure
$YBa_2Cu_3O_7$	123	92	2	Orthorhombic
$Bi_2Sr_2CuO_6$	Bi-2201	20	1	Tetragonal
$Bi_2Sr_2CaCu_2O_8$	Bi-2212	85	2	Tetragonal
$Bi_2Sr_2Ca_2Cu_3O_6$	Bi-2223	110	3	Tetragonal
$Tl_2Ba_2CuO_6$	Tl-2201	80	1	Tetragonal
$Tl_2Ba_2CaCu_2O_8$	Tl-2212	108	2	Tetragonal
$Tl_2Ba_2Ca_2Cu_3O_{10}$	Tl-2223	125	3	Tetragonal
$TlBa_2Ca_3Cu_4O_{11}$	Tl-1234	122	4	Tetragonal
$HgBa_2CuO_4$	Hg-1201	94	1	Tetragonal
$HgBa_2CaCu_2O_6$	Hg-1212	128	2	Tetragonal
$HgBa_2Ca_2Cu_3O_8$	Hg-1223	134	3	Tetragonal

Origin: Superconductivity, Wikipedia, Free Encyclopedia, 2010.

Table 6. [70] Critical temperature (T_c), crystal structure and lattice constants of some high-T_c superconductors

Here, any plans which will be put into practice to replace gasoline must be obviously non-CO_2 producing, and will include the ones already mentioned, e.g. wind, solar, and enhanced geothermal.
On the other hand, at a given time, and also the wind characteristics so that one need not worry about hours or days of irregularity but it is necessary to have stores for solar energy and wind energy for the big cities, these stores will have to be large.
Here, the virtues of hydrogen (for storage) are attractive. It is easy to produce from electricity, the form in which the solar and wind energy is most immediately available, and so large stores of hydrogen, at the moment, is the main way we hope to overcome the difficulty of transfer and storage of the cheapest of our renewable clean energies, no Global Warming.
A world which is set up to use solar and wind, together with appropriate storage for the big cities, would lead to a world without Global Warming by means of CO_2.
Of course, we look toward to a hope that we will be able to rely upon superconductivity. Here a breakthrough occurred in 1986, when, for the first time, it was possible to prove superconductivity in materials that retained this property above the boiling point of liquid nitrogen, 77 ^{o} K. (See Table 6).

4.7 Approximate estimate of the cost of changing to an inexhaustable energy from fossil fuels

It is when we look at the financial side of the big change, that resistance looms high in one's mind.
The first thing we could do to get over the great tax hump which confronts us in the near future is to reduce the energy per person which is used by American citizens.[10] Certainly,

[10] About twice that used by Europeans (as in e.g., England, France, Italy, et cetera.)

there are now countries in the Middle East where the citizen per person needs are more than 10 KW, the amount that Americans say they need.
In seeking some rationale for aiming our estimate of the renewable energy needed, 6kW is the equivalent power per person we shall assume.[11]

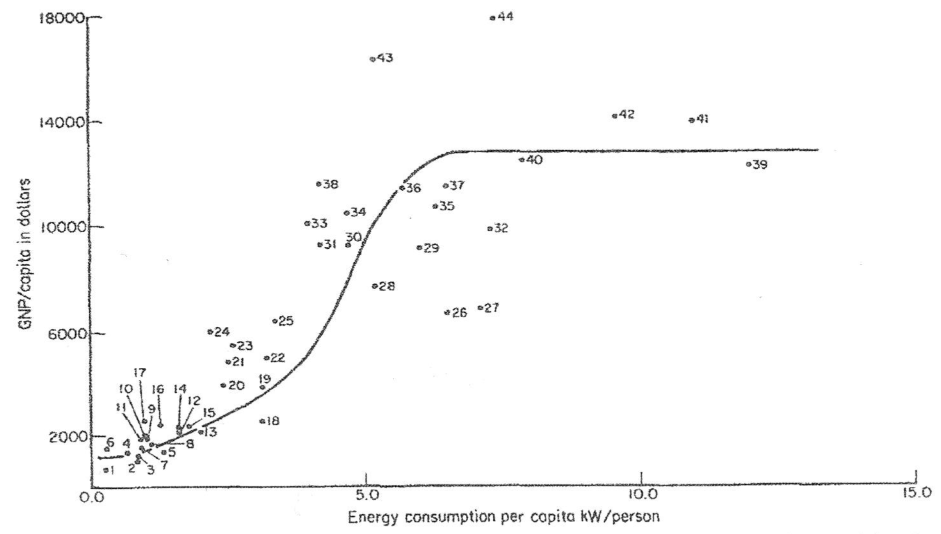

Fig. 10 Energy consumption *per capita* relationship with GNP *per capita*. 1. Ivory Coast, 2. Costa Rica, 3. Turkey, 4. Tunisia, 5. Jamaica, 6. Paraguay, 7. Ecuador, 8. Jordan, 9. Malaysia, 10. Chile, 11. Brazil, 12. South Korea, 13. Argentina, 14. Portugal, 15. Mexico, 16. Algeria, 17. Uruguay, 18. South Africa, 19. Venezuela, 20. Greece, 21. Spain, 22. Ireland, 23. Israel, 24. Hong Kong, 25. Italy, 26. Singapore, 27. Trinidad & Tobago, 28. New Zealand, 29. Belgium, 30. United Kingdom, 31. Austria, 32. Netherlands, 33. Japan, 34. France, 35. Finland, 36. West Germany, 37. Australia, 38. Denmark, 39. Canada, 40. Sweden, 41. Norway, 42. United States, 43. Switzerland, 44. Kuwait.

Fig. 25.{ P. Dandapani, 1987} [72]

With this limiting assumption, and conscious of the energy difficulties that face us, let us try for a very approximate 2010 cost estimate.
We start with a population of 300 million people, i.e. 3.10^8 and we are going towards a 6kW per person economy. This refers to the energy of all functions of the civilization, including for the USA, the heaviest items expenditure are on military operations, twice the per head expenditure of citizens in the main European powers.
What is the average cost per kw of wind or solar energy that, on average, would supply energy at the rate of 1kW. The amount varies from estimate to estimate, but on the whole, $5,000 per kW is a median value. Thus, the value for the USA would be: **3.10^8 6.5000.**
This is $9 trillion.

[11] Comparison with the income and living standard of other nations, an interesting result arises. It appears that until around 6kW per person, the increase in living standards increases exponentially with increase in income. However, around 6 kW, there is no further increase in living standard. This presents a big question in Sociology.

Over what time would we have to pay this very large cost? Here, it's going to only be possible to make an arbitrary assumption that we could pay it over fifty years. Taxation could be used to discourage the population from using CO_2 producing energy and encourage them in the direction in the new CO_2 free energy.
The cost of the 9 trillion will sink to 0.18 trillion per year or 180 billion per year if paid over fifty years.
Sums as large as this are difficult to comprehend, but it may be helpful to know that we spend $900 billion per year (four times more) operating our armed forces.

4.8 The cost of hydrogen as an energy storage medium

In some cases, sources of hydrogen will originate away from the place where the energy is needed. Further, if it comes from wind and solar, the sources will be from storage systems (although if we introduce enhanced geothermal the supply will be stable).
The principal ancillary costs of storage (1.70 / GJ) transportation of the energy (3.00 / GJ) and finally, distribution. By "distribution" Tappan Bose and Malbrunot charge 15.00 / GJ {2006} [73]. This latter cost seems high even if the main cost of distributing the hydrogen in the form of electricity is passing through a fuel cell and assuming an efficiency of 50 percent. This will cost around $9.60 / GJ to get the hydrogen after storage back to electricity.
To obtain the cost of raw hydrogen, the after costs of which we are discussing, let us start by taking $22 / GJ as the cost of hydrogen from wind energy by means of the electrolysis of water at room temperature.[12]
Thus, with this value for the raw hydrogen, the cost of electricity of stored hydrogen at distance from the source would be about $37.00 / GJ.

4.9 The cost of liquifying hydrogen

The attitude taken by most to liquefying gaseous hydrogen is that it will be too expensive, because liquefaction of such low temperature needed is inefficient in a Carnot sense. The hydrogen boiling point is 20.28°K {E. Wiberg, N. Wiberg, et al, 2001} [74].
Now, Tappan Bose and Malbrunot have come up with a different view. They point out that the cost of liquefying hydrogen is not so out of reach when one considers the comparison should be made not with raw hydrogen from the plant but delivered hydrogen which the French Canadians gives as $40-$48/GJ. Thus, using the liquid saves several things, and these are transportation costs, and of course, there is no need for compression, storage and use of the fuel cell.
There are several costs arrived at by Tapan Bose and Malbrunot {2006} [73] and the ones with which we are going to use as a benchmark is that for a GJ of gaseous hydrogen, - $48?GJ. (Compare the known cost of gaseous hydrogen, raw, at the electrolyzer of $20, - the range of cost goes from $16 to $26 depending on the temperature of the electrolysis.)

[12] The older means of obtaining hydrogen from this system reforming of natural gas is no longer admissible if we are going to ban CO_2 from entering our atmosphere, we cannot use these low cost methods of producing hydrogen and must resort to electricity. The cost of this is a longer story, but optimistic figures have been given by the wind energy association of America (.02c /kWh) and by the group that has sent helicopters up to 15,000 feet to milk the winds there. (.02c /kWh)

Conversion machinery producing electricity and hydrogen. Paid over fifty years.	9. 10^{12} 300 10^9/year	Conversion machinery assumed built over twenty-year period. If capital cost paid at same rate, cost would be \$250.10^9 per year (about ½ the cost of the U.S. Military budget).
Raw Hydrogen from Methane (CO_2-free).	\$9.50/GJ	Process described in reference [75].
Electrolysis, raw, at plant.	\$14.50/GJ	Cost of electricity assumed (2008) is \$.03 c.kWh, \$.02ckWh, tested. \$.04 c/kWh from Nano-Solar [76]
Electrolysis, raw, at plant, 1000°C	\$12.26/GJ	Uses $U_3O_8Y_2O_3$, membrane, Bevan, [77].
Ancillary costs of storage, transport, and delivery, (after electrolysis).	\$25.00/GJ	Involves storage, transfer, and delivery.
Liquid H_2 (including cost of electrolysis).	\$51.00/GJ	This is 25 percent increase in passing from gaseous to liquid is less than that imagined.

Table 7.

4.10 Hydrogen would be a dangerous fuel to handle

Hydrogen is a dangerous fuel, but the degree of danger has to be compared with that of a reasonable alternative, natural gas.

What is different with hydrogen that makes it more dangerous than natural gas that the mixture of hydrogen and air becomes explosive over a wider range of compositions than with natural gas.

Thus, one can imagine a practical example of hydrogen leaking out into an enclosed space, such as a garage, versus natural gas in the same situation. Here, the leaking hydrogen will be more dangerous than the leaking natural gas because, the garage atmosphere will become explosive, far more easily with natural gas.

These dangers may be lessened by the fact that the power of the hydrogen explosion is 4 times less than that of a natural gas.

Another aspect of the hydrogen versus natural gas comparison is that the burning of hydrogen in the air is a straightforward matter of the burning gas going upwards (see Figure 26). On the other hand, a car on fire with gasoline is extremely dangerous with the fire spreading and many dangerous vapors of organic compounds that are being consumed by the burning gas. The appearance of the car undergoing a natural gas explosion versus a car undergoing a hydrogen explosion, is impressively in favor of the hydrogen.

4.11 The so-called "liquid hydrogen" {G. Olah, et al, 2006} [78]

Hydrogen seemed the number one solution as a medium to some of our pollution problems and those who support this idea may be excited to know that Global Warming is attributed to automotive exhaust gases, another strong indication in favor of the use of hydrogen as an automotive fuel (with lower cost).

Fig 26. Tapan Bose and Pierre Malbrunot, et al, Hydrogen: Facing the Energy Challenge of the 21st Century, John Libby Eurotext, UK, December 2006, p.59.

A large-scale use of a Hydrogen Economy has grown as indicated by the size of the International Journal of Hydrogen Energy. In the early 1970's a single thin volume every two months, the journal is a signal of its use but now in 201, it is published twice per month in thick issues.

Although the cost of making a GJ of hydrogen from water by means of electrolysis from wind is reasonable and at room temperature is about $22.00 per GJ, this leaves out several steps that would have to be accepted by anyone who uses hydrogen in a practical situation.

For one thing, hydrogen is a gas and has to be stored, piped and transmitted and reconverted to electricity.

The total of these additional costs on top of what the electrolyzer gave, means as much as $40.00 / GJ, or in Tappan, Bose & Malbrunot, $48.00/GJ.

4.12 Should "liquid hydrogen" be cheaper?

Olah suggested [78] "Liquid Hydrogen" as a nickname for methanol, but this does not deal with the most important point of going to hydrogen. It does not form CO_2 pollution.

The content of a suggestion which may solve the hydrogen cost problem comes out of a development of Olah's idea of a methanol economy but has within it a significant difference and this is what I wish to represent here.

Thus, the methanol economy as written by Olah and colleagues {2006} [78] gives helpful information about the properties of methanol as a medium of energy (Table 8). Thus, storage and transport of methanol would be little different from what the world uses in its treatment of gasoline.

Transportation, too, would no longer need new cars or a new infrastructure!

In fact, replacing gasoline with methanol would allow us to continue our present economy with little difference. However, there is one thing missing: how can we use methanol as a medium of energy if it would still cause Global Warming?

Property	**Electricity**	**Methanol**	**H_2 Liquid**	**H_2 Gas**
Methods of preparation	Photovoltaic; or heat engine, et cetera.	Photosynthetic; or CO_2 from rocks + H_2 from water.	heat $H_2O \rightarrow H_2$ Elec Liq. N_2 $\rightarrow H_2$ (Low T) Expansion $\rightarrow H_2$ (liquid)	heat $H_2O \rightarrow H_2$ elec
Mixes with water	Not applicable	Complex; but in gasoline forms two immiscible layers if water present	Not applicable	Not applicable
Corrosion	Zero	Significant problem	Zero	Zero
Flame speed Flame temperature	Not applicable Not applicable	2900°C	306 cm sec^{-1} 2050°C	306 cm sec^{-1} 2050°C
Luminosity	Not applicable	Fair	Poor	Poor
Production of pollutants on combustion	Zero	CO + Aldehydes worse than gasoline ~ NOX worse than H_2	Zero	Zero
Use in fuel cell	Not applicable	Poor compared with H_2 better than oil	The best	The best
Compatible present IC Engine	Not applicable	Good. Some redesign necessary	Good. Fuel injection needed	Good. Gas storage >300 miles ok Li cells
Storage	Difficult in large amounts	Easy	Liquefaction costs $2-$3 per MBTU	Compressed gas in tank.
Transmission	Too expensive >1000 km	Costs slightly less than h2 in pipeline	Costs 25%> methanol	0.2 cents per 1000 km
Biological hazard	Safety preventions well practiced	Toxic; air pollution caused by large spills	Zero	Zero
Consumer acceptance before facts realized	Excellent	Very good	Poor	Poor

Table 8. electricity, methanol, and hydrogen compared as fuels [79]

Consider the formula of methanol, CH_3OH, it can be found from:

$$3H_2 + CO_2 \rightarrow CH_3OH + H_2O \quad (5)$$

Instead of making methanol with ordinary CO_2 and hydrogen we take the trouble to get the CO_2 we need firstly from the atmosphere. If we avoid momentarily the problem of how to get the CO_2 from the atmosphere in large amounts, then we can combine H_2 and CO_2 directly to form methanol {S. Ono, et al, 1986; I. Yasudaa, U. Shiraski, 2007}[80, 81, 82]. This is a process that has been worked on in Japan.
Methanol formed via CO_2 from the atmosphere produces no net greenhouse warming because although when we burn it to produce energy it does inject CO_2 into the atmosphere, we already got CO_2 in the methanol from the atmosphere so no extra CO_2 enters the atmosphere when we burn methanol created with in from the atmosphere.
Hence, there would be NO increase in Global Warming in a methanol economy if the one great exception towards what Olah said is made, that the CO_2 that is part of the makeup of the methanol, comes from the atmosphere itself.
Let us count the advantages that would occur if we did have $methanol_{at.}$
As far as transportation is concerned, we would go to a different gas tank and pour this special methanol into their cars rather than gasoline. Over a period of, say, fifteen years, the whole country would be converted and methanol would become a general medium of energy, and the problem of Global Warming would have been solved.
Another advantage is that we would not necessarily have to change our manufacturing. We could go on with our present fleet of cars, but now run them on methanol made from the atmosphere. There would be no rebuilding of the infrastructure. Of course, we firstly have to obtain CO_2 from the atmosphere.

4.13 Methods for obtaining CO_2 from the atmosphere

So far, in this account of "liquid hydrogen" we have stated the virtues of what would happen, were we to have methanol formed with CO_2 from the atmosphere. A Methanol Economy with the methanol from the atmosphere now will be like having hydrogen with the difference of no longer dealing with a gas, having to store it, transport it, reconvert it to methanol and use that more or less as we use gasoline.
The first problem, then, is to collect the wind and devise how to bring a large stream of air to the machine, and one of the answers which comes to mind is to figure on (admittedly a supposition) that there will be a good deal of energy made from wind in our future.
The next thing is to suggest that the wind that you wish to collect will come from a stream of wind to electricity generator in a wind belt.
Now then, suppose the wind sweeps through the wind generator, does its kinetic work there, and sweeps on at present it's just allowed to dissipate itself in the air behind it and has no further purpose.

WIND GENERATOR METHOD

We would collect the wind behind the wind generator in a wide mouthed tube, decreasing the diameter of the tube, until it's down to say 5'.
In this still a very wide tube, into which we put highly powdered magnesium oxide. We heat this MgO at 350° C in the tube containing the oxide. We keep the powdered magnesium oxide in small particles, not filling the tube, but when the wind comes through it, there will

be good contact between the magnesium oxide particles and the wind. At this appropriate temperature, a combination will occur and magnesium carbonate will result.
Of course, we have to do experiments and find out how long the tube has to be to get say 90% reaction of CO_2 and MgO, and furthermore, what should be the minimal temperature for a 95 percent dissociation (with catalyst).
What we are planning is a batch process and at the end of the first period, the flow of the air is suspended, and the magnesium carbonate now in the tube, is heated to more than 700° C. The magnesium carbonate breaks down and goes back to oxide, and a result of this is that CO_2 is produced, and is in a stream which is what we need, and can be piped off to a side circuit where it is brought into contact with a storchiometric amount of hydrogen.

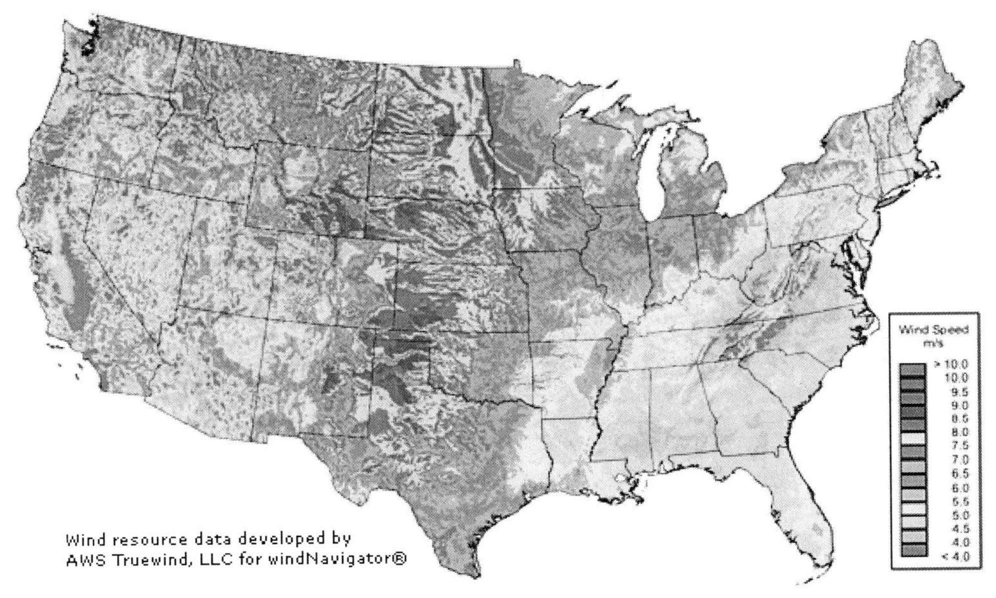

Fig. 27. WIND RESOURCE MAPT OF USA [83] United States and State – 80-Meter Wind Resource Maps

4.14 Zeroeth aproximation calculations by Dr. Rey Sidik [84]: Methanol from the Atmosphere

"I followed your guidelines in carrying out the following calculations.
So, the question is;
How does the cost of 1Gj of CH3OH per Eq. [1] compare to the cost of 1GJ of H_2 (including storage +transportation+delivery costs) ?
Let's collect the thermodynamic data for the chemicals [CRC Handbook of Physics & Chemistry, 1991]:
CO_2 + 3 H_2 = CH3OH(liq.) + H_2O(liq.) [1]
at standard state:
Kcal/mol
del.G -94.25 0 -39.76 -56.68

del.H -94.05 0 -57.04 -68.31
del.G for reaction = (-56.68 -39.76) - (-94.25) = - 2.19 Kcal/mol = - 9 Kj/mol
del.H for reaction = (-68.31 -57.04) - (-94.05) = - 31.3 Kcal/mol = - 131 Kj/mol
So the CO_2 conversion reaction is exothermic and spontaneous at room temp. This heat was Not used in the following calculation.
Now, let's find out how many moles of CH3OH gives us 1GJ of heat energy:
CH3OH(liq.) + 3/2(O2) = CO_2 + 2 H_2O(liq.) [2]
del.H -57.04 0 -94.05 -68.31
del.H for reaction = 2(-68.31) -94.05 + 57.04 = - 173.63 Kcal/mol = - 726 Kj/mol
1 GJ /[726x10^(-6)] = 1377 moles of CH3OH
But, to produce 1 mole of CH3OH we need 3 moles of H_2.
Thus, 1 GJ of methanol needs 3x1377 = 4132 moles of H_2.
Since 1GJ of H_2 is equivalent to 3499 moles of H_2 {1GJ/[285.81 kJ/mol x10^(-6) = 3499 moles H_2},
to produce 1GJ of methanol, we need 4132/3499 = 1.18 GJ of H_2.
Thus, as a zeroth approximation, 1Gj of methanol needs 1.2 GJ of H_2 and 1377 moles of CO_2.
By the way, 1 GJ of methanol = 1377 moles x 32 g/mol = 44 Kg/density = 56 liter = 15 gallon
Now, let's calculate the air volume and diameter of the cylinder (cawl) just after the windmill that are required to CAPTURE 1377 moles of CO_2 if the wind blows at 20 mph:
CO_2 concentration in the air is 0.037%v, using PV=nRT, n=1.5x10^(-5) moles/liter,
at 100% capture efficiency, we need an air volume of 1377/n ~= 92000 cubic meter.
A wind of 20 mph travels 20x1.6/12 = 2.7 km/5min, which means this wind can form an air column of 2.7 km in 5 min, so the radius of this column is what we need to find out:
Air volume = h x pi x r^2, where r is the radius of column,
92000 = 2.7x1000 x 3.14 x r^2, r= 10.85 ~ 11 meter.
Hence, the diameter of column or cawl that is needed to supply enough CO_2 to produce 1GJ of methanol in 5 minutes is 22 meter. This seems to be the size of a typical windmill ?!
The minimum energy required to capture CO_2 with MgO absorption is calculated as you suggested:
Cp [cal/K, mole]: 8.9 (CO_2), 9.0 (MgO), 18.0 (MgCO3)
del.H = sum of Cp x (700 - 300 degree) = (18 + 9 + 8.9) x 400 = 14.36 Kcal/mole = 60 Kj/mole
to capture 1377 moles of CO_2, we need 1377x60=83 Mj = 23 KW.hr ~ 1$ worth of electricity @ 4cents/Kw.hr.
Thus, CO_2 capture at least cost $1 per 1GJ of methanol production, once the capital cost of equipment is paid for.
The final answer to the question of if 1GJ of methanol obtained as in reaction [1] is cheaper than 1GJ H_2 plus its storage+transportation+delivery cost:
CO_2 + 3 H_2 = CH3OH(liq.) + H_2O(liq.) [1]
1$/GJ methanol 20$/GJ 1.2x20+1=25$/GJ
My conclusion from this exercise is, at the zeroth approximations of
a. CO_2 capture efficiency is 100% and energy use is also close to 100% efficiency
b. CO_2 conversion to methanol is 100 % efficient
c. capital cost of the equipment can be recouped within short period time, say 1-2 years
d. the cost of H_2 storage+transportation+delivery is about 20$/GJ H_2 per your note"

4.15 Useful quantities: calculations of distinguished professor Jerry North [85]

"The current concentration of carbon dioxide is 380 ppm. I start with the air pressure that is the weight of air per square meter (100,000 Pascals). The mass is then this number divided by g=10m/s^2 or 10,000 kg/m^2.

The area of the Earth is 4.5 10^14 m^2. So the total mass of the atmosphere is 4.5 10^18kg. The molecular weight of air is 28 kg/kmol, giving us 1.6 10^17 kmol of air.
The number of kmoles of CO_2 is then 380 10^(-6) 1.6 10^17+6.1 10^13 kmol.
The molecular weight of CO_2 is 44. So we have at last 2.7 10^15 kg of CO_2 in the global air. And it is rising at about 0.5 percent per year.
Hence, we have about 2.7 10^12 metric Tonnes (1000kg) of CO_2 in the air (nearly 3 trillion metric tonnes).
On the mixing of CO_2: it is well mixed vertically in the atmosphere up to many tens of km (the so-called turbopause). The mixing time from Northern to Southern Hemisphere is one to two years."
[85} These calculations came (with his permission) from Distinguished Professor Jerry North (July 2007) in the Meteorology Department of Texas A&M University, College Station, Texas, 77843.

4.16 A cryogenic approach

The second suggested method is called cryogenic. In principle, it's possible to extract the CO_2 from the air by passing it through a cold zone kept at temperatures a little less than that at which CO_2 becomes a solid, and drops down out of the air. (77°K)
Now, the positive side is that it's a simple process thermally but it has a negative aspect to it as in trying to get the 0.03% CO_2 out of the air, we should have to cool down the air which contains it.
It's simple to make an analysis of how much this costs. I was able to come up with a zeroeth approximation for the cost of the resulting final methanol. (1 GJ).

4.17 The electrochemical method

The suggestion is to start off with potassium hydroxide solution, KOH. When we pass air into KOH, it will extract CO_2 By making K_2CO_3.
Now then, we would have a solution of potassium carbonate, and if we go on passing the air at low cost (winds) through the solution until the potassium carbonate has got up to the saturation limit at which no more can be formed by passing air.
If we connect these up to a suitable power source, we will then get hydrogen off at the cathode, and CO_2 off at the anode. (Oxygen will evolve with CO_2 if the anode potential is positive.
This is what we need, CO_2 and hydrogen. These are the elements from which we can make methanol. And so, for the moment, neglecting the fact that one has to have 3 molecules of H_2 one can now proceed and make methanol by a well-known chemical process:

$$3H_2 + CO_2 \rightarrow CH_3OH + H_2O \text{ (with CuZn catalyst)} \quad (6)$$

I think that we are at the beginning of the choice of methods of removing CO_2 from the atmosphere. There is a scientist at Columbia University in New York, Professor Lackner [86], who has been devoting much time to this process.
Whatever the final method, the process, direct $3H_2 + CO_2 \rightarrow CH_3OH + H_2O$ is waiting for it, and we know we can directly go then, to the reliable electrolysis of water, producing a stream of hydrogen, and the CO_2 as we have been describing. CO_2 off [13]

---

[13] The other important thing is that we have to guard the potential of the anode because if two positive (or the current density are high) there will be an emission of O_2 with the CO_2.

Under these circumstances, then, the "hydrogen economy" can be made with "liquid hydrogen" (i.e., methanol made with CO_2 from the atmosphere.)

4.18 Politics

There will be much politics in the ratification of what we have suggested. In fact, as far as the US and the UK is concerned, both these countries house the main oil companies in the world, and it's difficult to see that there could be a change over to a new energy system without collaboration with the giant companies. There will never be a battle between two different systems. Eventually their profit-making future is to become energy companies, we may hope for collaboration and use of the minimum cost ideas that we present here.

5. Summary

The problem of Global Warming comes down to changing the energy production from those that produce CO_2 to those that do not. (It cannot be sequestered for more than ten years.)
One cannot put a date on the time when untreated carbonaceous fuels will make it impractical to live here because this depends upon the place on the earth we are considering.
Inhabitants of the huge cold country of Canada and the enormous space left open in Siberia, would welcome unbridled Global Warming for twenty or thirty years, with the flood of people trying to escape the heat and come to somewhere where it was easier to live, they would be inundated with new inhabitants and for a while it would look as though they had made an acceptable change. However, it's obvious that finally Global Warming, if unchecked, would invade the northern areas too, and even these lands would become too hot to hold us.
Apart from the original energy which one needs, one has to think of the medium and this is where hydrogen might be regarded as the solution to a problem - no Global Warming. The problem is how to store electricity in large amounts e.g. for a city. Now, we have discussed "liquid hydrogen" and this would solve that problem too, because storing liquid hydrogen would be no different than storing oil or gasoline. Hence, methanol from the atmosphere might be the answer to hydrogen high cost.
Time is a pressing issue. The Chinese government has made the announcement that it is going to convert the transport system in China to renewables within eleven years from 2010. Meanwhile, democracies are not well known for quick decisions. This is one of the problems we shall have to face as we move closer to the vast Changes which are being made by the Chinese and which will overtake the United States within ten years, as we will have to battle not only the problem of exhausting energy and Global Warming but the competitive power of a China, rapidly building an energy supply (plus a 300 mph train system.)

6. References

[1] The Hydrogen Economy. Opportunities, Costs, Barriers and R&D Needs. National Research Council and National Academy of Engineering, National Academies Press, Washington DC, 2004.
[2] Patrick Coffey, Cathedrals of Science: The Personalities and Rivalries That Made Modern Chemistry, Oxford University Press, 2008. ISBN 978-0-19-532134-0

[3] T.N. Veziroglu, I. Gurkan, M.M. Padki, International Journal of Hydrogen Energy, 14, 1989, 257.
[4] CO_2 over 1000 years. The Hydrogen Economy. Opportunities, Costs, Barriers and R&D Needs. National Research Council and National Academy of Engineering, National Academies Press, Washington DC, 2004; John O'M. Bockris, "Renewable Energies: Feasibility, Time, and Cost Options, Nova Science Publishers, New York, 2009.
[5] "Climate change could be accelerated by 'methane time bomb'", The Telegraph, Heidi Blake, February 22, 2010.
[6] "Methane and Carbon Monoxide in the Troposphere". http://www.atmosp.physics.utoronto.ca/people/loic/chemistry.html. Retrieved 2008-07-18.
[7] "Methane bubbles climate trouble". BBC News. 2006-09-07. http://news.bbc.co.uk/2/hi/science/nature/5321046.stm. Retrieved 2006-09-07
[8] Shakhova, Natalia; Semiletov, Igor (2007), "Methane release and coastal environment in the East Siberian Arctic shelf", *Journal of Marine Systems* 66 (1-4): 227–243, doi:10.1016/j.jmarsys.2006.06.006
[9] Climate Change 2001: The Scientific Basis (Cambridge Univ. Press, Cambridge, 2001)
[10] N. E. Shakhova, I. P. Semiletov, A. N. Salyuk, N. N. Bel'cheva, and D. A. Kosmach, (2007). "Methane Anomalies in the Near-Water Atmospheric Layer above the Shelf of East Siberian Arctic Shelf". *Doklady Earth Sciences* 415 (5): 764–768. doi:10.1134/S1028334X07050236.
[11] Walter, Km; Zimov, Sa; Chanton, Jp; Verbyla, D; Chapin, Fs, 3Rd (Sep 2006). "Methane bubbling from Siberian thaw lakes as a positive feedback to climate warming.". *Nature* 443 (7107): 71–5. doi:10.1038/nature05040. ISSN 0028-0836. PMID 16957728.
[12] http://globalwarmingcycles.info/, 2010
[13] Private communications in 2009 with JOMB.
[14] Private communications with DOE and JOMB, 2009.
[15] Mapping methane emissions from a marine geological seep source using imaging spectrometry Dar A. Roberts a, □, Eliza S. Bradley a, Ross Cheung b, Ira Leifer c, Philip E. Dennison d, Jack S. Margolis, Remote Sensing of Environment, 114, (2007) 592-606.
[16] T.N. Veziroglu, I. Gurkan, and M.M. Padki, International Journal of Hydrogen Energy, 14, 1989, 257.
[17] T.N. Veziroglu, I. Gurkan, and M.M. Padki, International Journal of Hydrogen Energy, 14, 1989
[18] Anthropogenic Global Warming is Nonsense, *by Edward Townes* (libertarian) Sunday, December 30, 2007 http://www.nolanchart.com/article805.html
[19] Pelham, Brett (2009-04-22). "Awareness, Opinions About Global Warming Vary Worldwide". Gallup. http://www.gallup.com/poll/117772/Awareness-Opinions-Global-Warming-Vary-Worldwide.aspx. Retrieved 2009-07-14.
[20] Julio Usaola and Edgardo D. Castronuovo, Wind Energy In Electricity Markets with High Wind Penetration, Nova Science Publishers, New York, 2009.
[21] Cedrick N. Osphey (Ed), Wind Power: Technology, Economics and Policies, Nova Sciences Publishers, New York, 2009.

[22] The Destructive Power of Wind: Turbine Disintigrates, *Hank Green 25/02/08,* http://www.ecogeek.org/wind-power/1396
[23] John O'M. Bockris, "Renewable Energies: Feasibility, Time, and Cost Options, Nova Science Publishers, New York, 2009.
[24] Bryan W. Roberts, David H. Shepard, Ken Caldeira, M. Elizabeth Cannon, David G. Eccles, Albert J. Grenier, and Jonathan F. Freidin, "Harnessing High Altitude Wind Power," IEEE Transactions on Energy Conversion, Vol. 22, No. 1, March 2007.
[25] John O'M. Bockris, Renewable Energies: Feasibility, Time and Cost Options, Nova Science Publishers, New York, 2009, pgs 16-21.
[26] Energy: The Solar-Hydrogen Alternative, J. O'M. Bockris, John Wiley and Sons, New York, 1975, pages 151-153.
[27] Muradov and Veziroglu private communications with them 2009. Muradov, N.Z. and T.N. Veziroğlu (2008) "Green" path from fossil-based to hydrogen economy: An overview of carbon-neutral technologies. *International Journal of Hydrogen Energy* 33, 6804- 6839.
[28] John O'M. Bockris, Renewable Energies: Feasibility, Time and Cost Options, Nova Science Publishers, New York, 2009, pgs 21-23.
[29] Bryan W. Roberts, David H. Shepard, Ken Caldeira, M. Elizabeth Cannon, David G. Eccles, Albert J. Grenier, and Jonathan F. Freidin, "Harnessing High Altitude Wind Power," IEEE Transactions on Energy Conversion, Vol. 22, No. 1, March 2007.
[30] Wind Energy Fact Sheet, American Wind Energy Association 2009.
[31] Environmental Conservation, John O'M. Bockris, No. 4, Vol 2, 1975
[32] http://meted.ucar.edu/hurrican/strike/text/htc_desc.htm
[33]Wind Energy Resource Atlas, http://rredc.nrel.gov/wind/pubs/atlas/chp2.html.
[33a] Wind Energy Resource Atlas http://rredc.nrel.gov/wind/pubs/atlas/chp3.html
[34] John O'M. Bockris, Renewable Energies: Feasibility, Time and Cost Options, Nova Science Publishers, New York, 2009, pgs 22-23.
[35] Wind Energy Fact Sheet, American Wind Energy Association, 2001.
[36] Richard Heinberg, The Party's Over, New Society Publishers, Gabriola Island, 2005.
[36a] Richard Heinberg, The Party's Over, New Society Publications, 2006, p 154.
[36b] Richard Heinberg, The Party's Over, New Society Publications, 2006, p 156.
[37] Kenneth S. Deffeyes. *Hubbert's Peak : The Impending World Oil Shortage,* Princeton University Press (August 11, 2003), ISBN 0-691-11625-3.
[38] Richard Heinberg. *The Party's Over: Oil, War, and the Fate of Industrial Societies,* New Society Press ISBN 0-86571-482-7
[39] Mathew R. Simmons. *Twilight in the Desert: The Coming Saudi Oil Shock and the World Economy,* Wiley (June 10, 2005), ISBN 0-471-73876-X
[40] Muradov, N.Z. and T.N. Veziroğlu (2008) "Green" path from fossil-based to hydrogen economy: An overview of carbon-neutral technologies. *International Journal of Hydrogen Energy* 33, 6804-6839.
[41] Kraft A. Ehricke, "The Power Relay Satellite (PRS) Concept in the Framework of the Overall Energy Picture, North American Aerospace Group, Rockwell International, December 1973.

[42] Kraft A. Ehricke, "The Power Relay Satellite (PRS) Concept in the Framework of the Overall Energy Picture, North American Aerospace Group, Rockwell International, December 1973, pg
[43] Kraft A. Ehricke, "The Power Relay Satellite (PRS) Concept in the Framework of the Overall Energy Picture, North American Aerospace Group, Rockwell International, December 1973, pg 141, Fig 8.1.
[44] Kraft A. Ehricke, "The Power Relay Satellite (PRS) Concept in the Framework of the Overall Energy Picture, North American Aerospace Group, Rockwell International, December 1973, pg 147-149, figs 8.5, 8.6, Tables 8.3 and 8.4.
[45] Kraft A. Ehricke, "The Power Relay Satellite (PRS) Concept in the Framework of the Overall Energy Picture, North American Aerospace Group, Rockwell International, December 1973, pg 150, figs 8.7, Table 8.5.
[46] Energy: The Solar-Hydrogen Alternative, J. O'M. Bockris, John Wiley and Sons, New York, 1975, pages 151-153.
[47] AArthur Cristian, Bakken Oil Formation In Dakota/Montana Provides USA 8 Times As Much Oil As Saudi Arabia @ $16 Per Barrel x 500 Billion Barrels Thu,, 09/11/2008http://www.loveforlife.com.au/node/5492; and Rod Nickel, Harnessing the Boom, The StarPhoenix, Friday, May 09, 2008
[48] "Branson launches $25m climate bid". BBC.co.uk. http://news.bbc.co.uk/1/hi/sci/tech/6345557.stm. Retrieved 2008-04-30.
[49] "US Department of Energy on greenhouse gases". http://www.eia.doe.gov/oiaf/1605/ggccebro/chapter1.html. Retrieved 2007-10-04.
[50] "Race for millions". Cosmos Magazine. http://www.cosmosmagazine.com/features/online/1075/the-race-bransons-millions. Retrieved 2008-11-05.
[51] From hydrocarbon to hydrogen-carbon to hydrogen economy Authors: Muradov, NZ; Veziroglu, TN, Journal: INT J HYDROGEN ENERG, 30 (3): 225-237 MAR 2005.
[52] Department of Energy Cost of Energy Section, 2010.
[53] Kenneth S. Deffeyes. *Hubbert's Peak : The Impending World Oil Shortage*, Princeton University Press (August 11, 2003), ISBN 0-691-11625-3.
[54] Arthur Cristian, Bakken Oil Formation In Dakota/Montana Provides USA 8 Times As Much Oil As Saudi Arabia @ $16 Per Barrel x 500 Billion Barrels Thu,, 09/11/2008http://www.loveforlife.com.au/node/5492; and Rod Nickel, Harnessing the Boom, The Star Phoenix, Friday, May 09, 2008
[55] "Branson launches $25m climate bid". BBC.co.uk. http://news.bbc.co.uk/1/hi/sci/tech/6345557.stm. Retrieved 2008-04-30.
[56] "US Department of Energy on greenhouse gases". http://www.eia.doe.gov/oiaf/1605/ggccebro/chapter1.html. Retrieved 2007-10-04.
[57] "Race for millions". Cosmos Magazine. http://www.cosmosmagazine.com/features/online/1075/the-race-bransons-millions. Retrieved 2008-11-05.
[58] Richard Heinberg. *The Party's Over: Oil, War, and the Fate of Industrial Societies*, New Society Press ISBN 0-86571-482-7
[59] Mathew R. Simmons. *Twilight in the Desert: The Coming Saudi Oil Shock and the World Economy*, Wiley (June 10, 2005), ISBN 0-471-73876-X

[60] From hydrocarbon to hydrogen-carbon to hydrogen economy Authors: Muradov, NZ; Veziroglu, TN, Journal: INT J HYDROGEN ENERG, 30 (3): 225-237 MAR 2005.
[61] Renewable Energy Resources, John Tidwell, Tony Weir, 2005, Taylor & Francis Publishers.
[62] NREL Photograph and Information Exchange
[63] Council of Australian Governments, July 2006
[64] High Temperature Solar Thermal Technology, Wyld Group, February 4, 2009 (Australian Library Collection)
[65] Popular Science, November 2007
[66] Closed cycle OTEC Schematic-Ocean Thermal Energy Conversion: Possible application in Certain Pacific Island Nations, Alicia Altagracia Aponte.
[67] 2009-10-24 13:49 (UTC) Geothermie_Prinzip01.jpg Geothermie_Prinzip.svg: Geothermie_Prinzip01.jpg: "Siemens Pressebild" http://www.siemens.com derivative work: FischX (talk) Geothermie_Prinzip01.jpg: "Siemens Pressebild" http://www.siemens.com derivative work: Ytrottier This file is licensed under the Creative Commons Attribution-Share Alike 3.0 Unported license. You are free: to share - to copy, distribute and transmit the work; to remix - to adapt the work Under the following conditions: attribution - You must attribute the work in the manner specified by the author or licensor (but not in any way that suggests that they endorse you or your use of the work). share alike - If you alter, transform, or build upon this work, you may distribute the resulting work only under the same or similar license to this one.
[68] http://en.wikipedia.org/wiki/Enhanced_Geothermal_System http://altarockenergy.com/AltaRockEnergy.2009-03-19.pdf ; MIT Report, 2009, Earth Science Deparment.
[69] http://en.wikipedia.org/wiki/Tidal_power, October 2007.
[70] :Superconductivity, Wikipedia, Free Encyclopedia, 2010.
[71] J.O'M. Bockris, "A Hydrogen Economy", Environment, 13, 1971, 51; and History of Hydrogen Fact Sheet, The National Hydrogen Energy Association http://www.hydrogenassociation.org/general/factSheet_history.pdf
[72] P. Dandapani, Personal Income and Energy, Int. J. Hydrogen Energy, 12, 1987, 439.
[73] Tapan Bose and Pierre Malbrunot, et al, Hydrogen: Facing the Energy Challenge of the21st Century, John Libby Eurotext, UK, December 2006
[74] Wiberg, Egon; Wiberg, Nils; Holleman, Arnold Frederick (2001). *Inorganic chemistry*. Academic Press. p. 240. ISBN 0123526515. http://books.google.com/books?id=vEwj1WZKThEC&pg=PA240.
[75] Nazim Muradov, ""Hydrogen via methane decomposition: an application for decarbonization of fossil fuels"", International Journal of Hydrogen Energy 26 (2001) pp. 1165-1175.
[76] Popular Science, November 12, 2007.
[77] Judge Bevan, S. Badwell, and J.O'M. Bockris, Evolution and Dissolution of Oxygen on Urania-Yttria, Acta Electrochimica, 25, 1980
[78] "The Methanol Economy: Beyond Oil and Gas", Olah, Goeppert & Prakash, Wiley, 2006.
[79] http://www.iea.org/work/2002/stavanger/mhi.pdf

[80] S. Ono, et al, "The Effect of CO2, CH4, H20, and N2 on Mg—MI Alloy as hydrogen Transportation" IJHE, 11, 6, 1986, 381-387.
[81] http://www.brain-c-jcoal.info/cctinjapan-files/english/2_5A1.pdf
[82] Isamu Yasudaa and Yoshinori Shirasaki (Tokyo Gas) Development and Demonstration of Membrane Reformer System for Highly efficient Hydrogen Production from Natural Gas, *Materials Science Forum* Vols. 539-543 (2007) pp 1403-1408
[83] Wind Powering America, U.S. Department of Energy, http://www.windpoweringamerica.gov/wind_maps.asp
[84] Private communications with JOMB and Rey Sidik, calculations.
[85] Private communication between JOMB and Jerry North (July 2007)
[86] http://www.columbia.edu/~kl2010/members_lackner.htm

TABLES

Table 1. Wind classes and wind speed.J. O'M. Bockris original and in John O'M. Bockris, Renewable Energies: Feasibility, Time and Cost Options, Nova Science Publishers, New York, 2009

Table 2. Power Relay Satellite Concept, Kraft A. Ehricke, "The Power Relay Satellite (PRS) Concept in the Framework of the Overall Energy Picture, North American Aerospace Group, Rockwell International, December 1973, pg 147-149, figs 8.5, 8.6, Tables 8.3 and 8.4.

Table 3. Character of microwave PRS energy systems. Kraft A. Ehricke, "The Power Relay Satellite (PRS) Concept in the Framework of the Overall Energy Picture, North American Aerospace Group, Rockwell International, December 1973, pg 147-149, figs 8.5, 8.6, Tables 8.3 and 8.4.

Table 4. http://en.wikipedia.org/wiki/Enhanced_Geothermal_System http://altarockenergy.com/AltaRockEnergy.2009-03-19.pdf

Table 5. SOME LARGER TIDAL POWER SCHEMES UNDER CONSIDERATION AROUND THE WORLD http://en.wikipedia.org/wiki/Tidal_power, October 2007.

Table 6. Critical Temperature (Tc), crystal structure and lattice constants of some high-Tc http://en.wikipedia.org/wiki/Superconductivity 2010

Table 7. Cost of Raw Hydrogen, Various conditions.

Table 8. Electricity, Methanol, and Hydrogen Compared as Fuels.

FIGURES

Fig. 1. This figure was created by Robert A. Rohde from published data and is part of the Global Warming Art project.Original image: http://www.globalwarmingart.com/wiki/Image:Greenhouse_Effect_png It was converted to SVG by User:Rugby471.Permission is granted to copy, distribute and/or modify this document under the terms of the **GNU Free Documentation License**, Version 1.2 only as published by the Free Software Foundation; with no Invariant Sections, no Front-Cover Texts, and no Back-Cover Texts. A copy of the license is included in the section entitled "Text of the GNU Free Documentation License."; 1997.

Fig. 2. The Keeling Curve of atmospheric CO_2 concentrations measured at Mauna Loa Observatory. Work, from Image:Mauna Loa Carbon Dioxide.png, uploaded in Commons by Nils Simon under licence GFDL & CC-NC-SA ; itself created by Robert A. Rohde from NOAA published data and is incorporated into the Global

Warming Art project. *Permission is granted to copy, distribute and/or modify this document under the terms of the **GNU Free Documentation License**, Version 1.2 or any later version published by the Free Software Foundation; with no Invariant Sections, no Front-Cover Texts, and no Back-Cover Texts. A copy of the license is included in the section entitled "GNU Free Documentation License";*

Fig. 3. From Tapan Bose and Pierre Malbrunot, et al, Hydrogen: Facing the Energy Challenge of the 21st Century, John Libby Eurotext, UK, December 2006, page 17.

Fig. 4. This image shows the instrumental record of global average temperatures as compiled by the Climatic Research Unit of the University of East Anglia and the Hadley Centre of the0 UK Meteorological Office. Data set TaveGL2v was used. The most recent documentation for this data set is Jones, P.D. and Moberg, A. (2003) "Hemispheric and large-scale surface air temperature variations: An extensive revision and an update to 2001". Journal of Climate, 16,206-223.

Fig. 5. The Hydrogen Economy. Opportunities, Costs, Barriers and R&D Needs. National Research Council and National Academy of Engineering, National Academies Press, Washington DC, 2004.

Fig. 6. Mapping methane emissions from a marine geological seep source using imaging spectrometry Dar A. Roberts a,□, Eliza S. Bradley a, Ross Cheung b, Ira Leifer c, Philip E. Dennison d, Jack S. Margolis, Remote Sensing of Environment, 114, 2010, pg 600.

Fig. 7a. http://www.cnsm.csulb.edu/departments/geology/people/bperry/geology303/_derived/geol303text.html_txt_atmoscell_big.gif

Fig. 7b. http://mabryonline.org/blogs/woolsey/images/global%20winds%202-1.jpg

Fig. 8. Energy Center, John O'M. Bockris Original.

Fig. 9. A & B: Energy Manual, Iowa Energy Center, 2006, www.energy.iastate.edu/renewable/wind/wem/wem-08_power.html

Fig. 10. A possible arrangement for a sea-borne generator. {J.Bockris, 1975} Environmental Conservation John O'M. Bockris, No. 4, Vol 2, 1975; John O'M. Bockris, Renewable Energies: Feasibility, Time and Cost Options, Nova Science Publishers, New York, 2009, pg 14.

Fig. 11. An alternative arrangement for a sea-borne generator. {J.Bockris, 1975} Environmental Conservation John O'M. Bockris, No. 4, Vol 2, 1975; John O'M. Bockris, Renewable Energies: Feasibility, Time and Cost Options, Nova Science Publishers, New York, 2009, pg 15.

Fig. 12. Diagram of the FEG in flight, showing the craft's nose-up angle, _, which is identical to the control axis angle, _c, as no cyclic pitch use is planned. The rotors fore and aft flapping angle, a1, is shown as the angle between the normal to the tip-path plane and the control axis. The total rotor thrust component along the control axis is T, and normal to this axis is the component force H. If T and H forces are combined vectorally the total rotor force is almost normal to the tip-path plane {J. Bockris, 2009} [25] John O'M. Bockris, Renewable Energies: Feasibility, Time and Cost Options, Nova Science Publishers, New York, 2009, pgs 16-21 and.Bryan W. Roberts, et al., "Harnessing High Altitude Wind Power, IEEE http://www.jp-petit.org/ENERGIES_DOUCES/eolienne_cerf_volant/eolienne_cerf_volant.pdf pg 4.

Fig. 13. Rendering of Sky Wind Power Corp.'s planned 240 kW, four-rotor demonstration craft {B. Roberts, D. Shepherd, et al, 2007: [29]. Bryan W. Roberts, et al.,

"Harnessing High Altitude Wind Power, IEEE http://www.jp-petit.org/ENERGIES_DOUCES/eolienne_cerf_volant/eolienne_cerf_volant.pdf, pg 2.

Fig. 14. "World map showing two mid latitude westerly wind belts (shown in pink). The northern belt blows from west to east across North America, the North Atlantic Ocean, Europe, and Asia. The southern belt blows from west to east across the South Pacific Ocean, Chile, Argentina, the South Atlantic Ocean, South Africa, the South Indian Ocean, Southern Australia, and New Zealand. The yellow arrows in the picture also show two tropical easterly wind belts blowing from east to west on either side of the equator. The northern tropical easterly belt blows across the Pacific Ocean, Southeast Asia, India, the North Indian Ocean, the Arabian Peninsula, Saharan Africa, the Atlantic Ocean, the Caribbean Sea, Southern Mexico, and Central America. The southern belt blows from east to west across Northern Australia, the Indian Ocean, Southern Africa, the South Atlantic Ocean, the middle of South America, and the South Pacific Ocean." [32]

Fig. 15. Source: GWEC, Worldwatch. Figure 15. From Wind Energy Fact Sheet, American Wind Energy Association, 2004.{AWEA, 2001} [35].

Fig. 16. Figure shows the great distances between areas of high insolation; and those of high concentration of affluent groups with manufacture {Kraft-Ehricke, 1973} [41].

Fig. 17. Power relay satellite concept {Kraft-Ehricke, 1973} [44].

Fig. 18. Range of a number of Primary Energy Power Plant Systems {Kraft-Ehricke, 1973} [44].

Fig. 19. Linear array of waveguide-fed helix elements {J. Bockris, 1975} [46].

Fig. 20. New presentation of data in figure 20 of http://www.hubbertpeak.com/hubbert/1956/1956.pdf. Meant as replacement for non-free en::Image:Hubbert-fig-20.png 2007-03-04 (original upload date) Transferred from en.wikipedia; transferred to Commons by User:Pline using CommonsHelper Original uploader was Hankwang at en.wikipedia CC-BY-2.5; Released under the GNU Free Documentation License http://en.wikipedia.org/wiki/File:Hubbert_peak_oil_plot.svg

Fig. 21. (a) Schematic of a power tower. Image adapted from Energy Efficiency Renewable Energy Network {J. Tidwell, 2005} [61];

Fig. 21. (b): Solar Two, power tower. Image courtesy of NREL's Photographic Information Exchange [62].

Fig. 22a. Schematic of a parabolic trough concentrator. Image adapted from Energy Efficiency Renewable Energy Network {Council of Australian Governments, 2006} [63].

Fig. 22b. Trough concentrator system at the Australian National University, which is designed to incorporate photovoltaic power generation or water heating and steam production. (Image courtesy of the Centre for Sustainable Energy systems, Australian National University, {Wyld Group, 2009} [64].

Fig. 23. Schematic of Closed-cycle OTEC system. Closed cycle OTEC Schematic-Ocean Thermal Energy Conversion: Possible application in Certain Pacific Island Nations, Alicia Altagracia Aponte [66].

Fig. 24. Diagram of EGS with numeric labels. 1:Reservoir 2:Pump house 3:Heat exchanger 4:Turbine hall 5:Production well 6:Injection well 7:Hot water to district heating

8:Porous rock 9:Well 10:Solid bedrock. 2009-10-24 13:49 (UTC) Geothermie_Prinzip01.jpg Geothermie_Prinzip.svg: Geothermie_Prinzip01.jpg: "Siemens Pressebild" http://www.siemens.com derivative work: FischX (talk) Geothermie_Prinzip01.jpg: "Siemens Pressebild" http://www.siemens.com derivative work: Ytrottier This file is licensed under the Creative Commons Attribution-Share Alike 3.0 Unported license. You are free: **to share** - to copy, distribute and transmit the work; **to remix** - to adapt the work Under the following conditions: **attribution** - You must attribute the work in the manner specified by the author or licensor (but not in any way that suggests that they endorse you or your use of the work). **share alike** - If you alter, transform, or build upon this work, you may distribute the resulting work only under the same or similar license to this one [67].

Fig. 25. P. Dandapani, Personal Income and Energy, Int. J. Hydrogen Energy, 12, 1987, 439.

Fig. 26. Tapan Bose and Pierre Malbrunot, et al, Hydrogen: Facing the Energy Challenge of the21st Century, John Libby Eurotext, UK, December 2006, p.59.

Fig. 27. Wind Powering America, U.S. Department of Energy, http://www.windpoweringamerica.gov/wind_maps.asp

Section 5

11

Simulating Alpine Tundra Vegetation Dynamics in Response to Global Warming in China

Yanqing A. Zhang[1], Minghua Song[2], and Jeffery M. Welker[3]
[1]*Department of Geography, and School of Computing Science, Simon Fraser University, BC,*
[2]*Institute of Geographic Sciences and Natural Resources Research, the Chinese Academy of Sciences, Beijing,*
[3]*Environment and Natural Resources Institute, University of Alaska Anchorage, AK,*
[1]*Canada*
[2]*P.R. China*
[3]*USA*

1. Introduction

Global temperatures are increasing due to the effects of greenhouse gases emission. It is projected that climate changes will have profound biological effects, including the changes in species distributions as well as vegetation patterns (Walther et al., 2002; Klanderud & Birks, 2003; Pauli et al., 2003; Tape et al., 2006). Many results from observations and experiments (Parmesan, 1996; Molau & Alatalo, 1998; Parmesan et al., 1999; Welker et al., 2000, 2005; Schimel et al., 2004; Sullivan & Welker, 2005), and simulation studies (Cramer & Leemans, 1991; Harras & Prentice, 2003) have depicted alterations in C and N cycling, trace gas exchanges and shifts in the distribution of vegetation boundary and the mixture of shrubs and grasses.

The Tibetan Plateau covers approximately 2.5 million km^2 with an average altitude of more than 4000 m dominated by alpine tundra (Zheng, 2000). Alpine tundra vegetation is predicted to be one of the most sensitive terrestrial ecosystems to changing climate (Korner, 1992; Grabherr et al., 1994; Chapin et al., 1992, 2000). This type of ecosystem is composed of slow-glowing plants and are dominated by the soils which can be concentrated with high organic matter near surface soil that undergo frost heave and cryoturbation (Billings, 1987; Xia, 1988). Both plant growth and possible organic matter decomposition are predicted to increase under warmer climates, which may cause alpine ecosystem carbon flux and energy flow changes (Chapin et al., 1997; Kato et al., 2006). Simultaneously, warmer weather may increase plant growth, and primary production (Bowman et al., 1993; Wookey et al., 1995) as well as changes in species dominance (Walker et al., 1994; Klein et al., 2007). We report findings that are derived from a short-term responses to simulated environmental warming, focusing on aboveground biomass of three dominated life forms and community compositional attributes.

Based on 38 years (1959-1996) of climate observations and statistical analysis, the annual mean temperature increased during this period ranged from 0.4 to 0.6°C in the area of

Haibei Alpine Tundra Ecosystem Research Station (Li et al., 2004), that is located on north-eastern part of Qinghai-Tibetan Plateau (37°N, 101°E). In order to study alpine tundra vegetation changes at the regional scale, we model alpine tundra vegetation spatial and temporal dynamics in response to global warming by integrating a raster-based cellular automata and a Geographic Information System (Zhang et al., 2008). Temperature changes across the study area are not only due to elevation, but also to aspect and distance from the nearest stream channel. The liner regression model provided a temperature spatial distribution based on elevation alone, which is the primary step. The normalized temperature surface created by the Multi-Criteria Evaluation (MCE) is highly representative of the potential temperature distribution in a normalized fuzzy format. Assuming each vegetation type in the raster cell unit reacts as homogeneous entity, we conduct a spatial and temporal simulation by combining cellular automata and MCE provided in the IDRISI software (Eastman, 2003).

Global changes have strong effects on terrestrial ecosystems but with significant regional differences. The Tibetan Plateau is currently experiencing rapid changes in temperature (Zhang et al., 1993). Fluctuations in temperature have had significant effects on alpine tundra ecosystem, which produces the important changes in the global energy balance and carbon budget (Cao & Woodward, 1998; Zhou, 2001; Kato et al., 2006). The Qinghai-Tibetan Plateau is situated in southwestern China (Fig. 1), and is the highest continental

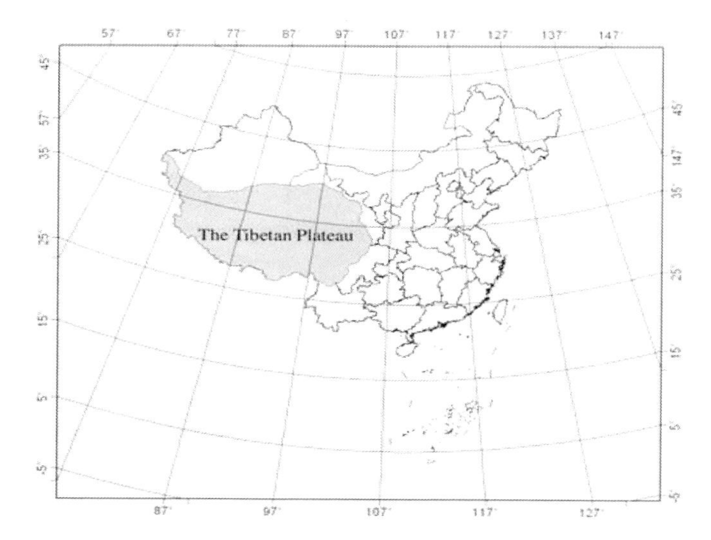

Fig. 1. The location of the Tibetan Plateau in China.

landmass in the world. Elevation ranges from 2500 to 8000 m with an average altitude of more than 4000 m. Uplifting of the plateau created and then strengthened the South Asia Monsoon, and affects terrestrial ecosystems in China owing to its unique location and high elevation (topography) (Zhang, 1993; Thompson et al., 1989). The development and evolution of species and vegetation on the Qinghai-Tibetan Plateau were influenced significantly by a fluctuating climate during the uplift. Ni (2000) simulated biomes on the Tibetan Plateau using the improved BIOME3 model (BIOME3-China) under the present climate conditions, as well as under a scenario with a CO_2 concentration of 500 ppmv. A combined biogeography biochemistry model, BIOME4 (Kaplan et al., 2003) was improved to simulate the alpine vegetation changes at the biome level (Song, et al., 2005).

In this chapter, we review the important ecological findings from simulated environmental changes on the alpine tundra vegetation (Zhang & Welker, 1996). We present a changing alpine tundra vegetaion using Vegetation Dynamic Simulation Model (VDSM) integrated with scenarios of global temperature increase of 1 to 3°C (Zhang et al., 2008). With BIOME4 model (Song et al., 2005), we illustrate the vegetation biomass changes and the vegetation distribution dynamics in the region of Qinghai-Tibetan Plateau in responses to global warming.

2. Tibetan alpine tundra above ground biomass and community responses to simulated changes in climate

A suite of abiotic conditions may be modified as weather patterns and regional climates change altering biospheric and atmospheric processes in tundra ecosystems (Maxwell, 1992; Shaver et al., 1992; Jonasson et al., 1993; Grabherr et al., 1994; Larigauderie & Korner, 1995). For instance, warmer air temperatures will likely alter the flux of water from these ecosystems to the atmosphere drying soils and contributing to increased cloud formation. Simultaneously, warmer conditions may increase plant growth, primary production and carbon sequestration, so long as cloud cover is not affected and other factors such as water or nutrients do not limit photosynthesis and growth (Haag, 1974; Bowman et al., 1993; Wookey et al., 1995).

The ecological consequences of changes in tundra environmental conditions will be manifested in a host of processesincluding shifts in primary production (Bowman et al., 1993; Walker et al., 1994), trace gas fluxes (Brooks et al., 1995), plant and soil mineral nutrition (Nadelhoffer et al., 1991; Shaver & Chapin 1991), reproductive plant biology (Wookey et al., 1993, 1994), leaf carbon isotope discrimination (Welker et al., 1993), as well as changes in species dominance (Walker et al., 1994). However, it is unclear whether all these processes are sensitive to short-term changes in environmental conditions in all tundra habitats or whether multiple years of climate change are necessary to elicit detectable alterations in plant performance and species abundance. To date, most studies of alpine tundra responses to in situ changes in climate, using field manipulations, have been confined to sites in North America and in Western Europe (Kmrner, 1992; Chapin et al., 1995; Kennedy, 1995) without the consideration of the extensive alpine tundra in Asia, and in particular, western China.

2.1 Experimental treatments and obervations

Our research site is located near Haibei Alpine Meadow Ecosystem Station (37°N, 101°E) at an elevation of 3250 m (Xia, 1989; Cincotta et al., 1992). The vegetation of our frield site is typical of a *Kobresia humilis* meadow (Zhou et al., 1987, Zhang & Zhou, 1992). Our field experiment was initiated in June 1991 and the first season was complated in October 1991. Four treatments were implemented as (1) Minigreenhouses (G) (2) Shade (S) (3) Side Fences (SF) (4) Control plot (C). The size of experimental plot is 2 m x 5 m. A completely randomized design was used to establish the 16 treatment plots consisting of four treatments (G, S, SF, C) replicated four times. The detail site setup, microclimate monitoring and frield observation were described by Zhang and Welker (1996).

The greenhouse treatment increased mean air temperature by 20% from 12.4 to 17.8° Cover the course of the growing season (Table 1). Warmer air temperature subsequently caused higher soil temperatures at 5, 10, and 15 cm under greenhouse (G) as opposed to ambient (C) conditions (Table 1). The mean vapor density was significantly increased under

Treatments		Control	Green-house	Shade	Side fence
Mean air temperature (°C)		12.38	17.78	14.33	12.91
Mean soil temperature (°C)	5 cm	12.79	16.07	13.85	12.44
	10 cm	12.29	15.14	12.76	12.06
	15 cm	12.19	13.95	12.74	12.10
Vapor density (g m^{-3})		4.00	12.00	6.00	5.80
Soil suction (Kpa)	10 cm	14.80	21.07	30.39	14.94

Table 1. Abiotic conditions from the four treatments between July and October 1991

warmer temperatures of the greenhouse (G) from 4 to 12 g m^{-3}. The soil suction was essentially the same between all treatment plots, except for under shaded (S) conditions, and the soil suction was consistently higher indicating a very lower soil water content for a dryer envrimental condition. The shade treatment (S), while reducing irradiance, also resulted in a slight increase in air temperature and soil temperature at 5 cm. The shade treatment (S) had no effect on soil temperatures at 10 cm or 15 cm nor did the shade treatment alter the vapor densities. Side fences (SF) had no effect on ambient air temperatures and subsequently no effect on soil temperatures.

2.2 Results and discussions

Total community aboveground biomass in all four treatments was not significantly different in July (Table 2). The peak aboveground biomass between Greenhouse (G), occurred in September 351.36 g m^{-2}, and ambient (C) condition, occurred in October 346.19 g m^{-2} have no significant difference at Haibei Apline Meadow Ecosystem Research Satation. However, lowered irradiance (S) resulted in a 23% decrease in total community biomass within 5 wk of treatment applications. Total biomass under reduced irradiance (S) continued to be the lowest over the course of the season reaching a maximum of only 80% of the peak biomass under ambient (C) conditions.

Total maximum aboveground biomass at our Tibetan alpine tundra site ranged from 161 to 351 g m^{-2} under ambient conditions (Table 2). These ranges in biomass are similar to the peak aboveground biomass at other alpine tundra sites such as on Niwot Ridge, Colorado, U.S.A., where the intercommunity aboveground biomass in different vegetation types ranges from 71 to 309 g m^{-2} (Walker et al., 1994). Our environmental manipulations simulating climate warming resulted in warmer air and soil temperatures between 1 and 5°C, which is within the ranges of increase reported for higher elevations in Western Europe over the past 15 years (Rozanski et al., 1992; Grabherr et al., 1994) and is within the ranges predicted for tundra habitats under a doubling of CO_2 over the next 50 yr (Maxwell et al., 1992). The season long average increases are also similar to those accomplished in other tundra experimental warming treatments though our lack of nighttime measurements means our averages are slightly higher than those actually experienced by plants and soil in these treatment plots (Chapin & Shaver, 1985; Wookey et al., 1993; Parsons et al., 1994; Kennedy, 1995). However, most importantly, higher temperatures were maintained in our warmed plots into October and may partially explain the extended growing season observed for grasses.

Date	3 Jul.	1 Aug.	2 Sept.	2 Oct.
Control	161.16 ± 10.23^a	269.36 ± 17.57^a	351.36 ± 15.55^a	285.68 ± 5.49^b
Greenhouse	157.52 ± 7.13^a	252.37 ± 16.57^a	334.61 ± 13.97^a	346.19 ± 11.81^a
Shade	145.86 ± 7.51^a	206.69 ± 17.63^b	278.93 ± 13.78^b	266.21 ± 10.63^b
Side fence	164.00 ± 9.25^a	247.30 ± 10.80^a	370.08 ± 6.45^a	300.80 ± 5.07^b

Differences between the treatments within each month at $p < 0.05$ are noted by different letters.

Table 2. Total aboveground biomass (g m^{-2}) from the four treatments in July, August, September, and October 1991

Aboveground biomass was initially similar among all treatments for forbs, sedges and grasses (Fig. 2a). Within 5 weeks after the warming treatments were implemented, grass biomass was significantly higher in the warmed as compared to control conditions (Fig. 2b). Conversely, grass biomass was significantly reduced during this same period under shaded conditions (Fig. 2b). Reductions of wind using side fences (SF) had no significant effect on grass, sedge or forb biomass (Fig. 2b). By September, grass biomass differences between control and warmed plots were nonsignificant though forb biomass was significantly ($p < 0.05$) lower in the greenhouses (G) as opposed to control conditions (C) (Fig. 2c). Lower irradiance had a significant effect on grass growth and in September, grass biomass was 36% less in shaded (S) as opposed to control conditions. Forb biomass was slightly higher in side-fenced areas as compared to control conditions. Between September and October grass in control plots started to senesce and biomass began to decline (Fig. 2c, 1d). However, under warmed (G) conditions, grass biomass was significantly ($p < 0.01$) higher in warmed as opposed to control conditions in October which postponed community senescence (Fig. 2d). This prolonged growth, or postponed senescence during the fall in warmed plots occurred as the greenhouses maintained warmer air and soil temperatures than ambient conditions. Biomass of grasses and forbs were slightly lower under shaded (S) conditions in October, while sedge biomass was significantly ($p < 0.05$) higher under these same reduced irradiance conditions (Fig. 2d).
Species importance values as a measure of community level responses are presented in Table 3. Under reduced radiation (S) reductions in *Elymus* and *Festuca* were associated with increases in *Stipa* and *Scirpus* which dramatically altered the composition and structure of these plant communities. Changes in community composition and structure under warmer conditions (G) were manifested by lower importance values for *Poa* and *Kobresia* with corresponding increases in importance values for *Stipa* and *Oxytropis* (Table 3).
Grass and forb biomass production was especially sensitive to warmer conditions (Fig. 2). Grass aboveground biomass was 25% greater under warmer conditions after only 5 week of warming while forb biomass decreased by 30% (Fig. 2b). Differences in aboveground grass

biomass between warmer and control conditions were diminished by September when grass biomasses were not significantly different (Fig. 2c).
However, it appears that community senescence, which usually starts in September, was postponed until sometime in October under warmer (G) conditions as evidenced by no decline in aboveground community biomass between September and October (Table 2). This postponing of senescence and subsequently an extension of the growing season under warmed conditions, resulted in part because peak grass biomass was not realized until early October amounting to 177 g m-2 (Fig. 2d). The ability of the grass life form at our site to exhibit a rapid, positive response to warmer conditions and to extend the season of growth is likely the result of (1) the existence of a large leaf area at the time of treatment application, (2) the inherent physiological capacity of grasses to alter patterns of resource allocation (Welker et al., 1985, 1987; Welker & Briske, 1992), (3) their morphological and demographic capacity to elongate fall tillers (Briske & Butler, 1989), and (4) the ability to grow when environmental constraints are temporally removed (Sala et al., 1992). Grasses at other tundra sites have also exhibited an ability to respond rapidly to simulated changes in climate as exemplified by *Calamagrostis* biomass increases in the sub-arctic at Abisko, Sweden under warmer conditions (Parsons et al., 1995). The grass growth response reported by Parsons et al. (1995), in what is typically a dwarf shrub dominated ecosystem, was due in large part to

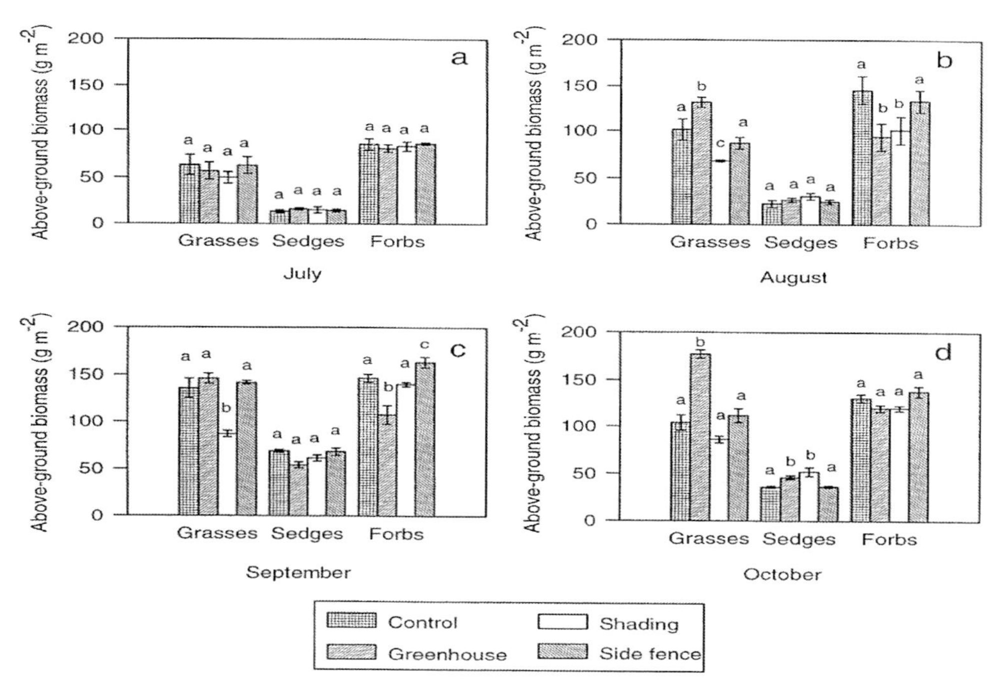

Fig. 2. The aboveground biomass of grasses, sedges, and forbs in control, greenhouses, shaded, and side fenced treatment plots sampled in July, August, September, and October 1991. Superscripts of the different letters denote biomasses which were significantly different ($p < 0.05$) for each individual sampling date.

Plant species	C	G	S	SF
Elymus nutans	52.05	54.07	23.80	56.09
Festuca ovina	35.05	39.08	23.64	33.42
Poa pratensis	21.12	14.53	27.99	20.04
Koeleria cristata	26.32	17.63	18.22	23.86
Stipa aliena	22.29	29.01	31.74	21.21
Kobresia humilis	19.65	6.55	18.66	23.85
Carex scabriostris	12.62	10.14	14.04	13.32
Scirpus distigmaticus	15.10	10.72	35.22	16.54
Saussurea superba	24.16	30.61	27.18	21.13
Gentiana aristata	8.15	10.11	22.13	10.01
Oxytropis ochrocephala	16.17	34.95	15.82	26.33
Trigonella ruthenica	8.69	16.33	8.01	7.75
Taraxacum mongolicum	8.67	10.93	9.48	9.19
Potentilla bifurca	10.50	10.20	9.02	13.94
Aster flaccidus	15.34	9.25	8.62	5.71
Oxytropis caerulea	11.87	13.55	11.23	8.40
Potentilla anserina	8.83	8.74	8.67	16.91
Gentiana straminea	14.65	8.86	21.62	14.88

[a] C—control; G—greenhouse; S—shade; SF—side fence.

Table 3. The Important value of domiant plant species between four treatment plots[a]

an extensive, preexisting network of underground *Calamagrostis* meristems, capable of rapid shoot extension and leaf development up through the dwarf shrub understory.

The shift in alpine tundra community biomass characteristics whereby maximum biomass is maintained into the autumn is different from what might be observed in arctic tundra dominated by deciduous dwarf shrubs. Prolonged growth of many arctic plants in autumn is unlikely due to photoperiodic cues which control senescence (Murry & Miller, 1982). Thus, even if conditions in arctic tundra were warmer in fall, the ability of many dominant life forms to either produce new fall foliage or continue expansion of existing leaf and shoot biomass is limited by life history traits. And while graminoids, such as *Eriophorum* may constitute a large fraction of the biomass in these systems (Shaver et al., 1992), extended growth in fall under warmer temperatures may be unlikely due to the low solar angles in autumn.

The ability of grasses to utilize favorable conditions at the end of the season is a trait similar to that observed for other tundra lifeforms such as evergreen shrub species (Karlsson, 1985; Welker et al., 1995). For instance, Welker et al. (1995) have found evidence that *Dryas octopetala*, a wintergreen species, has the capacity to exhibit net carbon assimilation at the end of the season under warmer, wetter, and fertilized conditions when plants in control conditions have ceased gaining carbon, which is made possible in part by its evergreen nature. In addition, Karlsson (1985) found that 20% of the carbon acquired by the evergreen dwarf shrub, *Vaccinium vitisidaea* occurred in spring and in autumn, before leaf emergence or after leaf senescence in the deciduous species, *Vaccinium uliginosum*. Thus, evergreen dwarf shrubs are also a tundra life form which due to their inherent life history characteristics can respond to changes in environmental conditions which occur in spring, and fall (Wookey et al., 1993; Welker et al., 1995).

The opportunistic behavior of grasses we observed was not evident for forbs. During the initial 5 weeks, forb biomass was reduced under warmer conditions while grass biomass was increasing (Fig. 2b). The opposite response for forbs may have been due in part to the grasses out-competing forbs for water, nutrients and or light. However, the overall community level response was that total biomass was not different between warmed (G) and control (C) conditions after 5 weeks of experimental applications (Table 2). This observation of similar community biomass under modified environmental conditions is consistent with the observations of Chapin and Shaver (1985). These authors found that arctic tundra total community production (current years growth) in perturbed and in control plots remained the same. This inherent buffering was achieved because some species or life forms increased growth while others exhibited reduced growth. They concluded that conditions favorable for one species or life form are less favorable for others, though the total community or ecosystem production changes annually very little (Chapin et al., 1995). This attribute of tundra ecosystems may be the result of the inherently low nutrient levels available to plants in tundra which constrains system level primary production (Shaver et al., 1992).

The one life form in our study which appeared to be the least responsive to simulated climate warming were the sedges, consisting primarily of Kobresia humillis. The lack of significant increases in biomass until the end of the first season under warmer or shaded conditions indicates that this life form has a relatively low sensitivity to temperature and irradiance. However, other sedges, such as Kobresia myosuroides on Niwot Ridge, Colorado, exhibits an increase in biomass under elevated nutrient availability (Bowman et al., 1993). This would suggest that while the warmer conditions in soils under our minigreenhouses may have elevated soil mineralization and increased nutrient pools available to plants (Jonasson et al., 1993; Robinson et al., 1995) the increases were either not sufficient to alter Kobresia growth, or that Kobresia root uptake rates are low, and its ability to compete for soil nutrients with grasses is low (Black et al., 1994; Falkengren-Grerup, 1995). Even though soil nutritionm may have been altered under warmed conditions, the ability of sedges at our site to acquire these resources in a competitive setting appears to be limited, in part due possibly to resource capture by soil microbes (Jackson et al., 1989). However, in future years changes in rooting patterns may enable this species to capitalize on changes in soil resources.

In conclusion, our findings suggest that Tibetan alpine grasses are predisposed to rapid increases in biomass under simulated climate warming due in part to their inherent life historytraits. In addition, the ability of grasses to produce tillers late in the season under warmer conditions extends the period of carbon gain and extends the period in which the community exhibits maximum aboveground biomass. We find that sedges at our site are insensitive in the short term to changes in environmental conditions, while forbs may decrease at the expense of grass biomass. Increases in cloudiness over the Tibetan alpine tundra would likely result in lower aboveground biomass, but if accompanied by higher rainfall the effects may be counter-acting. The extension of peak community biomass into the autumn may in the long term have cascading effects on net ecosystem CO_2 fluxes, nutrient cycling, and forage availability to grazers (Welker et al., 2004).

3. Cellular automata: simulating alpine tundra vegetation dynamics in response to global warming

Spatial modeling processes are available in current GIS software such as IDRISI, which is capable of dealing with a large set of raster data and manipulating the data via operations in a series of discrete time steps, where single raster cells can be influenced by their neighborhood or other data in an overlay. All map layers are imposed on the same grid system. This type of GIS environment provides a sophisticated tool to help us target the real problem in a complex system (Wolfram, 1984; Coulelis, 1985; Itami, 1994; White & Engelen, 2000; Giles, 2002). In our study, we use GIS analysis, linear regression, MCE, cellular automata (CA), and a raster image calculator to build a unique Vegetation Dynamic Simulation Model (VDSM). Global warming scenarios are interpreted as inputs of the spatial parameters. Large processing tasks are completed by the computer system.

The predicted outcome of this study is that individual vegetation types will respond to a global mean temperature increase (GMTI) in 2100 of 1 or 3°C by either expanding or shrinking their range because of plant species' suitability to the warmer and drier climate conditions. This corresponds to 0.1 or 0.3°C per decade, respectively (Johnes & Briffa, 1992; Leemans, 2004).

3.1 Methods and simulation model

3.1.1 Data set

Our study area is located near the Haibei Alpine Meadow Ecosystem Resaerch Station, Qinghai Province, China (37°29'-37°45'N, 101°12'-101°33'E) (Zhang & Zhou, 1992). The elevation in the study area varies from 3000 to 4500 m a.s.l. The model uses the following data:

- A 90 m x 90 m resolution DEM;
- Temperature derived from the DEM using an empirical linear regression model (Zhang, 2005);
- Land surface parameters (aspect, slope, stream channel density) derived from the DEM using GIS analysis tool provided in the IDRISI software (Eastman, 2003);
- A raster vegetation map (30 m x 30 m pixels) produced in 1988 (Zhang, 2005). The vegetation map with a total of 10 vegetation classes is resampled to match the DEM resolution (90 m x 90 m) (Fig. 3).

3.1.2 Multi-criteria evaluation

Constrains are defined as the limited area that are not considered to be natural vegetation, such as water bodies, glaciers, gravel slpoes and artificial grasslands. They represent area where the natural vegetaion cannot grow or are otherwise constrained. A Boolean image is created to display inclusion and exclusion of the constraint conditions.

$$y = MCE\ (F\,1, F\,2, F\,3, ...Fn) \qquad (1)$$

The factors (F1, F2, ...Fn) used in the MCE are selected based on the most important variables that determine the output y in Equation (1).

We use MCE in step 1 (Fig. 4) to determine a normalized surface temperature, which calibrates the temperature by these spatial parameters: aspect, suitable surface temperature, and distance to the nearest stream channel.

The temperature varies along with these spatial parameters: (1) It increases from the north (a=0°) to the southeast (b=145°), with the highest values from the southeast to the southwest (c =275°), then decreasing to the north (d =360). The change pattern can be described as a sigmoidal fuzzy function type, in a symmetric shape with specific values (a, b, c, d) in Table 4. (2) In the lower valley of our study area, the monthly mean temperature in July is 10.1°C

(Li et al., 2004). The temperature decreases with increasing elevation. We define the temperature less than a=0°C as unsuitable for alpine plant growth. The temperature from 0°C to b=5°C is defined as less suitable for alpine plant growth; the temperature from 5°C to c=13°C is defined as most suitable. The temperature from 15.5°C to d=18°C reduces the

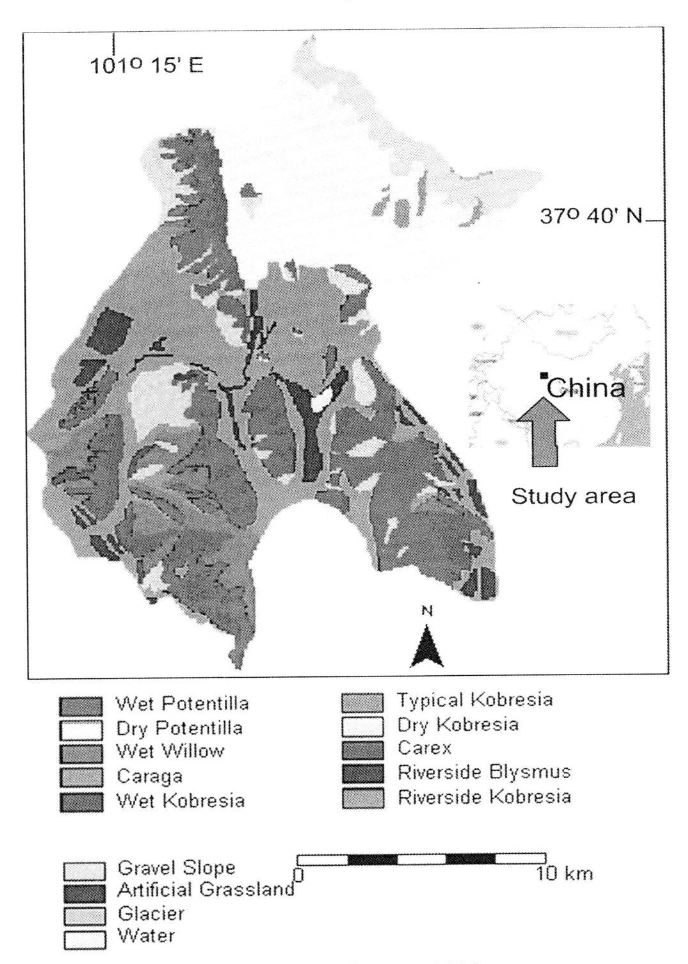

Fig. 3. Haibei Alpine Tundra Vegetation Distribution, 1988

suitability for alpine plant growth, representing the dry, south-facing areas (Table 4). In the Analytical Hierarchy Process (AHP), the most important temperature is in the range of b=5°C to c=13°C, which present the sigmoidal fuzzy function type and symmetrical shape. (3) The distance to a stream affects the temperature. If the area is within 10 m of a stream or water body, its temperature is closer to the stream or water temperature; beyond 600 m from a stream or water body, the temperature is minimally affected by the nearest water bodies. The values are defined as a = 10 m, b = 600 m, with a sigmoidal fuzzy function type and monotonically deceasing shape.

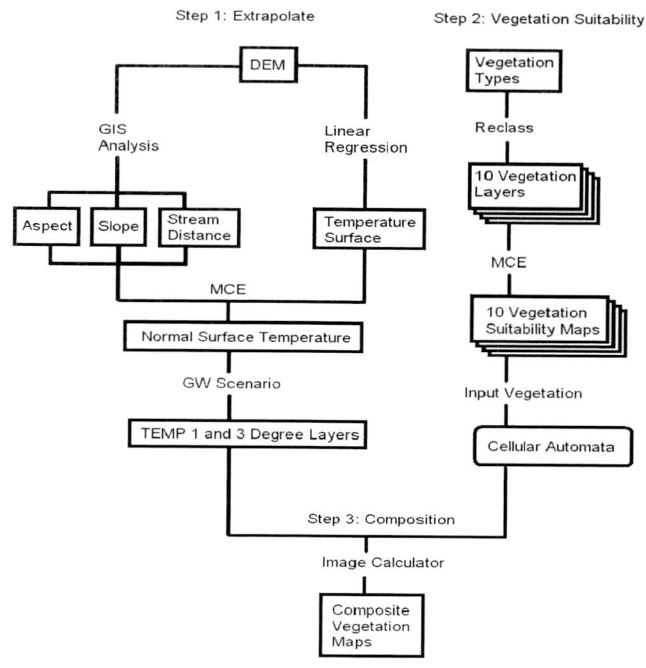

Fig. 4. Vegetation Dynamic Simulation Model (VDSM)

	Factors to standardize with fuzzy	Membership function type	Membership function shape	a	b	c	d	AHP weights
Temperature	aspect	Sigmoidal	Symmetric	0	145	275	360	0.2631
	temperature suitability	Sigmoidal	Symmetric	–2	5	13	16	0.5472
	distance to stream	Sigmoidal	Monotonically decreasing	10	600	—	—	0.1897

Notes: aspect: 0–360u; temperature suitability: uC; distance to stream: meters; a, b, c, d: suitabilities for the factors; AHP weights derived from pairwise comparison up to an acceptable level.

Table 4. Normalized surface temperature's fuzzy membeship function types and shapes, factor's suitability values (a,b,c,d) and AHP weights

In step 2 (Fig. 4), we use the MCE method and create a suitability map for each vegetation type using the aspect, slope, and the distance to stream channels. The suitability values (a, b, c, d) in Table 5 are defined based on the vegetation distribution in the study area (Zhang & Zhou, 1992). The fuzzy membership function type and shape and AHP weights for each vegetation are calculated and reported in Table 5 (Eastman, 2003).

3.1.3 Macro modeler

The macro model is created to simulate changes in each vegetation type through time and space, integrating operations such as overlay, scalar, fuzzy module, and cellatom (Fig. 5). These operations are available in the IDRISI software package (Eastman, 2003) and have to be built in Macro Modeler with an initial scalar value (0.0–1.0) and fuzzy set values (0–255). The GMTI scenarios are implemented in the Macro Modeler by adjusting the scalar operation to increase the temperature by 0.1°C in a discrete time period. The same logic is applicable to 0.3°C in a discrete time period (Leemans, 2004). Running the simulation for 10 iterations, the effects of increasing temperature on each vegetation type are accumulated in the output image.

Vegetation type	Factors to standardize with fuzzy	Membership function type	Membership function shape	a	b	c	d	AHP weights
Wet Potentilla Shrub	aspect	Sigmoidal	Monotonically Increasing	0	3	—	—	0.5842
	Slope	Sigmoidal	Symmetric	0	10	20	48	0.2318
	distance to stream	Sigmoidal	Symmetric	0	1500	1800	3700	0.184
Dry Potentilla Shrub	aspect	Sigmoidal	Symmetric	0	160	240	360	0.2627
	Slope	Sigmoidal	Monotonically Increasing	0	25	—	—	0.1591
	distance to stream	Sigmoidal	Monotonically Increasing	0	1800	—	—	0.5782
Wet Willow Shrub	aspect	Sigmoidal	Symmetric	0	250	300	360	0.4434
	slope	Sigmoidal	Symmetric	0	15	22	48	0.3874
	distance to stream	Sigmoidal	Symmetric	0	500	1900	3700	0.1692
Caragana Shrub	aspect	Sigmoidal	Symmetric	0	240	255	351	0.4836
	slope	Sigmoidal	Symmetric	0	6	12	20	0.1677
	distance to stream	Sigmoidal	Symmetric	0	500	600	1000	0.3487
Wet Kobresia Meadow	aspect	Sigmoidal	Symmetric	0	240	315	360	0.625
	slope	Sigmoidal	Symmetric	0	10	16	30	0.2385
	distance to stream	Sigmoidal	Monotonically decreasing	—	—	200	600	0.1365
Typical Kobresia Meadow	DEM	Sigmoidal	Monotonically deceasing	—	—	3100	3300	0.2599
	slope	J-shape	Monotonically deceasing	—	—	5	10	0.3275
	distance to stream	Sigmoidal	Symmetric	0	50	1900	3000	0.4126
Dry Kobresia Meadow	aspect	Sigmoidal	Symmetric	0	200	270	300	0.5936
	slope	Sigmoidal	Symmetric	0	1	8	18	0.2493
	distance to stream	Sigmoidal	Symmetric	0	200	1200	1800	0.1571
Carex Meadow	aspect	Sigmoidal	Symmetric	0	70	120	200	0.4934
	slope	Sigmoidal	Symmetric	0	10	15	20	0.3108
	distance to stream	Sigmoidal	Symmetric	0	500	900	1500	0.1958
Riverside Blysmus Meadow	DEM	Sigmoidal	Monotonically decreasing	—	—	3200	3300	0.6483
	slope	Sigmoidal	Monotonically decreasing	—	—	0	3	0.2297
	distance to stream	Sigmoidal	Monotonically decreasing	—	—	0	600	0.122
Riverside Kobresia Meadow	DEM	Sigmoidal	Monotonically decreasing	—	—	3200	3300	0.5396
	slope	Sigmoidal	Monotonically decreasing	—	—	0	10	0.297
	distance to stream	Sigmoidal	Monotonically decreasing	—	—	0	600	0.1634

Note: Slope:0-90°; DEM: indicates elevation range for relatively flat area; for others, refers to the note on Table 4

Table 5. Ten vegetation types' Fuzzy membership function types and shape, factor's suitability values (a,b,c, and d), and AHP weights

The cellular automata module is implemented in the Macro Modeler (Fig. 5) and used to form a uniform raster image to represent global warming effects in a spatial context, which operates over discrete time steps (Coulelis, 1985; Giles, 2002; White & Engelen, 2000). The change in each cell depends on the parameters or requirements set by the user and the surrounding neighbors (Wolfram, 1984; Itami, 1994; Ruxton & Saravia, 1998). This project used the CELLATOM module with a 3*3 filter and reclassifies an output cell if at least 3 neighbors contain non- null values. We define the 10 iterations using the DynaLink module (Eastman, 2003). Winthin each iteration, each vegetation map is dynamically updated by running Cellatom, then it is overlaid with the vegetation suitability map altered by a GMTI of 0.1 or 0.3°C. Thus, after 10 iterations, a final suitable vegetation map is produced. The same dynamic processing is repeated for each vegetation type resulting in a total of 10 vegetation suitability maps in response to warmer weather.

3.1.4 Composite final vegetation map

Our objective is to create a composite vegetation map for each global warming scenario, GMTI of 1 or 3°C over time (Fig. 4). All of the 10 vegetation suitability maps with a GMTI of 1 or 3°C are combined in order to produce a composite map using the image calculator module in IDRISI (Eastman, 2003). The highest suitability among the vegetation types is selected to represent the successful vegetation type in that cell.

$$Veg\_dominant = MAX(Veg1, Veg2, ...Veg10) \quad (2)$$

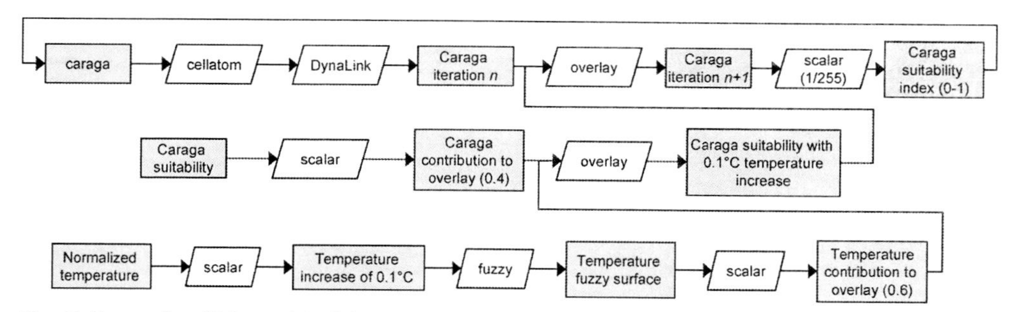

Fig. 5. Example of Macro Modeler incorporating Cellular Automata for Caragana Shrub

Equation (2) above creates a map where the value at each pixel corresponds to the vegetation type with the highest probability of thriving, where Veg1, Veg2 ... Veg10 correspond to each of the 10 vegetation types. The 10 resulting maps of dominant vegetation types are overlaid to produce a composite distribution of the most probable vegetation types.

3.2 Results and discussions

3.2.1 Normalized temperature spatial distribution

Temperature changes across the study area are not only due to elevation, but also due to aspect and distance from the nearest stream channel. The linear regression model provided a temperature spatial distribution based on elevation alone, which is our primary step. Furthermore, the normalized temperature surface created by the MCE is highly representative of the potential temperature distribution in a normalized fuzzy format (Fig. 6).

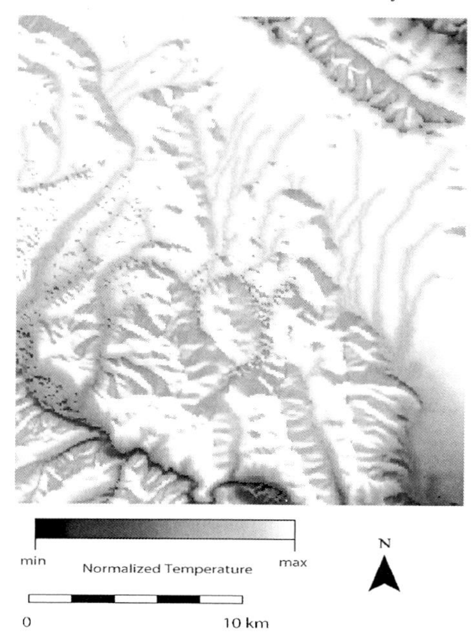

Fig. 6. Normalized temperature with values from 0 to 255.

Temperature distribution is correlated with and controlled primarily by elevation. Numerous spatial interpretation methods have been applied to estimate the spatial distribution of temperature (Li, 2005). The interpolation results do not always agree with the actual sample points, including using geo-statistical methods, and spatio-temporal spline. These methods are highly dependent on the distance to the sample points, and the surface equation. In our study, our first step is to create the primary temperature surface based a linear relation with elevation. The objective is to obtain a more accurate temperature map in terms of aspect, suitable temperature, and distance to the stream. We use the Multi-Criteria Evaluation with Weighted Linear Combination (MCE_WLC) to calibrate the spatial temperature distribution. The fuzzy memberships between the temperature and each factor (aspect, suitable temperature, distance to stream) are based on previous research works (Zhang & Zhou, 1992; Zhang & Welker, 1996; Zhang, 2005). The output, the normalized temperature surface is set into fuzzy format (0–255). Since the temperature is major factor on determining vegetation composition, structure, and distribution, the normalized temperature surface plays an important role when we simulate vegetation dynamics in spatial and temporal dimensions.

3.2.2 Vegetation change comparison

We calculate the percent area change for each vegetation type (Fig. 7) before we compose the final vegetation map. By increasing the global mean temperature (0.1 or 0.3°C per decade), the percent area change of each vegetation type indicates the potential expansion from their original ranges. For example, without competing with other types of vegetation, both Dry *Potentilla* Shrub and Dry *Kobresia* Meadow could expand their vegetation area by 23%. The difference between them is that Dry *Potentilla* Shrub responds more positively to a GMTI of 1°C, and Dry *Kobresia* Meadow responds more positively to a GMTI of 3°C. Similar phenomena are also observed in other vegetation types.

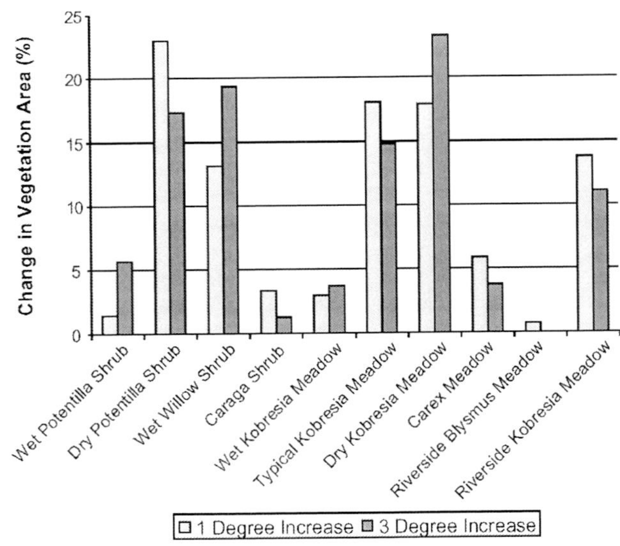

Fig. 7. Percent change in vegetation area with GMTI at 1 and 3°C.

After we compose the final vegetation map, the highest suitability among the vegetation types is finally selected to represent the successful vegetation type in every cell. For instance, Dry *Potentilla* shrub and Dry *Kobresia* meadow expand into areas previously occupied by wet types of vegetation (Fig. 8). The Riverside *Blysmus* meadow, which requires moist conditions, disappears completely with a 3°C temperature increase. In general, the dry vegetation types demonstrate significant expansion from their original ranges and tend to become more dominant in the study area.

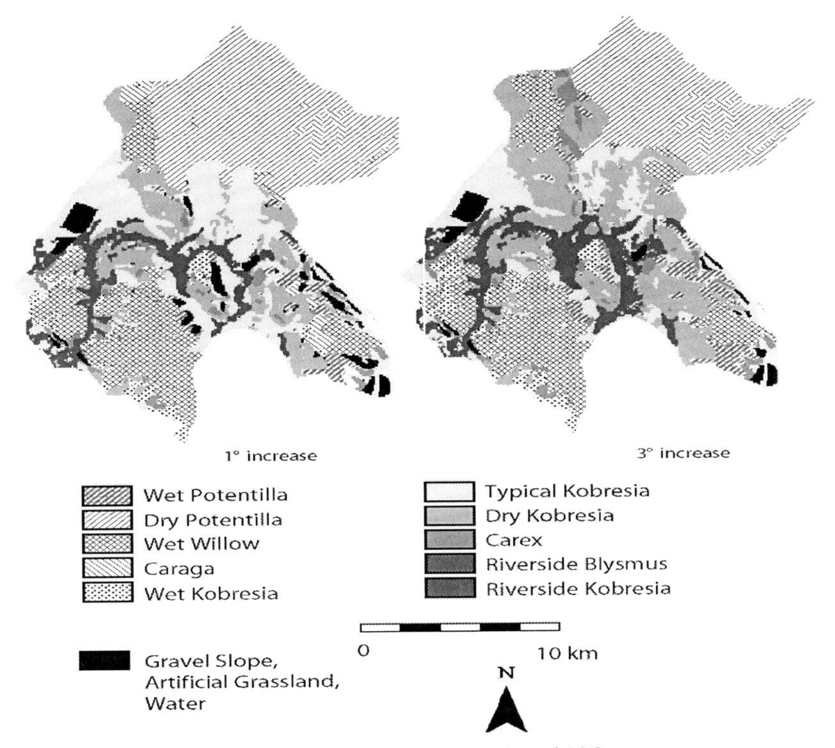

Fig. 8. Final Composite Vegetation Maps For GTMI at 1 and 3°C

3.2.3 Vegetation dynamics over time

The time dimension in the CA module of IDRISI is interpreted as a discrete time step, which corresponds to a temperature-time dimension. The CA module is integrated into the macro modeler, and uses the DynaLink module (Eastman, 2003) to simulate vegetation change within each iteration. The GMTI is defined as a spatial parameter in the deterministic model with a temperature increase of 0.1 or 0.3°C per decade.

3.2.4 Vegetation Dynamic Simulation Model (VDMS)

The VDSM is an example showing that spatial modeling can solve complex ecosystem problems in terms of having the capability to simulate spatial and temporal vegetation dynamics. In this case study, we model the spatial distribution of temperature and create

GMTI scenarios as a spatial grid image. The vegetation dynamics are simulated in discrete time by applying CA in a Macro Modeler. In future studies, this model will be capable of modeling the water-time dimension that makes the simulation more adaptable to global warming research. The VDSM could be potentially incorporated with a normal climate change model to assist in a long-term ecosystem simulation. Alternatively, the VDSM is capable of linking with a stochastic model of temperature change, in which we might be able to forecast an ecosystem disaster.

The VDSM starts by evaluating how to solve a complex vegetation dynamics problem using CA. The VDSM is built by combining the MCE, Macro Modeler, CA, image calculator, Scalar, and Fuzzy functions in IDRISI. Making a clear objective helps us to look into many available modelers and functions in order to solve the problem within the IDRISI software environment. For instance, using Decision Wizard, we create an objective and define a set of constraints, which eliminates the areas that are not natural vegetation. The spatial patterns of the factors (temperature, aspect, slope, and distance to stream) are created as continuous surfaces using Fuzzy functions. The transition rules in Macro Modeler are defined as the maximum potential suitable vegetation in each cell as well as over the study area. The frame work of VDSM (Fig. 4.) is a summary of the model structure and functionalities. VDSM is flexible enough to be integrated with other sub-models that are available from GIScience technology. Figure 3 provides an example of Macro Modeler incorporating Cellular Automata for the modeling of *Caragana* Shrub. In the case of a GMTI of 0.1°C, the suitability map for *Caragana* is weighted at 0.4 and the normalized temperature map is weighted at 0.6. These two factors are combined using the overlay module to produce a map of suitability for *Caragana* Shrub with an incremental temperature increase of 0.1°C. Ten iterations on one vegetation layer are simulated and updated after each iteration using the Cellatom and Dynalink modules. At the end of the iterations, we obtained an accumulated effect of GMTI of 1°C on the vegetation layer.

It would also be possible to incorporate the water layer with the vegetation layer and temperature layer, which can be linked by its weighted value in the VDSM. Thus, the VDSM not only provides a discrete time representation, but also demonstrates how we could develop our model for use with more complex scenarios (Chapin et al., 2006; McGuire et al., 2006).

Compositing a final vegetation map demonstrates the power of GIS analysis in IDRISI. The image calculator module in IDRISI successfully carries out interpolation of the image calculation presented in Equation (2) at the grid cell level. The VDSM illustrates how we could study vegetation dynamics and model many other spatio-temporal phenomena.

The VDSM integrates the suitability maps created from MCE, Macro-Modeler, CA, and spatial environmental factors. And the temperature-time dimension model is incorporated into the VDSM, which makes the temperature a spatial parameter that affects the vegetation dynamics over a discrete time step. The simulating processes conducted by Macao Modeler generate the temperature increase of 0.1 to 0.3°C per decade, which represents the influences of the different global warming scenarios. The results from Fig. 7 and 7 demonstrate that global temperature increase reduces moisture availability (Zhang & Welker, 1996) such that dry vegetation can invade areas previously occupied by vegetation adapted to moist conditions. The structure of the model is generally applicable to other situations, but the particular factors and constraints used in this model are unique to the Haibei alpine tundra ecosystem.

Global warming has strong effects on the alpine ecosystems in terms of altering the biomes and ecosystem biodiversity (Cao & Woodward, 1998; Ni, 2000; Song et al., 2005, Chapin et al., 2006). The alpine ecosystem in the region of the Qinghai-Tibetan plateau is sensitive and vulnerable to the changing climate (Zhang & Welker, 1996; Kato et al., 2006). The VDSM illustrates that altering global mean temperature changes the alpine vegetation dynamics in terms of having the capability to simulate spatial and temporal vegetation dynamics (Itami, 1994; Leemans, 2004). With the future integration of water condition (Sala et al., 1992; Hodkinson et al., 1999) and disturbance regimes (Zhang, 1990; Cincotta et al., 1992; Zhang & Liu, 2003; Chapin et al., 2006) into the VDSM, the simulation could model more detailed mechanisms and complex feedbacks (McGuire, 2006) of the alpine tundra ecosystem to the changing climate.

4. Simulating Tibetan Plateau alpine vegetation distribution in response to global warming

Vegetation patterns on the plateau were very sensitive and vulnerable to global change, where the growth and distribution of plants depended heavily on local climate conditions (Hou et al., 1982; Zhang et al., 1996). The undisturbed vegetation on the Tibetan Plateau provides an ideal natural laboratory for the research on the sensitivity and responses of alpine vegetation to climate changes. The distributions of the major dominant species and of the vegetation types on the Tibetan Plateau have been investigated since the early 1950s (Anon., 1985). Based on previous research, Zheng (1996) depicted the physiography of the Tibetan Plateau. Ni (2000) simulated biomes on the Tibetan Plateau using the improved BIOME3 model (BIOME3-China) under the present climate conditions, as well as under a scenario with a CO_2 concentration of 500 ppmv. The BIOME3-China used nine plant functional types (PFTs); it did not include the PFTs especially occurring in alpine vegetation, such as cold graminoid or forb, and cushion forb.
In this study, a combined biogeography biochemistry model, BIOME4 (Kaplan et al., 2003) was improved to simulate alpine vegetation at the biome level. We apply the model to the present day, and the end of the 21st century in a scenario with unchecked increase in atmospheric CO_2 concentra tion. We compare the modelled present vegetation to a map of present-day natural vegetation distribution. The future scenario allows us then to assess the sensitivity of alpine vegetation to changes in atmospheric CO_2 concentration and climate.

4.1 Methods

4.1.1 Model description

BIOME4, developed from the BIOME3 model (Haxeltine & Prentice, 1996), is a intergrated carbon and water flux model that predicts the global steady state of vegetation distribution, structure, and biogeochemistry, taking account of interactions among these aspects. The BIOME4 model followed most of the algorithms and rules of BIOME3. It is driven by long-term averages of monthly mean temperature, sunshine and precipitation. In addition, the model requires information on soil texture and soil depth in order to determine water holding capacity and percolation rates. CO_2 concentration is specified. For BIOME4, the improved model using 12 plant functional types (PFTs) that represent broad, physiologically distinct classes, ranging from alpine vegetation (e.g. cushion forbs) to tropical rain forest trees. The PFTs are tropical broad-leaved evergreen, tropical broad-leaved rain-green, temperate broad-leaved evergreen, temperate broad-leaved summergreen, temperate coniferous evergreen, boreal coniferous evergreen, temperate summergreen conifer, temperate grass, temperate xerophytic shrub, cold shrub, cold graminoid or forb, and cushion forb (Ni, 2000; Kaplan et al., 2003; Yu, 1999). Each PFT is assigned a small number of bioclimatic limits which determine whether it could be present in a given grid cell (Table 6).

Plant Functional Types	Min Tc	max	GDD Min	GDD_0 min	min Tw	max	α min	D
Tropical broad-leaved evergreen	16						0.85	1
Tropical broad-leaved raingreen	12	16					0.80	1
Temperate broad-leaved evergreen	7	12					0.70	2
Temperate broad-leaved summergreen	3	7	1500				0.65	3
Temperate coniferous-leaf evergreen	- 0	3	1200				0.60	3
Temperate summergreen conifer	- 5	7	1200				0.60	3
Boreal coniferous-leaf evergreen	- 8	0	350				0.75	3
Temperate xerophytic shrub	- 2	2					0.30	4
Cold shrub	- 8	- 1		1000	7	11	0.25	5
Temperate grass	- 2	1		1800			0.25	6
Cold graminoid or forb		- 13			5	7	0.40	7
Cushion forb				100			0.45	7

Table 6. Bioclimatic limits of each plant functional type in the model (Tc stands for mean temperature of the coldest month, Tw for mean temperature of the warmest month, GDD for growing degree-days on 5°C base, GDD0 for growing degree-days on 0°C base, α for Priestley-Taylor coefficient of annual moisture availability and D for dominance class).

The computational core of BIOME4 is a coupled carbon and water flux scheme, which determines the seasonal maximum leaf area index (LAI) and maximizes NPP for any given PFT, based on a daily time step simulation of soil water balance and monthly process-based calculations of canopy conductance, photosynthesis, respiration and phenological state (Haxeltine & Prentice, 1996). To identify the biome for a given grid cell, the model ranks the tree and non-tree PFTs that were calculated for that grid cell. The ranking is defined according to a set of rules based on the computed biogeochemical variables, which include NPP, LAI, and mean annual soil moisture. The resulting ranked combinations of PFTs lead to an assignment to one of the biomes (Table 7).

4.2 Climate scenarios

4.2.1 Modern climate data

A Chinese grid-based long-term mean climatology (temperature, precipitation and sunshine) database (1961-1990) was used for a modern vegetation simulation, and as a baseline for other modeling experiments. The climate data for China simulated by the PRISM climate model (Parameter-elevation Regressions on Independent Slopes Model) developed by the Oregon Space Climate Research Center. Daly (1994, 2000) simulated the Chinese 0.05° × 0.05° gridded long-term mean climatology based on 2450 station mean values for monthly temperature, monthly percentage of potential sunshine hours, and monthly total precipitation throughout China and its adjacent regions. An atmospheric CO_2 concentration of 340 ppmv was used to link BIOME4 to the present-day baseline simulation.

Biomes	Plant functional types
Sub-tropical broad-leaved forest	Tropical broad-leaved evergreen Tropical broad-leaved raingreen Temperate broad-leaved evergreen
Mantane broad-leaved forest	Temperate broad-leaved evergreen Temperate broad-leaved summergreen
Sub-alpine coniferous-leaf forest	Temperate coniferous-leaf evergreen Temperate summergreen conifer Boreal coniferous-leaf evergreen
Montane shrub steppe	Temperate xerophytic shrub Temperate grass
Montane steppe	Temperate grass Temperate xerophytic shrub
Alpine meadow	Cold graminoid or forb Cushion forb
Alpine steppe	Cold graminoid or forb Cold shrub
Montane desert	Cold shrub Cold graminoid or forb
Alpine desert	Cold shrub Cold graminoid or forb
Deciduous coniferous broad–leaf forest	Temperate broad-leaved summergreen Temperate coniferous-leaf evergreen Temperate summergreen conifer

Table 7. Biomes and plant functional types on the Tibetan Plateau at present

4.2.2 Future climate projection

The climatic conditions under increasing greenhousegas concentrations and sulfate aerosols have been simulated by atmospheric general circulation models (AGCMs). These models were commonly used in the construction and application of climate change scenarios for climate change impacts assessments (Neilson et al., 1998; Cramer et al., 2001). HadCM3 is a coupled atmosphere-ocean GCM developed at the Hadley Centre (Cox et al., 1999). The model was driven by computing the averages for 1931-1960 and for 2070-2099. We used the mean climate anomalies, and then interpolated the anomalies to the grid in high resolution (Fig. 9).

The anomalies were added to the baseline climatology to produce the climate fields used to drive improved BIOME4 to assess the sensitivity of alpine vegetation to possible future climate changes. The emissions scenario (Anon., 1996) included an increase in atmospheric CO_2 concentration from 340 to 500 ppmv and increase in sulphate aerosol concentration for the 21st century simulation. The simulation is not intended as a realistic forward projection and it was used to illustrate a possible course of climate change and thus to give an impression of the sensitivity of alpine ecosystems to climate change.

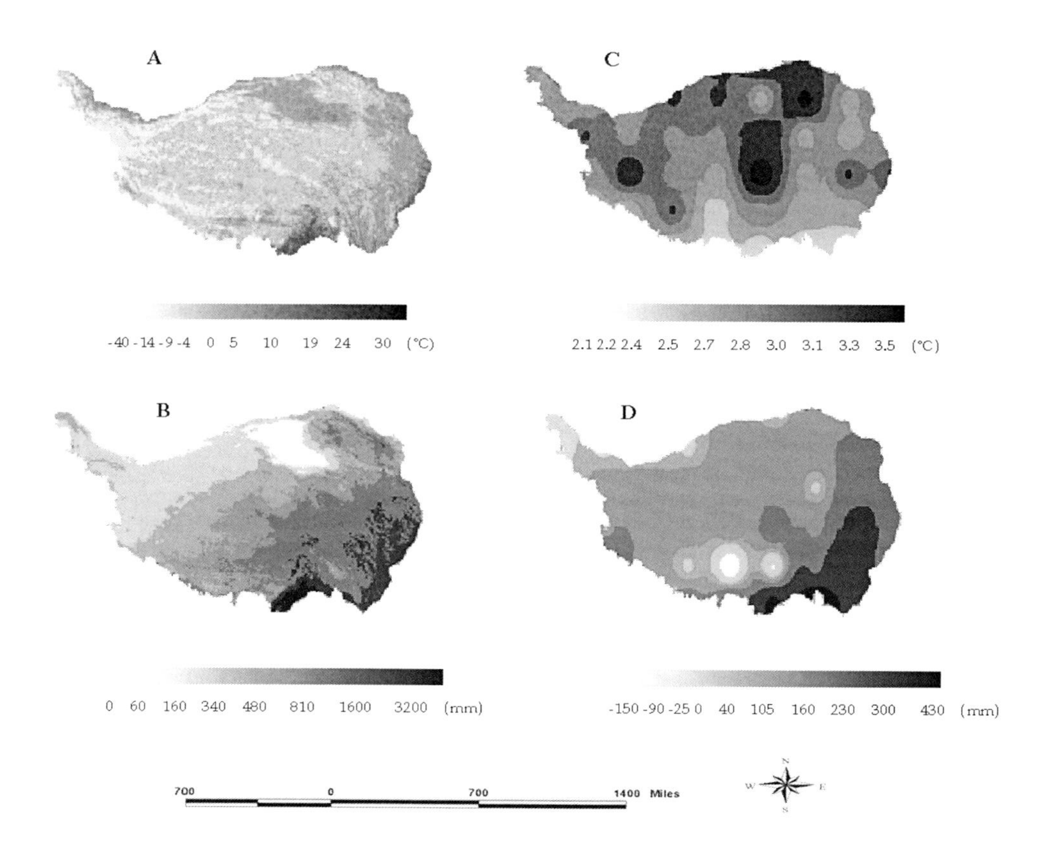

Fig. 9. Annual mean temperature (**A**) and annual precipitation (**B**) on the Tibetan Plateau, and anomalies in annual mean temperature (**C**) and annual precipitation (**D**) simulated by the Hadley Centre GCM (Johns et al., 1997; Mitchell et al., 1995).

4.2.3 Soil data

A digitized soil texture data set for the Tibetan Plateau was derived from Xiong & Li (1987). The soil texture information was interpolated to 0.05° × 0.05° grid cells. Eight soil types were classified.

4.2.4 Vegetation data

A map of potential natural vegetation of the Tibetan Plateau on 0.05° × 0.05° grid cells was derived from a digital vegetation map at a scale of 1 : 4 000 000 (Hou et al., 1982), which presents 113 vegetation units. These units were classified into nine categories based on the physical-geographical regions system of the Tibetan Plateau (Zheng, 1996). Each vegetation type was required to be floristically distinguishable to compare them with simulated vegetation maps (Fig. 10a, b).

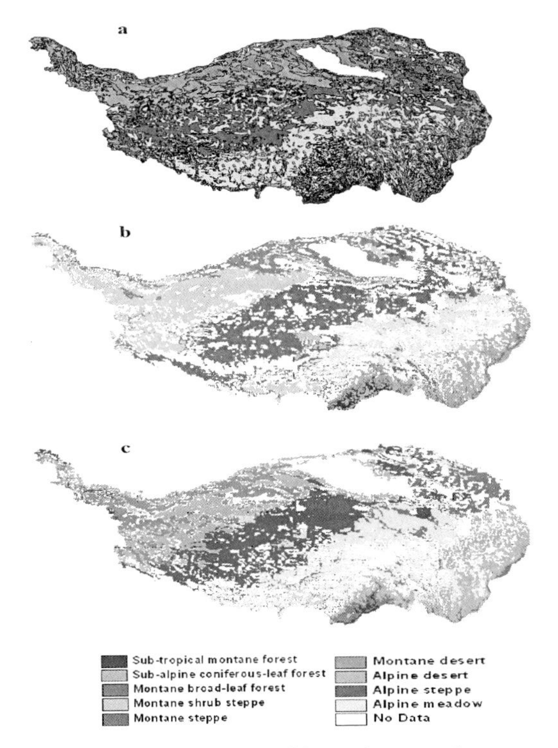

Fig. 10. Biomes on the Tibetan Plateau
a. Natural vegetation patterns
b. Biomes simulated by improved BIOME4
c. Biomes predicted by improved BIOME4

4.2.5 Assessment of the simulated results

The agreement between simulated and natural vegetation maps or reconstructed vegetation maps was quantified by the ΔV value. ΔV is a nontrivial and attribute-based measure of dissimilarity between biomes (Sykes et al., 1999). Dissimilarity between two maps (ΔV) was obtained by area-weighted averaging of ΔV over the model grid. The criterion of ΔV was cited (Sykes et al., 1999). ΔV values < 0.15 can be considered to point to excellent agreement between simulated and actual distributions, 0.15-0.30 is very good, 0.30-0.45 good, 0.45-0.60 fair, 0.60-0.80 poor, and > 0.80 very poor.

4.3 Results and discussions

4.3.1 Present day

In a quantitative comparison between the simulated vegetation map and the modern natural vegetation map, 80.1% of grid cells (80100 cells) showed the same biome (Fig. 10). Percentage agreement for grid cells assigned to specific biomes in the natural vegetation map were: sub-tropical montane forest 65.4%; sub-alpine coniferous forest 50.5%; montane broad-leaved forest 49.7%; montane shrub steppe 43.6%; montane steppe 55.0%; montane desert 77.9%; alpine desert 81.3%; alpine steppe 85.1%; alpine meadow 68.5%. The ΔV values of each biome suggest that it is in excellent agreement for montane broad-leaved forest, sub-alpine coniferous forest and montane desert, and very good agreement for sub-tropical montane forest and alpine desert, and a good agreement for montane steppe, and fair agreement for alpine meadow and alpine steppe, and poor agreement for montane shrub steppe (Table 8 and Fig. 10a,b).

Biome	Area (×1000 km²)		ΔV	
	A	B	C	D
Sub-tropical montane forest	86.09	79.91	0.21	0.65
Montane broad-leaved forest	200.13	210.93	0.13	0.27
Sub-alpine coniferous-leaf forest	320.05	332.16	0.14	0.33
Montane shrub steppe	120.91	130.42	0.61	0.67
Alpine meadow	263.10	194.09	0.55	0.58
Montane steppe	158.11	296.13	0.37	0.81
Alpine steppe	675.98	583.52	0.52	0.41
Montane desert	373.39	449.25	0.14	0.22
Alpine desert	244.96	169.64	0.22	0.39
No data (Lake)	79.24	75.91		
Total	2521.96	2521.96	0.32	0.51

Table 8. Area (× 1000 km²) and ΔV values for each biome of the Tibetan Plateau. A = areas of simulated biomes under the current climate with CO_2 concentration = 340 ppmv; B = areas of simulated biomes under a scenario at the end of the next century with CO_2 concentration = 500 ppmv; C = ΔV values for comparison between simulated biome under current climate and actual vegetation distribution; D = ΔV values for comparison between simulated biome under a scenario with CO_2 concentration of 500 ppmv and simulated biome under current climate with CO_2 concentration of 340 ppmv.

4.3.2 Sensitivity to future changes

In the illustrative simulation of a 'greenhouse climate', the potentially forested area of the Tibetan Plateau increased substantially (Fig. 10c). The area of sub-tropical montane forest is slightly reduced, with replacement by montane broad-leaved and sub-alpine coniferous forest. The simulated tree line is farther north in most sectors than at present. Trees potentially invade shrubland/ meadow types where only fragments of forest exist today. Thus the simulations indicate a great sensitivity of the forest limit to CO_2-induced warming (Lloyd & Rupp et al., 2003; Lloyd & Fastie, 2003). The 'greenhouse climate' simulation also indicates major northward shifts of the alpine meadow biomes and a future reduction in the areas occupied by shrub-dominated montane steppe. The boundary between montane desert and alpine desert is found farther south than today. Our model results indicate that the extension of alpine desert would be reduced, while the area of montane desert would increase under the future climate scenarios with an atmospheric CO_2 concentration of 500 ppmv (Fig. 10c).

The improved BIOME4 model captures the main features of vegetation distribution on the Tibetan Plateau, such as the position of the alpine forest limit, its species composition in vegetation, regional differentiation in vertical vegetation, and the extent of alpine meadow, alpine steppe, and alpine desert. The spatial differentiation of physical-geographical regions on the plateau is determined mainly by topographic configuration and atmospheric circulation. The climate is warm and humid in the southeast, and cold and arid in the northwest (Zheng, 1996). The reduction in temperature and precipitation toward the northwest is the most important reason for the simplification of species complexity in the vegetation (Zhang et al., 1996). The vegetation types in this region change gradually from marine humid montane (tropical seasonal and rain forest, warm-temperate broad leaved evergreen forest, temperate deciduous forest, and conifer forest) in the southeastern region to continental semi-arid montane (temperate shrubland/meadow, temperate steppe, alpine meadow/shrubland, and alpine steppe) in the middle region to continental arid montainous (temperate desert, alpine desert, and ice/polar desert) in the north- western region (Ni, 2000).

The improved BIOME4 model simulated the biome distribution with very good agreement for the central and northwestern regions of the Tibetan Plateau (DV = 0.26 for non-forests), and with a good agreement for the southeast (ΔV = 0.32 for forests). Altogether 13.8% of the forest cells were simulated as non-forest due to misclassification, i.e. cold needle-leaved evergreen or cold deciduous forest cells were simulated as low and high shrub meadow, and 7.1% of non-forest cells were simulated as forest due to low and high shrub meadow cells being simulated as the tree-line forming biome. Under the control of both climate and complex physiognomy, the actual vegetation pattern on the Tibetan Plateau is a mosaic, especially for forest types in flat regions (Anon., 1980). But in our simulation, the model produced vegetation types with continuous distribution leading to unrealistic patterns. The major mismatches (where > 20% of cells assigned to one biome in the natural vegetation map were assigned to a different biome in the simulation) were between adjacent biomes in climate space (Fig. 10a, b). The simulated boundary between alpine meadow and alpine steppe is somewhat too far south. The natural vegetation map shows the boundary between alpine steppe and alpine desert farther northwest than the simulation, apparently because of lower temperature and humidity. Our model results cannot distinguish ice/polar desert from alpine desert (Fig. 10a, b). Vegetation patterns simulated by improved BIOME4 are similar to those modelled by Ni (2000) using BIOME3-China. In our simulation, shrubland and meadow were distinguished using additional PFTs specifically occurring in alpine vegetation (cold shrub, cold graminoid or forb, and cushion forb). Therefore, areas of montane steppe and alpine meadow simulated by improved BIOME4 are more precise.

In the simulation of future developments triggered by increased atmospheric CO_2 concentration both winter and summer temperatures rise throughout the region (Fig. 9). Simulated temperature anomalies in winter are generally higher than in summer. This trend can be confirmed by the climate change on the Tibetan Plateau during recent years, i.e. from 1951 to 1990 (Tang et al., 1998). Thus the CO_2 increase causes a large, year-round warming which produces a stronger effect on vegetation shifts. For example, there would be a reduction in sub-tropical montane forest, alpine meadow, alpine steppe and alpine desert, and an extension of montane broad-leaved forest, sub-alpine coniferous forest, montane shrub steppe, montane steppe and montane desert. These results are consistent with other reports that suggest a northward shift of the vegetation on the Tibetan Plateau under a warming climate (Ni, 2000; Zheng, 1996; Zhang et al., 1996).

5. Acknowledgments

We thank all these people Dr. Preminda Jacob and Chen Bo for assistance in the field at Haibei Alpine Meadow Ecosystem Station; Dr. Suzana Dragicevic, Verda Kocabas, and the

Simon Fraser University Spatial Information Systems (SIS) lab for their support of this research project; Dr. Jian Ni for his help from Max-Planck Institute for Biogeochemistry, Jena, Germany. Research was supported by Haibei Alpine Meadow Ecosystem Station 90-0318, the Biosphere Program, U. S. State Department Grant 1753-900561, and in part by U.S. International Tundra Experiment (USITEX)(NSF/OPP-9321730), and was financially supported in part by The Key Project funded by the Chinese Academy of Sciences (KZCX3-SW-339), and The National Natural Science Foundation (40331066). We thank all these exports who participated in these projects, Prof. XingMin Zhou, Dr. Richard Cincotta, Dr. CaiPing Zhou, Prof. Hua Ouyang, Dr. Mechael Peterman, Dr. Dorin Aun, Prof. YanMing Zhang, and Dr. Andy Parson.

6. References

Anon. (The scientific expedition teams to the Tibetan Plateau, Chinese Academy of Sciences) (1980). *Vegetation of Tibet*. Science Press, Beijing, CN. (In Chinese.)

Anon. (The scientific expedition teams to the Tibetan Plateau, Chinese Academy of Sciences) (1985). *Forests of Xizang*. Science Press, Beijing, CN. (In Chinese.)

Anon. (Intergovernmental panel on climate change working group I). 1996. *Climate change 1995: The science of climate change*. Cambridge University Press, New York, NY, US.

Billings, W. D. (1987). Constraints to plant growth, reproduction and establishment in arctic environments. Arctic and Alpine Research, 19: 357-365.

Black, R. A., Richards, J. R., & Manwaring, J. H. (1994). Nutrient uptake from enriched microsites by three Great Basin perennials. *Ecology* 75: 110-122.

Bowman, W. D., Theodose, T. A., Schardt, J. C., & Conant, R. T. (1993). Constraints of nutrient availability on primary production in two alpine tundra community. *Ecology* 74: 2085-2097.

Briske, D. D. & Butler, J. L., (1989). Density-dependent regulation of ramet populations within the bunchgrass Schizachyrium scoparium: interclonal versus intraclonal interence. *Journal of Ecology* 77: 963-974.

Brooks, P. D., Williams, M. W., Walker, D. A., & Schmidt, S. K. (1995). The Niwot Ridge snow fence experiment: Biogeo-chemical responses to changes in the seasonal snowpack. In Tonnessen, K., Williams, M. W., and Tanter, M. (eds.), *Biogeochemistry of Seasonally Snow-Covered Catchments (Proceedings of a Boulder Symposium, July 1995)*. International Association of Hydrological Sciences Publication 228, 293-302.

Cao, M.K & Woodward, F.I. (1998). Net primary and ecosytem production and carbon stocks of terrestrial ecosytems and their responses to climate cxhange. *Global Change Biology*, 4:185-198.

Chapin, F S. & Shaver, G. R. (1985). Individualistic growth response of tundra plant species to environmental manipulations in the field. *Ecology* 66: 564-576.

Chapin, F S., Jefferies, R. L., Reynolds, J. E, and Svoboda, J., 1992: Arctic plant physiological ecology in an ecosystem context. In Chapin, E S., Jefferies, R. L., Reynolds, J. E, Shaver, G. R., & Svoboda, J. (eds), *Arctic Ecosystems in a Changing Climate: An Ecophysiological Perspective*. San Diego: Academic Press, 441-452.

Chapin, F. S., Shaver, G. R., Giblin, A. E., Nadeloffer, K. J., & Laundre, J. A. (1995). Responses of arctic tundra to experimental and observed changes in climate. *Ecology* 76: 694-711.

Chapin, F.S. III, Walker, B.H., Hobbs, R.J., Hooper, D,U., Lawton, J.H.,Sala, O.E. & Tilman, D., (1997). Biotic control over the functioning of Ecosytsem. Science. 277.5325: 500-504.

Chapin, F. S., III, McGuire, A. D., Randerson, J., Pielke, R. Sr, Baldocchi, D., Hobbie, S. E., Roulet, N., Eugster, W., Kasischke, E., Rastetter, E. B., Zimov, S. A., & Running, S. W. (2000). Arctic and boreal ecosystems of western North America as components of the climate system. *Global Change Biology* 6: 211–223.

Chapin, F. S., III, Robards, M. D., Huntington, H. P., Johnstone, J. F., Trainor, S. F., Kofinas, G. P., Ruess, R. W., Fresco, N., Natcher, D. C. & Naylor, R. L. (2006). Directional changes in ecological communities and social-ecological systems: a framework for prediction based on Alaskan examples. *The American Naturalist* 168: S36–S49.

Cincotta, R. P., Zhang, Y. Q., & Zhou, X. M. (1992). Transhumant alpine pastoralism in northwestern Qinghaip rovince: An evaluation of livestock population response during China's agrarian economic reform. *Nomadic People* 30: 3-25.

Coulelis, H. (1985). Cellular world: a framework for modeling micro-macro dynamics. *Environment and Planning* A 17: 585–596.

Cox, P., Betts, R., Bunton, C., Essery, R., Rowntree, P.R. & Smith, J. (1999). The impact of new land surface physics on the GCM simulation of climate and climate sensitivity. *Clim. Dynamics* 15: 183-203.

Cramer, W. & Leemans, R. (1991). Assessing impacts of climate change on vegetation using climate classification systems. In: Shugart, H.H. & Solomon, A.M. (eds.) *Vegetation dynamics and global change*, pp. 190-217. Chapman & Hall, New York, NY, US.

Cramer W., Bondeau A., Woodward F. I., Prentice I. C., Betts R. A., Brovkin V., Cox P., Fisher V., Foley J. A., F riend A. D., Kucharikch C., Lomas M. R., R amankutty N., Sitch S., Smith B., White A. & Molling C.Y. (2001). Global response of terrestrial ecosystem structure and function to CO2 and climate change: results from six dynamic global vegetation models. *Global Change Biol.* 7: 357-373.

Daly, C., Neilson, R.P. & Phillips, D.L. (1994). A statistical-topographic model for mapping climatological precipitation over mountainous terrain. *J. Appl. Meteorol.* 33: 140-158.

Daly, C., Gibson, W.P., Hannaway, D. & Taylor, G.H. (2000). Development of new climate and plant adaptation maps for China. In: *Proceedings of the 12th Conference on Applied Climatology, May 8-11, 2000.* American Meteorogical Society, Asheville, NC, US.

Eastman, J. R. (2003). *IDRISI Kilimanjaro Tutorial. Manual Version 14.0.* Worcester, Massachusetts: Clark Labs of Clark University, 61–123.

Falkengren-Grerup, U . (1995). Interspecies differences in the preference of ammonium and nitrate in vascular plants. *Oecologia* 102: 305-311.

Giles, J. (2002). What kind of science is this? *Nature* 417: 216–218.

Grabherr. G., Gottfried,M . & Pauli, H. (1994). Climate effects on mountain plants. *Nature* 369: 448-450.

Harrison, S.P. & Prentice, I.C. (2003). Climate and CO2 controls on global vegetation distribution at the last glacial maximum: analysis based on palaeovegetation data, biome modellingand palaeoclimate simulations. *Global Change Biology* 9: 983-989.

Haxeltine, A. & Prentice, I.C. (1996). BIOME3: an equilibrium terrestrial biosphere model based on ecophysiological constraints, resource availability and competition among plant functional types. *Global Biogeochem. Cycl.* 10: 693-709.

Hodkinson, I. D., Webb, N. R., Bale, J. S. & Block, W. (1999). Hydrology, water availability and tundra ecosystem function in a changing climate: the need for a closer integration of ideas? *Global Change Biology* 5(3): 359–369.

Hou, X.Y., Sun, S.Z., Zhang, J.W., He, M.G., Wang, Y.F., Kong, D.Z. & Wang, S.Q. (1982). *Vegetation map of the people's Republic of China. Map Press of China,* Beijing, CN.

Itami, R. M. (1994). Simulating spatial dynamics: cellular automata theory. *Landscape and Urban Planning* 30: 27–47.

Jackson, L. E., Schimel, J. P. & Firestone, M. K. (1989). Short-term partitioning of ammonium and nitrate between plants and microbes in an annual grassland. *Soil Biology and Biochemistry* 21: 409-415.

Johnes, P. D. & Briffa, K. R. (1992). Global surface air temperature variations during the twentieth century. Part I: spatial, temporal and seasonal details. *The Holocene* 2: 165–179.

Jonasson, S., Havstrom, M., Jensen, M. & Callaghan, T. V. (1993). In situ mineralization of nitrogen and phosphorus of arctic soils afterp erturbations imulatingc limate change. *Oecologia* 95: 179-186.

Kaplan, J.O., Bigelow, N.H., Prentice, I.C., Harrison, S.P., Bartlein, J., Christensen, T.R., Cramer, W., Matveyeva, N.V., McGuire, A.D., Murray, D.F., Razzhivin, V.Y., Smith, B., Walker, D.A., Anderson, P.M., Andreev, A.A., Brubaker, L.B., Edwards, M.E. & Lozhkin, A.V. (2003). Climate change and Arctic ecosystems: 2. Modeling, paleodata model comparison and future projections. J. *Geophys. Res.* 108 (D19):8171.

Karlsson, P. S. (1985). Effect of water and mineral nutrient supply on a deciduous and evergreen dwarf shrub: Vaccinium uliginosum L. and V. vitisidaea L. Holarctic. *Ecology* 8: 1-8.

Kato, T., Tang, Y., Gu, S.,Hirota, M., Du,M., Li, Y. & Zhao, X. (2006). Temperature and biomass influences on interannual changes in CO2 exchange in an alpine meadow on the Qinghai-Tibetan Plateau. *Global Change Biology* 12:1285-1298.

Kennedy, A. D. (1995), Simulated climate change: are passive greenhouse a valid microcosn for testing the biological effects of environmental perturbation? *Global Change Biology* 1: 29-42.

Klanderud, K. & Birks, H.J.B. (2003). Recent increases in species richness and shifts in altitudinal distributions of Norwegian mountain plants. *Holocene* 13: 1-6.

Klein, J.A., Harte J. & Zhao X.Q. (2007). Experimental warming, not grazing, decreases rangeland quality on the Tibetian Plateau. *Ecological Applications* 17(2):341-557.

Korner, Ch. (1992). Response of alpine vegetation to global climate change. *In: International Conference on Landscape Ecological Impact of Climate Change*. Lunteren, The Netherlands, Catena Verlag, Supplement, 22: 85-96.

Leemans, R.E. (2004). Anotherreason for concern: regional and global impacts on ecosytems for different levels of climate change. *Global Environmental Change* 14: 219–228.

Li, Y. N., Zhao, X. Q., Cao, G. M., Zhao, L. & Wang, Q. X. (2004). Analysis on climates and vegetation productivity background at Haibei Alpine Meadow Ecosystem Research Station. *Plateau* Meteorology 23(4): 558–567.

Li, X., Cheng, G. D. & Lu, L. (2005). Spatial analysis of air temperature in the Qinghai-Tibet Plateau. *Arctic, Antarctic, and Alpine Research* 37(2): 246–252.

Lloyd, A.H., T. S. Rupp, C.L. Fastie & A. M. Starfield. (2003). Patterns and dynamics of treeline advance on the Seward Peninsula, Alaska. *Journal of Geophysical Research Atmospheres.* 108 (D2): 8161, doi: 10.1029/2001JD000852.

Lloyd, A.H. & C.L. Fastie (2003). Recent changes in treeline forest distribution and structure in interior Alaska. *Ecoscience.* 10(2):176-185.

Maxwell, B., (1992). Arctic climate: Potential for change under global warming. In Chapin, F S., Jefferies, R. L., Reynolds, J. F, Shaver, G. R., and Svoboda, J. (eds), *Arctic Ecosystems in a Changing Climate: An Ecophysiological Perspective.* San Diego: Academic Press, 11-34.

McGuire, A. D., Chapin, F. S., III, Walsh, J. E. & Wirth, C. (2006). Integrated regional changes in arctic climate feedbacks: implications for the global climate system. *Annual Review of Environmental* Resources 31: 61–91.

Molau, U. & Alatalo, J.M. (1998). Responses of subarctic alpine plant communities to simulated environmental change: Biodiversity of bryophytes, lichens, and vascular plants. *Ambio* 27: 322-329.

Murry, C. & Miller, P. C. (1982). Phenological observations of major plant growth forms and species in montane and Eriophorum vaginatum tussock tundra in central Alaska. *Holarctic Ecology* 5: 109-116.

Nadelhoffer, K. J., Giblin, A. E., Shaver, G. R. & Laundre, J. A. (1991). Effects of temperature and substrate quality on element mineralization in six arctic soils. *Ecology* 72: 242-253.

Neilson, R., Prentice, I.C. & Smith, B. (1998). Simulated changes in vegetation distribution under global warming. In: Watson, R.T. et al. (eds.) *The regional impacts of climate change*, pp. 439-456. Cambridge University Press, New York, NY, US.

Ni, J. (2000). A simulation of biomes on the Tibetan Plateau and their responses to global climate change. *Mount. Res. Devel.* 20: 80-89.

Parmesan, C. (1996). Climate and species' range. *Nature* 382: 765-766.

Parmesan, C., Ryholm, N., Stefanescu, C., Hill, J. K. et al. (1999). Poleward shifts in geographical ranges of butterfly species associated with regional warming. *Nature* 399: 579-583.

Parsons, A. N., Welker, J. M., Wookey, P. A., Press, M. C., Callaghan, T. V. & Lee, J. A. (1994). Growth responses of four sub-arctic dwarf shrubs to simulated climate change. *Journal of Ecology*, 82: 307-318.

Parsons, A. N., Press, M. C., Wookey, P. A., Welker, J. M., Robinson, C. H., Callaghan T. V. & Lee, J. A. (1995). Growth and reproductive output of Calamagrostis lapponica in response to simulated environmental change in the subarctic. *Oikos* 72:61-66.

Pauli, H., Gottfried, M. & Grabherr, G. (2003). The Piz Linard (3411m), the Grisons, Switzerland - Europe's oldest mountain vegetation study site. In: Nagy, L., Grabherr, G., Körner, C. & Thompson, D.B.A. (eds.) *Alpine biodiversity in Europe – A Europe-wide assessment of biological richness and change*, pp. 443-448. Springer-Verlag, Berlin, DE.

Robinson, C. H., Wookey, P. A., Parsons, A. N., Potter, J. A., Callaghan, T. V., Lee, J. A., Press, M. C. & Welker, J. M. (1995). Responses of plant litter decomposition and nitrogen mineralisation to simulated environmental change in a high arctic polar semi-desert and a subarctic dwarf shrub heath. *Oikos*, 74: 503-512.

Rozanski, K., Araguas-Araguas, L. & Gonfiantini,R. (1992). Relation between long-term trends of oxygen-18 isotope composition of precipitation and climate. *Science* 258: 981-985.

Ruxton, G. D., and Saravia, L. A. (1998). The need for biological realism in the updating of cellular automata models. *Ecological Modeling* 107(2–3): 105–112.

Sala, O. E., and Lauenroth, W. K. and Parton, W. J. (1992). Long-term soil water dynamics in the shortgrass steppe. *Ecology* 73: 1175-1181.

Schimel, J. S., Bilbrough, C. B. and Welker, J. M. 2004. The effect of changing snow cover on yearround soil nitrogen dynamics in Arctic tundra ecosystems. *Soil Biology and Biochemistry* 36: 217-227.

Shaver, G. R., Billings, W. D., Chapin, E S., Giblin, A. E., Na-delhoffer, K. J., Oechel, W. C. & Rastetter, E. B. (1992). Global change and the carbon balance of arctic ecosystems. *Bioscience* 42: 433-441.

Shaver, G. R. & Chapin, E S. (1991). Production:biomass relationships and element cycling in contrasting arctic vegetation types. *Ecological Monographs* 61: 1-31.

Song, M., Zhou, C. & Ouyang, H. (2005). Simulated distribution of vegetation types in response to climate chang on the Tibetan Plateau. *Journal of Vegetation Science* 16:341-350.

Sullivan, P. F. & Welker, J. M. 2005. Warming chambers stimulate early season growth of an arctic sedge: results of a minirhizotron field study. *Oecologia* 142: 616-626.

Sykes, M.T., Prentice, I.C. & Laarlf, F. (1999). Quantifying the impact of global climate change on potential natural vegetation. *Clim. Change* 41: 37-52.

Tang, M.C., Cheng, G.D. & Lin, Z.Y. (1998*). Contemporary climatic variations over Tibetan Plateau and their influences on environments*. Guangdong Science & Technology Press, Guangzhou, CN.

Tape, K., Sturm, M. & Racine, C. (2006). The evidence for shrub expansion in Northern Alaska and the Pan-Arctic. *Global Change Biol.* 12, 686–702.

Thompson, L., G., Mosley-Thompson, E., Bolzan, J. F. et al. (1989). Holocene-Late Pleistocene climatic ice core records from Qinghai-Tibetan Plateau. *Science* 246: 474-477.

Walker, M. D., Webber, P. J., Arnold, E. H. & Ebert-May, D. (1994). Effects of interannual climate variation on aboveground phytomass in alpine vegetation. *Ecology* 75:393-408.

Walther, G.R., Post, E., Convey, P., Menzel, A., Parmesan, C., Beebee, T.J.C., Fromentin, J.M., Hoegh-Guldberg, O. & Bairlein, F. (2002). Ecology responses to recent climate change. *Nature* 416: 389-395.

Welker, J. M., Rykiel, E. J., Briske, D. D. & Goeschel, J. D. (1985). Carbon import among vegetative tillers within two bunchgrasses: assessment with carbon-11 labelling. *Oecologia* 67: 209-212.

Welker, J. M., Briske, D. D. & Weaver, R. W. (1987). Nitrogen-15 partitioning within a three generation tiller sequence of the bunchgras Schizachyrium scoparium: response to selective defoliation. *Oecologia* 74: 330-334.

Welker, J. M. & Briske, D. D. (1992). Clonal biology of the temperate caespitose graminoid Schizachyrium scoparium: A synthesis with reference to climate change. *Oikos* 56:357- 365.

Welker, J. M., Wookey, P., Parsons, A. P., Callaghan, T. V., Press, M. C. & Lee, J. A. (1993). Leaf carbon isotope discrimination and demographic responses of Dryas octopetala to water and temperature manipulations in a high arctic polar semi-desert, Svalbard. *Oeclogia* 95: 463-749.

Welker, J. M., Svoboda, J., Henry, G., Molau, U., Parsons, A. N. & Wookey, P. A. (1995). Response of two Dryas species to ITEX environmental manipulations: A synthesis with circumpolar comparisions. In: *Proceedings from the 6th International Tundra Experiment (ITEX)*. Ottawa, Canada. April 1995. Abstract.

Welker, J. M., Fahnestock, J. T. & Jones, M. H. 2000. Annual CO2 flux from dry and moist arctic tundra: field responses to increases in summer temperature and winter snow depth. Climatic Change 44: 139-150.

Welker, J.M., Fahnestock, J.T., Povirk, K., Bilbrough, C. & Piper, R. 2004. Carbon and nitrogen dynamics in a long-term grazed alpine grassland. *Arctic, Antarctic and Alpine Research* 36: 10-19.

Welker, J. M., Fahnestock, J. T., Sullivan, P. & Chimner, R. A. (2005). Leaf mineral nutrition of arctic plants in response to long-term warming and deeper snow in N. Alaska. *Oikos* 109: 167-177.

White, R. & Engelen, G. (2000). High-resolution integrated modeling of the spatial dynamics of urban and regional system. *Environment and Urban Systems* 24: 383–400.

Wolfram, S. (1984). Cellular automata as models of complexity. Nature 311(4): 419–424.

Woodward, F.I. 1987. *Climate and plant distribution*. Cambridge University Press, Cambridge, UK.

Wookey, P. A., Parsons, A. N., Welker, J. M., Potter, J., Callaghan, T. V., Lee, J. A. & Press, M. C. (1993). Comparative responses of phenology and reproductive development to sim-ulated environmental change in sub-arctic and high arctic plants. *Oikos* 67: 490-502.

Wookey, P. A., Welker, J. M., Parsons, A. N., Press, M. C., Callaghan, T. V. & Lee, J. A. (1994). Differential growth, allocation and photosynthetic responses of *Polygonum viviparum* to simulated environmental change at a high arctic polar semi-desert. Oikos 70: 131-139.

Wookey, P. A., Robinson, C. H., Parsons, A. N., Welker, J. M., Press, M. C., Callaghan, T. V. & Lee, J. A. (1995). Environmental constraints on the growth, photosynthesis and reproductive development of Dryas octopetala at a high arctic polar semi-desert, Svalbard. *Oecologia* 102: 478-489.

Xiong Y. & Li, Q.K. (1987). *Soils in China.* Science Press, Beijing.

Xia, W. P. (1988). A brief introduction to the fundamental characteristics and the work in Haibei Research Station of Alpine Meadow Ecosystem. Proceedings of the *International Symposium of an Alpine Meadow Ecosystem*. Bejing: Academic Sinica, 1-10.

Yu, G. (1999). Studies on biomization and the global palaeovegetation project. *Adv. Earth Sci.* 14: 306-311.

Yu, G., Sun, X.J., Qin, B.Q., Song, C.Q., Li, H.Y., Prentice, I.C. & Harrison, S.P. (1998). Pollenbased reconstruction of vegetation patterns of China in mid-Holocene. *Sci. China (Ser. D)* 41: 130-136.

Zhang, X.S. (1993). The Tibetan Plateau in relation to the vegetation of China. *Ann. Missouri Bot. Garden* 70: 564-570.

Zhang, X.S., Yang, D.A., Zhou, G.S., Liu, C.Y. & Zhang, J. (1996). Model expectation of impacts of global climate change on biomes of the Tibetan Plateau. In: Omasa, K.,

Kai, K., Taoda, H., Uchijima, Z. & Yoshino, M. (eds.) *Climate change and plants in East Asia* pp. 25-38. Springer-Verlag, Tokyo, JP.

Zhang, Y. M., and Liu, J. K. (2003). Effects of plateau zokors (Myospalax *fontanierii*) effects on plant community and soil in an alpine meadow. *Journal of Mammalogy* 84(2): 644–651.

Zhang, Y. Q. (1990). A quantitative study on the characteristics and succession pattern of alpine shrublands under the different grazing intensities. *Acta Phytoecologia and Geobotanica Sinica* 14(4): 358–365.

Zhang, Y. Q. & Zhou, X. M. (1992). The quantitative classification and ordination of Haibei alpine meadow. *Acta Phytoecological ET Geobotanica Sinica* 16(1): 36–42.

Zhang, Y. Q. & Welker, J. M. (1996). Tibetan alpine tundra responses to simulated changes in climate: aboveground biomass and community responses. *Arctic and Alpine Research* 28(2): 203–209.

Zhang, Y. Q. (2005). Raster multi-criteria evaluation for experimental design with aggregating weighed factors. *GeoTec Event Proceeding*. Poster Presentation in Vancouver Canada. (http:// www.sfu.ca/geog/geog355fall04/yqz/index.htm).

Zhang, Y.Q.A., Peterman, M.R.,Aun, D.L. & Zhang Y.M. (2008). Cellular Automata:Simulating alpine tundra vegetation dynamics in Response to global warming. *Arctic, Antarctic and Alpine Research* 40(1):256-263.

Zheng, D. (1996). The system of physico-geographical regions of the Tibet Plateau. *Sci. China (Ser. D)* 39: 410-417.

Zhou, X. M., Wang, Zh. B. & Du, Q. (1987). *Qinghai Vegetation*. Qinghai People Press.

Zhou, X.M. (2001). *Chinese kobresia meadows*. Science, Beijing.

Corresponding author: Yanqing A. Zhang, instca@yahoo.com